# Producing Mus

During the last two decades, the field of music production has attracted considerable interest from the academic community, more recently becoming established as an important and flourishing research discipline in its own right.

*Producing Music* presents cutting-edge research across topics that both strengthen and broaden the range of the discipline as it currently stands. Bringing together the academic study of music production and practical techniques, this book illustrates the latest research on producing music.

Focusing on areas such as genre, technology, concepts, and contexts of production, Hepworth-Sawyer, Hodgson, and Marrington have compiled key research from practitioners and academics to present a comprehensive view of how music production has established itself and changed over the years.

**Russ Hepworth-Sawyer** started life as a sound engineer and occasional producer. He has over two decades' experience of all things audio. Russ is a member of the Audio Engineering Society and co-founder of the UK Mastering Section there. A former board member of the Music Producers Guild, Russ helped form their Mastering Group. Through MOTTOsound (www.mottosound.co.uk), Russ now works freelance in the industry as a mastering engineer, writer, and consultant. Russ currently lectures part-time for York St John University and has taught extensively in higher education at institutions including Leeds College of Music, London College of Music, and Rose Bruford College. He contributes from time to time in magazines such as *MusicTech, Pro Sound News Europe*, and *Sound On Sound*. He has also written many titles for Focal Press and Routledge. Russ is the co-founder of the Innovation In Music conference series (www.musicinnovation.co.uk) and also of the Perspectives On Music Production series of academic books.

**Jay Hodgson** is on faculty at Western University, where he primarily teaches courses on songwriting and project paradigm record production. He is also one of two mastering engineers at MOTTOsound, a boutique audio services house situated in England. In the last few years, Dr. Hodgson's masters have twice been nominated for Juno Awards and topped Beatport's global techno and house charts, and he has contributed music to films recognized by the likes of *Rolling Stone* magazine and which screened at the United Nations General Assembly. He was awarded a Governor General's academic medal in 2006, primarily in recognition of his research on audio recording, and his second book, *Understanding Records* (2010), was recently acquired by the Reading Room of the Rock and Roll Hall of Fame. He has other books published and forthcoming from Oxford University Press, Bloomsbury, Continuum, Wilfrid Laurier University Press, Focal Press, and Routledge.

**Mark Marrington** trained in composition and musicology at the University of Leeds (MMus, PhD) and is currently a senior lecturer in music production at York St John University. He has previously held teaching positions at Leeds College of Music and the University of Leeds (School of Electronic and Electrical Engineering). Mark has published chapters with Cambridge University Press, Bloomsbury Academic, Routledge, and Future Technology Press and has contributed articles to *British Music*, *Soundboard*, the *Musical Times*, and the *Journal on the Art of Record Production*. Since 2010, his research has been focused on a range of music production topics with a particular emphasis on the role of digital technologies in music creation and production. Other interests include songwriting, music technology pedagogy, the contemporary classical guitar, and British classical music in 20th century. His most recent research has been concerned with the aesthetics of classical music recording and a forthcoming monograph on the role of recordings in shaping the identity of the classical guitar in the 20th century.

# Perspectives on Music Production Series

**Series Editors**
Russ Hepworth-Sawyer
Jay Hodgson
Mark Marrington

**Titles in the Series**

**Mixing Music**
*Edited by Russ Hepworth-Sawyer and Jay Hodgson*

**Audio Mastering: The Artists**
Discussions from Pre-Production to Mastering
*Edited by Russ Hepworth-Sawyer and Jay Hodgson*

**Producing Music**
*Edited by Russ Hepworth-Sawyer, Jay Hodgson, and Mark Marrington*

# Producing Music

**Edited by Russ Hepworth-Sawyer,
Jay Hodgson, and Mark Marrington**

Routledge
Taylor & Francis Group

NEW YORK AND LONDON

First published 2019
by Routledge
52 Vanderbilt Avenue, New York, NY 10017

and by Routledge
2 Park Square, Milton Park, Abingdon, Oxon, OX14 4RN

*Routledge is an imprint of the Taylor & Francis Group, an informa business*

*Library of Congress Cataloging-in-Publication Data*
Names: Hepworth-Sawyer, Russ. | Hodgson, Jay. | Marrington, Mark.
Title: Producing music / edited by Russ Hepworth-Sawyer, Jay Hodgson, and Mark Marrington.
Description: New York, NY : Routledge, 2019. | Series: Perspectives on music production series | Includes index.
Identifiers: LCCN 2018050743 (print) | LCCN 2018051167 (ebook) | ISBN 9781351815109 (pdf) | ISBN 9781351815086 (mobi) | ISBN 9781351815093 (epub) | ISBN 9780415789219 (hardback : alk. paper) | ISBN 9780415789226 (pbk. : alk. paper) | ISBN 9781315212241 (ebook)
Subjects: LCSH: Sound recordings—Production and direction. | Popular music—Production and direction.
Classification: LCC ML3790 (ebook) | LCC ML3790 .P757 2019 (print) | DDC 781.49—dc23
LC record available at https://lccn.loc.gov/2018050743

ISBN: 978-0-415-78921-9 (hbk)
ISBN: 978-0-415-78922-6 (pbk)
ISBN: 978-1-315-21224-1 (ebk)

Typeset in Times New Roman
by Apex CoVantage, LLC

Printed and bound in Great Britain by
TJ International Ltd, Padstow, Cornwall

# Contents

# Contributor Biographies

**Alex Baxter** is a networked audio and audio electronics specialist who builds audio hardware and software. He also has wide-ranging freelance experience in audio mastering, restoration, and forensic audio. In addition to being the program leader for music technology at the University of Hertfordshire, where he leads the undergraduate degree awards in music production, music and sound design technology, audio recording and production, and live sound and lighting technology, Alex also works as an educational consultant for all things involving technology, music, and sound and has provided master classes and consultancy in a variety of settings, including for the University of Cambridge and teacher training providers, and delivered county-wide training in Hertfordshire and Bedfordshire. Alex is very interested in how the performance interface affects engagement with music, and this interest informs both his hardware and software development work.

**Thomas Brett** is a Broadway percussionist who holds a PhD in ethnomusicology from New York University. He has written essays for *Popular Music*, *Popular Music and Society*, the *Oxford Handbook of Music and Virtuality*, and the *Cambridge Companion to Percussion*. He writes about music at brettworks.com.

**Stace Constantinou** is a musician who composes for solo instruments, ensembles, and electronic music (live and fixed media). Having studied piano, composition, and electronic music at Huddersfield University, he went on to Kingston University, where he took a PhD in composition (awarded in 2015). Aged nine, he won a scholarship to attend music school, where he received grounding in the classical music tradition. Later, his electronic piece "Desert Storm" won the 1993 Ricordi Prize for Best Composition. And notable performers, including Jane Manning OBE, Christopher Redgate, Rhodri Davies, and Kate Ryder, have premiered his solo pieces. Influenced by music of the Western lineage from ancient through to classical and contemporary times, his composing has also been guided by mathematical ideas, for example, numerical, geometrical, and statistical models of pattern types, as well as a broad range of philosophical ideas from Plato, Aristotle, Rameau, Kant, Russell, and P. Schaeffer to contemporary writings on chaos theory. Having held positions at Morley College and Kingston University, he now works as a freelance musician, teacher, and producer. His works have been performed in concerts and festivals in Britain and on BBC national radio and Resonance FM, as well as on the European continent and in Australia.

**Michail Exarchos** (a.k.a. Stereo Mike) is a hip-hop musicologist and award-winning rap artist (MTV Best Greek Act), with nominations for multiple VMAs and an MTV Europe Music Award. He is the course leader for music mixing and mastering at London College of Music (University of West London), and his publications include articles for *Popular Music* journal (Cambridge University Press), Routledge, Cambridge Scholars, and the *Journal of Popular Music Education*. His self-produced/engineered album *Xli3h* has been included in the 30 Best Greek Hip-Hop albums of all time.

**Paul Ferguson** is an associate professor of audio engineering in the Applied Music Research Centre within the School of Arts and Creative Industries at Edinburgh Napier University. Since 2012, his research has focused on the use of high-speed research and education networks to allow artists separated by long distances to perform together in real-time. This research brings together two separate strands from Paul's industry background: ten years as an electronics/embedded software R&D engineer in the defense industry, where he led a 12-person team specializing in deterministic network communications, and 20 years as an audio engineer with a technical specialist role, supporting studio and live projects by artists such as Emerson, Lake and Palmer, Marillion, Bad Company, and Robert Plant. Since joining Edinburgh Napier University in 2001, Paul has maintained his music industry links and provides project support and consultancy. He is currently under NDA with audio companies AVID, Audinate, and Focusrite.

**Alexander C. Harden** is a researcher in narratology and popular music analysis and recent graduate of the University of Surrey. In 2018, he completed his doctoral thesis entitled "Narrativity, Worldmaking, and Recorded Popular Song" under the supervision of Professor Allan F. Moore. Alexander has also received an International Association for the Study of Popular Music (UK & Ireland Branch) Andrew Goodwin Memorial Prize for his essay "A World of My Own". His research interests include the hermeneutics of popular song, the art of record production, and narrativity across media.

**Russ Hepworth-Sawyer** started life as a sound engineer and occasional producer. He has over two decades' experience of all things audio. Russ is a member of the Audio Engineering Society and co-founder of the UK Mastering Section there. A former board member of the Music Producers Guild, Russ helped form their Mastering Group. Through MOTTOsound (www.mottosound.co.uk), Russ now works freelance in the industry as a mastering engineer, writer, and consultant. Russ currently lectures part-time for York St John University and has taught extensively in higher education at institutions including Leeds College of Music, London College of Music, and Rose Bruford College. He contributes from time to time in magazines such as *MusicTech*, *Pro Sound News Europe*, and *Sound On Sound*. He has also written many titles for Focal Press and Routledge. Russ is the co-founder of the Innovation In Music conference series (www. musicinnovation.co.uk) and of the Perspectives On Music Production series of academic books.

**Jay Hodgson** is on faculty at Western University, where he primarily teaches courses on songwriting and project paradigm record production. He is also one of two mastering engineers at MOTTOsound, a boutique audio services house situated in England. In the last few years, Dr. Hodgson's masters have twice been nominated for Juno Awards and topped Beatport's global techno and house charts, and he has contributed music to films recognized by the likes of *Rolling Stone* magazine and which screened at the United Nations General Assembly. He was awarded a Governor General's academic medal in 2006, primarily in recognition of his research on audio recording, and his second book, *Understanding Records* (2010), was recently acquired by the Reading Room of the Rock and Roll Hall of Fame. He has other books published and forthcoming from Oxford University Press, Bloomsbury, Continuum, Wilfrid Laurier University Press, Focal Press, and Routledge.

**Rob Lawrence** is a full-time audio producer, coach, and business consultant who specializes in immersive audio experiences. Having composed and produced audio for theater, film, music, and broadcast applications worldwide, he was awarded a scholarship, earning a first-class BSc (Hons) in audio production at SAE in Oxford, England.

**Gareth Llewellyn** began his sound career in film audio post before moving to Galaxy Studios, Belgium where he was heavily involved with the pioneering 3D audio work that led to the Auro-3D sound format. Gareth helped mix the world's first commercial 3D audio feature (Red Tails, Lucasfilm) with a team from Skywalker Sound, and went on to remix dozens of Hollywood features – pioneering new creative and technical approaches to deliver audience immersion through channel and object-based sound systems. Gareth has since showcased new kinds of audience immersion through his creative production house *Mixed Immersion* – delivering interactive 3D audio to brands, immersive theatre and artists. Gareth is currently working on mixed reality audio technology in his new startup, *MagicBeans*.

**Kallie Marie** is a music producer, composer, recording engineer, author, and educator. She received her MA in music production from the Leeds College of Music and has a background in women's studies. Highlights of her work include writing music for Big Foote Music + Sound, Nylon Studios, Warren Adams, and Loni Landon, and she was a featured composer in the NYC Electroacoustic Music Festival. As a producer and recording engineer,

Kallie has worked with Jeff Derringer, January Jane, RainMKERS, and Makes My Blood Dance. Additionally, Kallie is a contributing author at *Sonic Scoop*, and has worked as a product specialist with Tracktion Software, and is an adjunct faculty at New York University. She spoke at Barnard College's Gender Amplified Festival on the topic of audio fundamentals. Previously she has worked for Apple Computers and was formerly faculty at the Art Institute of New York City.

**Mark Marrington** trained in composition and musicology at the University of Leeds (MMus, PhD) and is currently a senior lecturer in music production at York St John University. He has previously held teaching positions at Leeds College of Music and the University of Leeds (School of Electronic and Electrical Engineering). Mark has published chapters with Cambridge University Press, Bloomsbury Academic, Routledge, and Future Technology Press and has contributed articles to *British Music*, *Soundboard*, the *Musical Times*, and the *Journal on the Art of Record Production*. Since 2010, his research has been focused on a range of music production topics with a particular emphasis on the role of digital technologies in music creation and production. Other interests include songwriting, music technology pedagogy, the contemporary classical guitar, and British classical music in the 20th century. His most recent research has been concerned with the aesthetics of classical music recording and a forthcoming monograph on the role of recordings in shaping the identity of the classical guitar in the 20th century.

**Zack Moir** is a lecturer in popular music at Edinburgh Napier University, UK. His research interests are in popular music in higher education, popular music composition pedagogy, the teaching and learning of improvisation, and real-time networked audio performance. He has published on the topics of popular music pedagogy, popular music making and leisure, popular music songwriting/composition, school-university transitions, and composition/production pedagogy. Zack is also an active composer/musician performing as a soloist and in ensembles internationally. Recent compositions include pieces for saxophone and tape, solo cello, and a reactive generative sound art installation for the Edinburgh International Science Festival.

**Austin Moore** is a senior lecturer and course leader of sound engineering and music production courses at the University of Huddersfield UK. He recently completed a PhD in music production and technology, which investigated the use of non-linearity in music production with a focus on dynamic range compression and the 1176 compressor. His research interests include music production sonic signatures, the perceptual effects of dynamic range compression, and semantic audio. He has a background in the music industry and spent many years producing and remixing various forms of electronic dance music under numerous artist names. Also, he is an experienced recording engineer and has worked in several professional recording studios.

**Pat O'Grady** has a PhD in music from Macquarie University. He works as a professional musician and teaches music and media subjects at Macquarie. His research examines the relationships between technologies and the politics of pop music production and consumption.

**Josh Ottum** holds an MFA in integrated composition improvisation and technology from UC Irvine and a PhD in interdisciplinary arts from Ohio University. His research focuses on sound studies, environmental humanities, and popular music studies. Ottum has released multiple albums with labels in the United States and Europe and has toured extensively. His music has appeared on television shows such as *The Voice* (NBC) and *Mad Men* (AMC) and the podcast *Freakonomics*. He is currently a professor of commercial music at Bakersfield College.

**Justin Paterson** is an academic, music producer, and author. His research has an international profile and has ranged from transient enhancement in multi-mic recordings through various papers on the musicology of record production to an AHRC-funded project developing interactive music playback via an album app with patent pending. Justin has exhibited computer art at the Victoria and Albert museum in London. He is currently the AHRC-funded principal investigator for the project The Commercialisation of Interactive Music. Commercial research bid partners have included BBC, Abbey Road Studios, Ninja Tune, MelodyVR, Warner Music Group, Digital Jam, Mixed Immersion, Blue Studios, Swedish Museum of Performing Arts, Vitensenteret (including its Planetarium), and halo Post. Justin is author of *The Drum Programming Handbook*. He is co-chair of the Innovation in Music conference series and an editor of the journal *Perspectives on Music Production*. He leads the MA Advanced Music Technology at University of West London.

**Marcio Pinho** is a graduate and master of popular music at the State University of Campinas (São Paulo, Brazil) and is currently a PhD candidate at Goethe University of Frankfurt am Main, Germany, under the supervision of Dr. Julio Mendívil. As a musician, he released two albums of Brazilian instrumental music with Trio Pulo do Gato.

**Matt Shelvock** is the Business Affairs Manager at bitbird records and Heroic Management, where he works with internationally celebrated artists such as San Holo, WRLD, DROELOE, Taska Black, Fytch, Duskus, Flaws, EASTGHOST, and numerous others. He is also a researcher, mastering engineer, and producer, having performed audio services on over 300 records, including Juno and CCMA nominated records, and other chart-topping releases (i.e., Runaway Angel, Arthur Oskan). During Matthew's previous career as a professional musician (2003–2012), he worked with Grammy/Juno/Billboard award-winning personnel such as Skip Prokop (Lighthouse, Janis Joplin), Josh Leo (LeAnn Rimes, Lynyrd Skynyrd, Glen Frey), and many others. Matthew's research investigates musical competencies that are required to complete and exploit recordings, such as audio engineering and music production skills, as well as global licensing strategies for rights holders. He is currently completing two manuscripts for Routledge/Focal press on audio mastering and cloud-based music production.

**Gareth Dylan Smith** is manager of program effectiveness at Little Kids Rock, president of the Association for Popular Music Education, a board member of the International Society for Music Education, and visiting research professor of music at New York University. Gareth's performance career extends from punk, hard rock, and psycho-ceilidh through jazz and musical theater. He has written for magazines including *Rolling Stone*, *In Tune Monthly*, and *Rhythm*. He has authored numerous scholarly books and articles, is a founding editor of the *Journal of Popular Music Education*, and contributes to several encyclopedias. Gareth's first love is to play drums.

**Robert Toft** is a researcher, vocal coach, and producer who specializes in historically informed performances of vocal music written between the 16th and 19th centuries. He has given master classes on the old bel canto style of singing at leading conservatories and music schools in Europe, North America, and Australia, and he has published five books on the history of singing and a monograph on 1960s popular music. He runs his own record label, Talbot Records, and he is currently completing a new book on recording classical music for Focal Press. His home base is in the Faculty of Music at Western University in London, Canada.

**Rodrigo Vicente** graduated in popular music and earned his doctorate in music at the State University of Campinas. Nowadays, Vicente works as a researcher in the Institute of Brazilian Studies – University of São Paulo, where he develops work on the composer Noel Rosa and the samba of Rio de Janeiro in the 1930s. Vicente is also a guitarist and music teacher.

**Robert Wilsmore** studied music at Bath College and then postgraduate composition at Nottingham University with composer Nicholas Sackman. He gained his doctorate in 1994. Prior to starting at York St John University in 2007, Wilsmore was assistant head of music at Leeds College of Music. Wilsmore has also led on national pedagogical research projects, including Collaborative Art Practices in HE: Mapping and Developing Pedagogical Models, A HEA PALATINE project (2011) with Liz Dobson (Huddersfield) and Christophe Alix (Hull, Scarborough). He is currently writing a book on coproduction with co-author Chris Johnson.

# 1

## Introduction

### Russ Hepworth-Sawyer, Jay Hodgson, and Mark Marrington

Welcome to Volume 2 of the *Perspectives on Music Production* series, *Producing Music*. As with our first book of the series, *Perspectives on Music Production: Mixing Music*, this volume explores a specific topic pertinent to the Music Production Studies field, in this instance, production itself. In sourcing *Producing Music* we have gathered a wide range of material concerning the nature of "music production" at the present juncture. As editors, we have not necessarily aimed to align its content with any particular position as regards how music production should be discussed within the academic arena. For example, the book does not assert that a "musicological" perspective should exclusively define the terms of the debate, even though some of the chapters herein might be deemed to draw upon such ideas as they have been represented in recent literature. While our call for chapters for the present volume obviously had to defer to existing frameworks and contexts that have defined the discipline of Music Production Studies over the last decade or so, we have been keen to remain open to fresh perspectives on what an academically-situated discussion of a music production topic might potentially consist of.

For this new volume, we have also taken a different approach to that of our first book (*Mixing Music*) in that we have grouped together chapters using four specific headings designed to reflect an observable thematic unity amongst particular author contributions. As such, Part 1 contains chapters which discuss music production in relation to specific stylistic and genre-focused contexts—hip hop, metal, classical music, bossa nova and so on. Part 2, by contrast, draws together chapters on the basis of their tendency to address specific technological contexts of music production, both old and new. In Part 3 we have placed three chapters together that engage in theoretically informed discussions of music production practice, including inter-disciplinary perspectives deriving from the field of narrative studies, theories relating to collaborative practice (co-production) and philosophical perspectives concerning the ontology of the audio object. Finally, in Part 4 we present three chapters that are concerned with the wider context of contemporary production practice, exploring, for example, recent debates concerning women in music production and the oft-raised questions concerning educationalists' needs to represent and reflect the changing nature of music production in the commercial arena. To further assist the reader in navigating the book's content we also present the following chapter summaries:

Matt Shelvock's chapter, which launches the book, offers an ethnographically couched discussion of the practice of hip hop beat-making through the lens of ANT (Actor Network Theory) and SCOT (Social Constuction of Technology) with a view to providing a corrective to the tendency to use such frameworks in discussions of production practice without giving due attention to the operational strategies adopted in relation to specific technologies by recordists themselves. As such Shelvock's chapter provides a succinct model for interrogating specific technologically driven aspects of creative practice to build upon in future research.

The debates raised in Shelvock's work are also alluded to by Mike Exarchos (Chapter 3), who sustains the genre-specific focus on hip hop, but this time homes in on a technology central to the genre—the Akai MPC—and the role it has played in shaping the "Boom Bap" aesthetic. In particular Exarchos's concern is to explore the relationship between subgenre aesthetics and MPC functionality, with a view to tracing stylistic changes within the hip hop genre in relation to the technology's evolution.

In Chapter 4, Mark Marrington's discussion of digital aesthetics in the production of contemporary metal music builds upon his earlier work exploring the impact of DAWs on creative production practice. Here metal music—and specifically certain of its more recent subgenres, cybergrind, djent and djent-step—provides a vehicle for observing the process of genre transformation that occurs when particular electronic music aesthetics engendered by DAWs are adopted within metal music practice. This general theme of the DAW's capacity to structure creative practice is also taken up in Thomas Brett's chapter discussing Live.

In Chapter 5, Rodrigo Vicente and Marcio Pinho move the genre focus to the Brazilian bossa nova and its development through the innovations of producer Tom Jobim and João Gilberto. What emerges from this informative discussion is the sense of the bossa nova as a musical form (particularly where its instrumentation and arrangement are concerned) whose stylistic evolution is inextricably bound up with the technical concerns of studio-based recording and the production process during this period in Brazil.

In Chapter 6 Josh Ottum presents a stylishly written critique of a seminal work in the *vaporwave* genre—James Ferraro's *Far Side Virtual (FSV)*—which takes in (amongst other things) retro-digital production aesthetics, postmodern theory and ideas concerning record production, place, art and environment. Like djent (discussed in Mark Marrington's chapter), *vaporwave* is a relatively recently evolved "cyber" genre whose development was fueled by the activities of the online production community. The music's relative intangibility has meant that it has garnered little attention in conventional accounts of record production aesthetics and Ottum's chapter thus constitutes a unique and valuable entry point for its consideration in the Music Production Studies arena.

Part 1's tour of genre-focused music production is completed by Robert Toft's engaging discussion (Chapter 7) of classical record production and the challenges facing performers who wish to record works from the past in an historically informed manner, utilizing acoustic spaces similar to those in which the music would have been heard originally. In particular, Toft discusses the approach taken by a group of recordists to creating historic soundscapes through modern studio techniques which create within the listener the sense that they are sitting in the same small room as the performers. With its detailed focus on pre-production preparation, production practices, and post-production editing and mixing, Toft's chapter offers much food for thought for the classical recordist.

In Part 2 the focus moves to issues concerning music production and technology, beginning with Pat O'Grady's exploration of the emergence of digital signal processing technologies designed to emulate famous analogue production tools. O'Grady's aim is to turn a critical lens on these recent shifts that have seen the proliferation of software simulations that look backward to earlier media for music production. In particular, O'Grady offers an insightful analysis (with particular reference to UAD) of industrial discourses within music production arguing that such technologies do not democratize the field of production practice, rather they play into problematic discourses of value and aesthetics, functioning to reinforce the social order of the field of recording.

Rob Lawrence's contribution, discussing the parameters and working practices around immersive audio applications, is of great interest in this volume. His chapter comes at a time when spatial audio has been gaining traction in many differing applications outside of the normal movie theatre or high specification recording context. New innovations, such as virtual reality, augmented reality and binaural headphones, now place an emphasis on the directionality of sound, so much so that the term "immersive" has replaced the once, rather trite term "spatial" when referring to surround technologies. Lawrence's work looks at the processes in production and mixing to best make use of these new emerging platforms.

Continuing the focus on this topic area in Chapter 10, Justin Paterson and Gareth Llewellyn identify key production practices, which will help define the way in which producers and engineers work within the 3D audio environment and challenge existing preconceptions. For example, the notion that the stereo field provides approximately 60 degrees of width has been with producers for generations and a set of standard procedures have become widely adopted, such as the phantom mono presentation of the vocal. In this chapter, the authors consider the paradigm of translating stereo techniques to 3D and the issues therein. This is a fascinating discussion of a form of production presentation that exemplifies the cutting-edge of innovation within the field.

Thomas Brett's chapter (11) focuses on creative practice within the DAW and in particular the workflows of electronic dance music producer-DJs based in Ableton Live. Brett views Live as a "self-contained techno-musical system" which reveals creativity in the context of electronic music production as a "recombinant art whose elements are modular and perpetually fungible." Such analyses of contemporary of DAWs in terms of the consequences of their constituent environments for musical thought are becoming increasingly common as their integration with music production practice becomes ever more seamless and transparent. Brett's chapter is of value not only in terms of its detailed evaluation of the workflows engendered by one particular DAW platform, here considered in terms of problem solving, but also in its revelations regarding the thinking of those high profile contemporary electronic music practitioners who work with it.

In Chapter 12 Paul Ferguson, Zack Moir and Gareth Dylan Smith offer a timely reflection on the position of research and development into solutions for real-time remote collaboration between musicians and producers. The chapter begins with an informative historical overview of the evolution of the technology and its current state of development, following which the authors examine the potential technical issues that may affect the quality and "usability" of recorded material garnered during remote recording sessions. They conclude with a discussion, drawing on their own practical experience, of the ways in which any technical and interpersonal issues might impact upon the musical experience of musicians and producers and the potential for meaningful creative collaboration using such technologies.

In the concluding chapter of Part 2, Austin Moore investigates processors and other audio equipment placed into the signal chain during the recording phase to print and impart a unique sonic signature. Through several key interviews, Moore's findings ground and display different techniques employed to add character to key recordings. Different instruments demand different processors and these are discussed in some detail through interviews. Rounding off this chapter, Moore relates interesting views on the mix from his interviewees.

Part 3's transition into conceptual ideas informing production practice begins with Stace Constantinou's thought-provoking examination of the nature of what he calls the "audio object." In particular his essay considers the manner in which this object is perceived in relation to the affordances of the DAW production environment (hence he may be profitably read alongside the chapters by Marrington and Brett in this volume) and situates his theoritizing in reference to ideas of Pierre Schaeffer. In addition to his drawing on commentary derived from interviews conducted with a number of contemporary electronic music practitioners, the author's thinking on the subject is also informed by his own extensive practice as a sound engineer, producer and composer in a variety of media.

Robert Wilsmore's essay on co-production in Chapter 15, which also anticipates his forthcoming monograph for the this series (with Chris Johnson), aims to establish a framework for the study of collaborative practices in music production through the identification of a series of types of practice that occur from the operations of a few individuals working closely together to the contribution of the whole world to the entirety of produced music. Following an initial discussion of the dominance of the singular producer in the literature and iconography of recorded music, Wilsmore offers four overarching types of collaborative practices (types of co-production) that break with this ideology of the singular producer and demonstrates the various ways in which joint authorship operates.

In Chapter 16 Alex Harden provides a valuable discussion of record production in support of narrative readings of recorded popular song. In particular he builds upon David Nicholls' (2007) work on narrative theory to support an assertion that narrative readings are formed from the combination of phonographic aspects such as sonic design and staging (after Lacasse, 2000); a persona enacted by the singer and lyrics (after Moore, 2012); and, a musical environment. In addition Harden draws upon Monika Fludernik's (2010) theory of "narrativisation" to propose that a track affords to the listener ways of narrative interpretation. With these theoretical perspectives in mind Harden then proceeds to focus on the role of record production through analyses of phonographic aspects in a range of contemporary popular repertoire in reference to four archetypal narrative parameters: setting, characterization, event-sequencing and point of view/mood.

Part 4 begins with Alex Baxter's chapter (17) discussing music production education and a potential change that could be necessary given the abundance of powerful and commonly available technology. Baxter discusses whether the approach to teaching technology and practice now needs to be reviewed in an age where the tools of the trade are literally ubiquitous. His chapter analyses the requirements in some detail and provides ideas for the future of music production education.

In Chapter 18 Russ Hepworth-Sawyer, having now reached a certain age himself, explores, via the roundtable format, the experiences in the late 2010s of the mid-career audio professional. Interviewing three professional engineers and producers, Hepworth-Sawyer explores the challenges such engineers have faced in establishing themselves in the modern music industry. The interviewees also comment on how they gained their skills and speculate on what the future holds for younger engineers coming up behind them.

Finally, Kallie Marie's chapter entitled "Conversations with Women in Music Production," brings to the fore the experiences of female producers and engineers in the field of music production. Through extensive interviews with some leading US figures, Marie seeks to begin a larger conversation on the topic for the benefit of the whole industry.

Hence, like *Mixing Music*, the first book in the *Perspectives on Music Production* series, the current volume fields a wide array of topics and practitioner perspectives with a view to providing a mosaic-like overview of the evolving field of Music Production Studies. Certain of the chapters presented herein cover conceptual and theoretical questions regarding the art form of production, while elsewhere others offer accounts of creative production practice that are both experiential and anecdotal. We are confident that, viewed as a whole, these various perspectives fulfill a key objective of the POMP series, which is namely to provide academics, professionals and other interested parties with a means for gaining a deeper understanding and appreciation of the situation of music production today.

Future books in the series, and calls for chapters can be viewed at www.hepworthhodgson.com

Part One

# Music Production and Genre

## 2

# Socio-Technical Networks and Music Production Competencies

## Towards an Actor Network Theory of Hip Hop Beat-Making

## Matt Shelvock

### INTRODUCTION

This article provides a first step towards using Actor Network Theory (ANT) to analyze the core competencies of hip-hop beat production. To do this, I address the genre's most foundational creative competencies and technologies, and I evaluate them as a network, that is, as a group of inter-related objects which engage in some type of back-and-forth exchange. As Bruno Latour (1996: 4), one of ANT's founders, states, the concept of a network is analytically useful because:

> A network notion implies a deeply different social theory: it has no a priori order relation; it is not tied to the axiological myth of a top and of a bottom of society; it makes absolutely no assumption whether a specific locus is macro- or micro- and does not modify the tools to study the element "a" or the element "b"; thus, it has no difficulty in following the transformation of a poorly connected element into a highly connected one and back.

As it stands, the majority of investigations of ANT and music production focus entirely on social contexts, or they combine actor-network theorization with insights derived from so-called *Social Construction of Technology* (SCOT) work, even though ANT and SCOT demonstrate conflicting methodological requirements. And neither of these approaches resembles actor-network theorization in other fields, such as in science and technology studies (STS) or human-computer interaction (HCI; Latour 2005: 1). Moreover, these studies often adopt an overly broad definition of the word *production*, a term which, for many scholars, means "everything done to create a recording of music" (Hepworth-Sawyer and Hodgson 2017: xii). As a result, formal analyses of *production* invoke discussions of anything from Greek mythology to critical theory,[1] but these investigations often neglect the various musical competencies actually required to produce records.

Fields that analyze music production still require a more comprehensive record of most studio activities in order to promote more accurate investigations. As mentioned in the past by Russ Hepworth-Sawyer and Jay Hodgson (2017: xiii), two of this volume's editors:

> The place of traditional academic and scholarly work on record production remains clear in the field. However, the place of research and reflection by professional recordists themselves remains less obvious. Though music production studies tend to include professional perspectives far more conscientiously than other areas of musical study, their contributions nonetheless are often bracketed in quiet ways.

Perhaps this is why the subfield of music production studies (MPS), announced in 2017 in *Perspectives Music Production*, has recently seen participation by so many recordists, scholars, and hybrid recordist/scholars, such as, to name only a few, myself, William Moylan, Jay Hodgson, Russ Hepworth-Sawyer, Mark Marrington, Martyn Phillips, Ruth Dockwray, Dean Nelson, Phil Harding, Gary Bromham, Rob Toulson, and Joshua D. Reiss (and many more, of course, in this volume). The common thread that unites these authors' contributions is their direct focus on socio-technical production competencies in the way that existing music-making cultures understand them, rather than the *arm's-length* approach adopted elsewhere in literature.

In order to address the many technical and creative misunderstandings currently propagated in other fields of music research, in this article, I provide increased clarity regarding the meaning of the term *production*, as used within existing music-making cultures. In this way, I hope to support increased accuracy in the future investigations of record production. Also, given the apparent constraints in other formal investigations of ANT and music production, and using hip hop as an example, the current article intends to clarify the following:

1.  the most pertinent interaction between technologies and humans in music production: the creative *operation* of these technologies (i.e., record creation techniques);
2.  the network characteristics of hip-hop beat production, in consideration of its primary *actants*, technologies, and competencies; and
3.  an improved model for future research into socio-technical dimensions of ANT and music production competencies.

## Background

Recently, a few authors have attempted to apply ANT in discussions of record production, yet, curiously, most authors minimize the role of music-making (and playback) technology in their analyses. This does not satisfy the requirements of actor network theorization, which was invented for analyzing network interactions between scientific discoveries, technologies, and humans. In order to sidestep this requirement, some researchers pair ANT with insights from a framework known as the social construction of technology (SCOT). However, in the field of science and technology studies (STS), where these methods were developed, SCOT and ANT are considered to have conflicting methodological stipulations. For instance, as Simon Zagorski-Thomas (2014: 4) writes:

> Actor-network theory (ANT), the social construction of technology (SCOT), and systems approach to creativity can be applied to various aspects of record production. This allows us to discuss both the technology, and the ways that humans engage with it.

Despite Zagorski-Thomas's comments, most available discussions of ANT and production provide sparse detail regarding human engagement with music-making technologies, and a more generalized sociological investigation typically replaces any substantive consideration of recording practice, that is, the specific competencies needed to produce a record. This approach, however, does not resemble how scholars use ANT in other fields, such as science and technology studies (STS). Bruno Latour (2005: 1), one of the founding scholars behind actor network theorization (along with Law and Callon), explains,

> What I want to do in the present work is to show why the social cannot be construed as a kind of material or domain and to dispute the project of providing a "social explanation" of some other state of affairs. Although this earlier project has been productive and probably necessary in the past, it has largely stopped being so thanks in part to the success of the social sciences. At the present stage of their development, it's no longer possible to inspect the precise ingredients that are entering into the composition of the social domain. What I want to do is to redefine the notion of social by going back to its original meaning and making it able to trace connections again.

Indeed, any study that purports to align itself with the goals of ANT must avoid assigning increased agency to the network's social factors, because technology also exerts its own agency within any socio-technical network. As noted by Strachan, a music scholar whose work more closely resembles the STS application of the theory (2017: 8),

> [D]igital technologies are more than mere tools; they are active and enmeshed within the creative process in a central way. In actor network theory terms, they are a key part within the socio-technological-human networks that go toward making music in the post digitization environment. Actor-network theory views both humans and objects as non-hierarchical actors (or actants) within sociotechnical networks.

This "dispersal of agency" between humans and technology is a central premise of ANT. However, as it stands, most published accounts of ANT and record production assign *increased* agency to humans, and social contexts, during record production. This approach contradicts the majority of published ANT research available in other fields of social and technological research.[2] As information systems researchers Tatnall and Gilding state:

> Actor-network theory extends ethnography to allow an analysis of both humans and technology by using a single register to analyse both, so avoiding the need to consider one as context for the other.
>
> (2005: 963)

Most scholarly accounts of ANT and music production seem to abruptly end at the point where recordists begin to *use* studio tools of any kind.[3] I find this curious, given that these accounts purport to study recordings, the production of recordings, and recording technology *per se*. To make matters worse, many of these scholars even go so far as to claim that they elucidate the usage of production technology by recordists within the abstracts of their articles and in the introductions they write, even though such analyses are nowhere to be found within the documents themselves.[4]

I do not believe this body of work is intentionally misleading, but many of their descriptions of recording practice do not resemble any standard studio operations I have ever encountered. I think the problem instead lies with a lack of available scholarly literature that explains production methods *per se*, even the most basic music production competencies, described by those who produce, engineer, and create music. As a result, ontological and practical misunderstandings about record production abound in scholarly literature. Recently, for instance, Hodgson described the need for more accurate research on record production (2014: 1):

> Understanding [recording practices] – even acknowledging that they exist – will provide a tremendous boon for recordists as they conceptualize their work, just as it will provide analysts with a more concrete and empirically accurate understanding of what they professionally research.

Already, only three years later, this lacuna is being addressed by researchers who contribute to the field known as *music production studies* (MPS), which officially formed in 2017.[5]

In order to satisfy the methodological requirements of ANT, scholars must resist the temptation to *socialize*, or personify, music-making technology. For instance, an LA2A compressor has no social history, as it is not sentient. In an ontological sense, LA2As do not congregate, they do not share ideas, and, thus, they cannot identify with any social construct.[6] In order to sidestep this problem, many students of music production combine ANT and SCOT methods. However, ANT and SCOT have opposing analytical requirements – the most obvious being that SCOT affords increased agency to social contexts, whereas ANT acknowledges the affordances of both technological and social *actors* – and in these specific cases, scholars use SCOT to contextualize studio technologies as social entities. This methodological misstep disqualifies these studies as actor-network theorizations (Tatnall and Gilding 2005: 963). In fact, Latour and his colleagues, Law and Callon, created ANT specifically to address the perceived shortcomings of SCOT, which has seen less scholarly attention since the early 2000s (apart from by the theory's founders, such as Bijker, et al.).

This research also does not consider the perspectives, let alone the day-to-day activities, of recordists themselves, whose point-of-view remains notably absent from most research on record production. This is also a critical error, as ANT is designed as a method for extending ethnographic research. Scholars must thus consider the working conditions of recordists themselves, to satisfy the methodological requirements of the theory (Tatnall and Gilding 2005: 963). In fact, until recently, most formal investigations of record production were primarily conducted by people who neither produced, engineered, nor otherwise contributed to the production of records, or they were undertaken by researchers with very limited, and historically confined, exposures to studio training (few researchers seem to have any release credits, whether amateur or professional).[7] Some may argue that this reflects the sociological tradition of *blackboxing* technology, that is, reducing a machine's complexities when discussing systemic socio-technical interactions. However, this method is clumsily applied in many publications on record production. One of the founding scholars of ANT, Bruno Latour (1999: 304) argues that *blackboxing* refers to

> the way scientific and technical work is made invisible by its own success. When a machine runs efficiently, when a matter of fact is settled, one need focus only on its inputs and outputs and not on its internal complexity. Thus, paradoxically, the more science and technology succeed, the more opaque and obscure they become.

Scholars often misrepresent the "internal complexity" Latour describes in the preceding passage, when they sidestep any discussion of how production technology is actually used, let alone how it works.[8] And, to be clear, standard music production competencies, such as mixing and mastering, are neither scientific nor purely technical (Shelvock 2017b: 10). As a result, the musical *inputs* and *outputs* of music production processes, and the musical competencies required to complete these tasks, are effectively ignored in this type of research (Shelvock 2017b: 1). Interestingly, many of these scholars attempt to reproduce the methods of Bijker et al. (2012: xxi), whose goal is to "focus on what social groups and actor networks actually say and do with technology". However, most available research on music production ignores the operational strategies recordists use and the ways these production competencies relate within a socio-technical network, such as techniques for using compressors, equalizers, stereo alteration devices, spatial effects, and so on. I argue that research which analyzes human interactions with these technologies, or what I call "production competencies", is more closely aligned with the majority of published ANT research than with work being done in any other fields (outside of music production studies, of course).

Available investigations of production spaces have certainly led to a number of thought-provoking social, cultural, and historical insights. However, many of these publications include titles which highlight the language of recordists without actually mentioning even the most basic record production techniques in the text. These investigations cannot benefit researchers (or practitioners) who focus on broad production competencies, such as tracking, mixing, and mastering, and they are of little use to those who study more specific competencies, such as sampling, *beat-slicing*, or arrange-window sequencing, for example. Given the amount of time it has taken scholars to recognize existing record creation techniques in a formal field of inquiry, it is safe to say that the field still has much work ahead of it.[9] In fact, in this article, I am suggesting that scholars adjust their analytic approach to studying socio-technical interactions so they include the most impactful area of human and technological interaction during the production of records: the *operation* of production equipment by recordists. As it stands, much more research which investigates how studio technologies become technological *partners* with recordists as they create recordings is needed. This requires scholars to elaborate music production competencies, as the newly formed field of music production studies (MPS) is beginning to.[10]

## Methods

In order to advance current research on music production and ANT, this article includes an ethnographic dimension. As stated earlier, a key feature of ANT is that it provides researchers an opportunity to extend the boundaries of ethnographic research. Like other research in MPS, I employ a practice-based method to inform my published work.[11] I have spent several years producing hip-hop tracks (as kingmobb and M'NK x Mobb), which are available on streaming services such as Apple Music, Spotify, Google Play, Tidal, and others, via ghosttape records (https://soundcloud.com/ghosttape-records), in addition to working as a mastering engineer. Indeed, beyond just for the purposes of this article, as a scholar of music production, my past experiences as a recordist comprise a crucial component of my research activity.

An important distinguishing characteristic of MPS, compared with other production literature, is that researchers are usually required to possess the production skills they critique.[12] This does not conflict with the methodological requirements of ANT, which

> unlike action research . . . is not concerned with emancipation of the researcher or practitioner and is not focused on making us better at developing information systems.
>
> (Tatnall and Gilding 2005: 963)

Beyond ANT, I would argue that not just MPS but music research in general requires input from those who produce records. I say this because a common argument against studying even routine music production competencies, for instance, is that the topic requires far too much specialist knowledge from researchers (Elsdon 2014). But the production techniques used to create a record are, of course, in no way obliged to avail themselves to music researchers who do not understand them. It would be far more logical instead for scholars to adapt their studies to prevalent musical practice, such as the collection of quasi-compositional techniques associated with music production.[13]

Of course, it should be noted that music production operations are not actually secret knowledge and have not been for a very long time. However, these operations were absent from formalized research until recently, causing many music scholars to believe the obverse. Now, at least 26 colleges and universities in the UK are accredited by professional recordist personnel in an initiative known as JAMS, or Joint Audio Media Education Support. The stated purpose of JAMS (2018) clarifies why this matters to the kind of music production study I'm modeling and proposing here:

> Assessed by industry professionals, we engage with education primarily through our Course Accreditation process which is designed to endorse relevance, quality and continuing innovation throughout the student curriculum.
>
> This not only equips students for a rapidly changing industry, but also ensures that many years of industry experience are not lost to future generations.
>
> JAMES continues to promote supportive links between industry and education.

Moreover, highly impactful conferences, such as *Innovation in Music* (2013, 2015, 2017), demonstrate that tenured and tenure-track researchers, music industry personnel, studio equipment manufacturers, and even *famous* recordists are all eager to contribute to music production research within higher education.

In the spirit of educational initiatives likes JAMES, I will now outline some of the crucial components of professional hip-hop beat-making using ANT terminology. In section 3, I provide a case study for the creation of a song available on Spotify, iTunes/Apple Music, Tidal, Google Play, and so on.

## HIP-HOP PRODUCTION AND THE ACTOR NETWORK: A TAXONOMY OF ACTANTS

In order to begin evaluating hip-hop beat-making as a network, I draw from the production session for a track I recently composed, called "Rinzai", under the kingmobb moniker for ghosttape records (GT004 2017).

This song is available on all major streaming platforms. Before I proceed with this, however, I should first provide a brief overview of the relevant "actants" existing within the network of hip-hop beat production, namely (i) sound sources; (ii) hardware controllers; (iii) virtual collaborators; and (iv) digital music creation environments, sound-shaping tools, and *sound* itself (i.e., psychophysical/Gestalt impressions); and (v) the mastering engineer. I explain each, in turn, in the following sections.

## Sound Sources: Samples and Synthesis

Hip-hop instrumentals usually rely on a technique known as sampling (Shelvock 2017a: 173). Historically, samples were taken from previously released recordings, and this technique always impinges on the intellectual property of other musicians without legal recourse for clearing their use. To this day, sometimes producers are taken to court for alleged copyright infringement.[14] However, depending on the nature of the sampling techniques used, and the source that musicians sample, many sample-based producers did not face litigation throughout the golden age of hip hop (c.1980s–1990s). This is because, in many cases, musicians acquired licenses to use samples through legal channels, known as sample clearance.

Musicians who did not follow sample-clearance protocols, however, would often utilize shorter samples (i.e., less than one second), which were much harder to trace to the original source. These samples are known as *one-shot*s.[15] One-shot samples are small audio files that typically contain a kick drum or snare drum being played only one time, for instance. Producers may also create one-shot samples from any conceivable musical sound in order to develop new textures – a practice reminiscent of electroacoustic compositional traditions. In addition, one-shots are commonly reversed, edited, or re-pitched in a number of ways before being loaded into some type of sample-triggering device, which I discuss later. Noah "Shebib" 40 (OVO Sound, Drake, Lil' Wayne, Action Bronson) remarks on his own use of samples, stating,

> I use 'em sort of as a tool I guess, you know? I'm a fan of publishing, so I try not to sample *too* much, but it's a tool in your arsenal. There is something to be said about not sampling, and if you say "I'm not going to take *that* from someone else", I say "great, congratulations – I'm happy for you". But, you *can* sample, and what if a sample makes the song better? What I try to do is grab something, take a loop out of it, then flip it, reverse it, distort it, or chop it up. Then I ask myself "Did I create something?"
>
> (Quoted in *Pensado's Place* 2014)

Sample-based techniques are even more prominent today than they were throughout the 1980s and 1990s due, in part, to the overwhelming prevalence of what are commonly known as "royalty-free" samples. In more precise terms, these are samples which are free of all licensing restrictions.

While most available scholarship discusses sampling as an act of social resistance, this is quickly changing as cloud-based sample databases become more popular. In fact, a number of dedicated recording labels now exclusively make recordings for producers to sample, without any fear of copyright litigation. Splice, for instance, has signed 71 of these labels, and some prominent examples include Loopmasters, CAPSUN Pro Audio, and Class A Samples. These companies provide users with audio samples in the form of one-shots, as well as longer musical passages that producers can legally incorporate within recordings as they please.

In addition, hip-hop producers use synthesis to create and manipulate sounds. For instance, the Roland TR-808 found common use in earlier hip-hop scenes, as it still does today.[16] These devices use an internal synthesizer to provide a number of synthesized percussion sounds, which could also be programmed into sophisticated musical sequences.

Synthesis is similarly common in G-Funk music, where producers often emulate Parliament Funkadelic's signature high-pitched synthesizers. In addition, producers from this scene, such as Dr. Dre, employ live instrumentation in their recordings. However, the prevalence of live instrumentation in G-Funk recordings is often overstated in scholarly literature. For example, Adam Krims (2000: 74) writes that G-Funk is:

> A style of generally West Coast rap whose musical tracks tend to deploy live instrumentation, heavy on bass and keyboards, with minimal (sometimes no) sampling and often highly conventional harmonic progressions and harmonies.

Yet, this is not the case. For instance, three of the genre's most famous tracks are based on a recurring sample, such as "Nuthin' but a 'G' Thang" by Snoop Dogg (1992), "Regulate" by Warren G and Nate Dogg (1994), and "Let Me Ride" by Dr. Dre (1992).[17]

Another hip-hop subgenre that commonly utilizes synthesis is Southern Rap, otherwise known as the *Dirty South* sound. Artists who rose to prominence in the early 2000s, such as TI, Ludacris, Rick Ross, Lil Wayne, and many others, chose to use sampling less frequently than previous artists (even though sampling still occurs). In particular, the sounds of Paul Wall, Z-RO, and the Screwed Up Click were particularly influential on current emcees outside of the American South, including A$AP Rocky and Lil Uzi Vert.

Today, the most common synthesized sounds in hip hop (and closely related subgenres, such as trap) are 808-style sub-basses, 808-style snare drums and claps, and 808-style kick drums, as well as plucks and pads (Shelvock 2017a: 176–177). "Plucks" and "pads" are also common. Equipment manufacturers and sound database designers (i.e., Splice, Noiize) use the term *pluck* to describe any synthesized sound with a staccato delivery. Pads, in contrast, are sustained sounds that typically provide background harmony. Interestingly, many sampled sounds are layered with synthesized sounds. For instance, producers commonly layer a saturated (and lo-fi-sounding) snare drum sample with a synthesized 808 snare drum. I have outlined some of these sounds, as heard on Travis Scott's "Butterfly Effect" (2017), in the following table:

| Travis Scott "Butterfly Effect" (2017) | |
| --- | --- |
| Time | Sound |
| 00:00 | 808 sub-bass |
| 00:00 | Synthesizer plucks* |
| 00:13 | Kick drum |
| 00:14 | 808 snare drum (layered with clap) |

*The producer applied a large amount of reverb and delay to these synthesizer plucks and also ensured that the notes sustained for a long duration. This also causes the main synthesizer melody (and counter-melodies) to sound like a pad.

## Hardware Controllers

In today's studios, producers make music using a number of hardware controllers. Some of these controllers are vintage synthesizers and samplers (such as the TR-808 or Korg EXS), but, more often, producers prefer to use modern hardware controllers. Even a simple, inexpensive device, such as Novation's popular Launchkey Mini, is capable of controlling plug-ins, synthesizer patches, and DAW operations while also providing an interface for MIDI input via 24 piano keys and 16 trigger pads.

For the most part, producers use these devices to control one-shot samples (as described in section 2.1). Producers can assign kick drum, snare, auxiliary percussion, and hi hat sounds to the keys on the keyboard, or a trigger pad matrix, for example, and trigger these samples manually. From this point, producers simply record themselves "playing" different drum parts on their controller, until they are satisfied with the result.

Another technique, known as *slicing*, involves taking a longer sound sample and assigning small segments to adjacent keys. Depending on the sample, and the desired result, producers may *slice* (i.e., cut) a sample at (i) a recurring rhythmic interval; (ii) at each transient; or (iii) by defining a custom time region within the sample. Once producers transform the original sample into discrete segments, the resulting segments are assigned to particular keys on a connected keystation or to trigger pads on a connected controller. At this point, the producer has effectively created a new musical instrument from the original sample. When used to produce beats, this process is reminiscent of the compositional techniques of Pierre Schaeffer, who is known for using tape recordings as an instrument in his innovative *musique concrète*.[18]

In addition to slicing, producers can also take one sample region and assign it to a keyboard (or another controller) so that the sample's pitch is adjusted by playing different keys. In fact, this process builds upon the design features of classic analogue sampling devices, such as the Mellotron, which used audio samples recorded onto tape, rather than digitally stored audio data.

With today's samplers, a chosen sample will typically play at its original pitch when middle C is played. When another key is played, the sample will sound at a higher or lower pitch, depending on which key is played. For instance, the C# above middle C will play the chosen sample at one semitone higher than it was recorded. This method allows producers to experiment with radically new timbres, because interesting sonic artifacts emerge when samples are sped up, slowed down, or time-stretched.

## Virtual Collaboration

Today, electronic, EDM, and hip-hop musicians often use cloud-based music production services, such as Splice, Noiiz, and Loopcloud, to source the raw materials they will use for their productions. These services provide access to large databases of high-quality license-free audio samples, synthesizer presets, plug-ins, and even complete, or incomplete, DAW session files. In effect, these services make professional music production materials readily available to anyone who has a subscription. Services such as Splice make high-quality audio samples available to all producers for a small monthly fee, and 71 record labels provide this database with royalty-free sounds.

As noted in previous literature, cloud-based services are even redefining the notion of *jamming* in the current digital music production scene. In a sense, those who use samples from cloud-based services participate in a form of technologically enabled collaboration (Koszolko 2015). When users adapt samples to their musical works in progress, they are also collaborating and effectively *remixing* the work of other recordists and musicians who create these samples.

Samples available on these services are even heard in chart-topping music productions. For instance, Kendrick Lamar's producer on the song "Feel", known as Sounwave, used a prominent Loopmasters sample on his April 2017 album, *Damn* (*COF_125_Am_LaidOut_Underwater.wav*). In fact, I unknowingly used the same sample in a song that debuted in a conference paper at the April 2017 meeting of the American Musicological Society in Toronto. This song was mastered, and distributed to, Spotify, Apple Music, Tidal, and other streaming services (please see "Cut Deep" *KNGMXBB x II*, 2017).[19]

## Digital Music Creation Environments, Sound-Shaping Tools, and Gestalt Impressions

Music producers use a variety of digital music-making tools that provide recording, editing, and MIDI capabilities. In hip hop, the most important of these tools is known as the "Digital Audio Workstation" (DAW), a type of modular software that allows the inclusion of sound-shaping *plugins* (Shelvock 2017a: 172). As noted in previous literature, DAWs establish a framework for creative action (Mooney 2011: 144; Marrington 2016: 55).

The most common DAWs for hip-hop production are Ableton and Logic, because they offer streamlined beat-slicing tools.[20] In Ableton, any audio segment may be selected, and distinct zones of this segment are assigned to sequential MIDI keys. A number of sample-slicing algorithms can be used, which gives users the ability to *cut* the original audio segment at a predefined time interval (i.e., 1 bar, ½ note, ¼ note, ⅛ note, and so on), at each transient, or by defining custom regions with Ableton's *warp markers*. Logic X offers a similar option (when flex-editing is enabled), but users are limited to shorter length samples. Once samples are assigned to MIDI notes, users typically connect a device capable of providing MIDI input, such as the hardware controllers discussed earlier.

In addition to recording, storing, sequencing, and exporting sound, DAWs offer a number of sound-shaping tools. Some of these tools alter the sequence and timing of musical events, such as tools associated with audio editing. Another set of tools causes measurable timbral, dynamic, or spatial alterations, such as tools associated with mixing or signal processing. Modern DAWs offer the ability to configure audio in both of these ways.

Another commonly used editing tool in hip hop involves the use of *groove templates*. This is because, as researchers have noted, hip hop and R&B use a peculiar microrhythmic strategy in comparison to other genres (Danielsen 2010; Shelvock 2017a). In both Ableton and Logic, for instance, users can apply an external groove to an audio segment or MIDI sequence. This requires producers to select a groove from a library, or extract a groove from an audio or MIDI segment. At this point, the groove may be assigned to any MIDI or audio event. This alters the timing to resemble the selected template. A common use for this technique includes extracting the groove from a famous piece, such as J-Dilla's "Last Donut of the Night" (2006). In this case, producers would create a musical segment, such as a drum sequence or melody, and apply the groove from the external track (i.e., J-Dilla's). In this case, the original audio would *shift* its rhythmic characteristics according to the transient structure of the groove template. And while groove templates are common, producers may also edit passages through mouse/keyboard interaction (i.e., by *hand*). This can be done to add, or remove, rhythmic *flaming* caused by microrhythmic discrepancies in the kick drum and bass instruments, depending on the desired aesthetic goal.[21]

DAWs also enable producers to engage in sophisticated sound-designing activities through the use of signal-processing devices. As noted in previous literature, sound design in hip hop, as well as EDM and electronic music, blurs the line between production and mixing (Shelvock 2017a: 177; Devine and Hodgson 2017: 157). This is because those who create hip-hop beats use mixing tools throughout the production process, and many create their own finalized mixdowns before mastering occurs.

The use of mixing tools alters the physical properties of audio signals in quantifiable ways, and as mentioned in my previous research, a record's sonic characteristics elicit psychophysical reactions in listeners (Shelvock 2015, 2016). In fact, Gestalt theory can explain the sonic impact of common mix techniques, such as layering, filter automation, spatial processing, and bus processing, insofar as these techniques engage a listener's psychoacoustic senses (Shelvock 2016, 2017a).[22]

In fact, the ways that other hip-hop *mixes* engage a listener's psychophysiology, for instance, are taken into consideration by producers during the production process. Often, engineers listen to reference recordings in comparable genres, and subgenres, in order to aid them in designing a record's sonic characteristics. Grammy-winning hip-hop mix engineer Matthew Weiss (Snoop Dogg, 9th Wonder, !llmind) states that when mixing hip hop, one must consider both

> [the] genre's aesthetic and the producer's intentions. Listen to the record and find the "leads". Vocals is a given. With the groove it's usually drums, but often the bass, and sometimes a rhythm guitar or keyboard. Figure out what's most important to be heard and make sure those elements shine. Everything else should work towards supporting the leads. Understanding genre can go a long way here.[23]

Indeed, producers and engineers routinely reference their in-progress productions to other available records in the genre. The *mix* mix characteristics heard in these diverse subgenres, as established by other recordists, is also a crucial component of the hip-hop production network. Even though a mix is neither *human*, nor *technology*, it is a preserved sonic design that exists through the performance of a technological competency, and, as a result, mixes should receive consideration within the actor-network framework discussed in this article.

## Mastering

Mastering engineers (and their chosen technologies) are another critical actant within the network of hip-hop beat production. Scholars have been slow to acknowledge the role of audio mastering in the music production process, even though mastering has been performed since the 1940s (Shelvock 2017b: 29).[24] Simply put, mastering is the final chance to address a record's mix characteristics, such as spectral balance, micro- and macro-dynamic levels, spatial characteristics, and stereo configuration (Shelvock 2017b: 1).

In the case of instrumental hip hop and beat creation, mastering engineers and producers typically build a strong working relationship. This is because beat producers often create and mix their own material. Once mastering engineers receive a record, they perform a type of *creative* and *technical* quality assurance. On the creative side, mastering engineers routinely reconfigure the sonic design of records, while, at the same time, they ensure the record meets the standards of various output formats (i.e., streaming or physical copy) and codecs (i.e., M4A, FLAC, and others). Mastering engineers also perform a quasi-curatorial duty insofar as they also ensure records sound like other available records (Shelvock 2017b: 127).

## A CASE STUDY IN INSTRUMENTAL HIP-HOP BEAT PRODUCTION

I will now describe the production process for "Rinzai", which is available on Spotify, iTunes, Tidal, Google Play, and others (via ghosttape records) in a case study format. As mentioned in 1.2, ANT extends ethnographic research methods (Tatnall and Gilding 2005: 968). Thus, ANT aligns with research from the field of MPS, where practice-based and practice-led research are common (companion audio examples are available on my personal website: www.mattshelvock.com/).

### Creating the Primary Hook/Theme from Samples

In order to create the main hook for "Rinzai", I imported a sample from Splice. Quite simply, my selection criterion was a melody or sequence that I enjoyed. I performed a keyword search using terms such as "classic hip hop", "Wu Tang", and "world music". After listening to a number of samples, I selected one called *Bpm86_G#_36Chambers_EthnicPluck* (on Splice/Prime Loops). This sample is a nod to the production style of the RZA of the Wu Tang clan, as the filename references their popular record (1993).

In Ableton, I used the "slice to new MIDI track" feature to convert the original sample to a number of discrete segments, which are then assigned to adjacent MIDI keys. Ableton provides the ability to automatically slice an audio file at the onset of each transient, at a predefined rhythmic interval, such as every 1/8th note, or, by defining customized regions of the audio sample. In this case, I used the transient slicing feature (Figure 2.1).

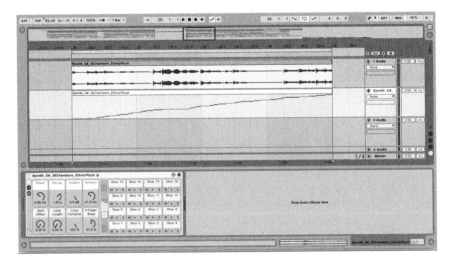

**Figure 2.1** Top row sample = original sample; Bottom row sample = original sample sliced at every transient and converted to MIDI. Sample: Bpm86_G#_ 36Chambers_EthnicPluck (on Splice/Prime Loops)

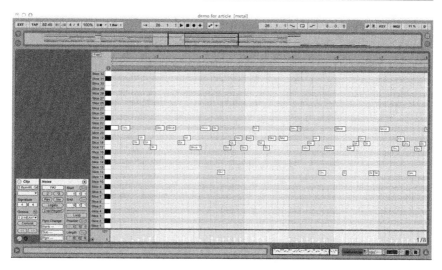

**Figure 2.2** New hook for "Rinzai", using the instrument created by slicing a sample called Bpm86_G#_36Chambers_EthnicPluck (on Splice/Prime Loops). Note how MIDI events do not always coincide with the Ableton's time grid.

Once these discrete audio segments were mapped to MIDI notes (via Ableton's Drum Rack), I created a new sequence by triggering these *slices* via a MIDI keyboard. To do so, I simply recorded myself *jamming* or *improvising* using slices from the original sample.[25] In addition, many of the MIDI events triggered do not coincide with Ableton's time grid. And although editing occurred at a later phase, many of these late rhythms are preserved in the final version of this song. The resultant sequence is pictured in Figure 2.2 (audio example 2.1).

## Establishing the Groove

Next, I began to record drum parts to accompany the sample. To do this, I recorded myself playing snare drum and kick drum passages separately. Ableton's loop-record feature also allowed me to audition a number of different kick and snare patterns, until I settled on a groove that I liked. In order to capture a classic 1990s' hip-hop vibe, as readers will also notice, occasionally the snare and kick drum stray from the digital time grid (Shelvock 2017a: 175, Figure 2.3, audio example 2.2).

After establishing a drum groove, I began to record bass parts using a MIDI keyboard (audio example 2.3). The keyboard controlled a popular synthesizer plugin known as *Massive* by Native Instruments. Listeners may notice that the bass occasionally occurs many milliseconds after the kick drum. In the following example, the bass occurs approximately 8.25 ms after the kick drum (Figure 2.4). This technique creates more sonic separation between low-frequency dominant instruments by causing them to occur at slightly different points in time. However, the delay is also short enough (i.e., much faster than a 64th note) so that listeners hear both

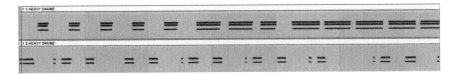

**Figure 2.3** Layered kicks and snares, with microrhythmic variation. The top track is the snare drum (SWUM_snare_14.wav; PL_BBERNB_Snare_5.wav; and one of Ableton's prepackaged 808 snares). And the bottom track is the kick drum (010_-_AT_Elastic_-_Kick_D.wav; RP6_Kick_1.wav).

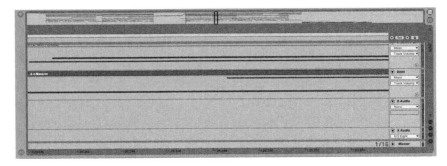

**Figure 2.4** An image that demonstrates microrhythmic discrepancy between the kick and bass. The bass occurs approximately 8.25 ms after the kick drum.

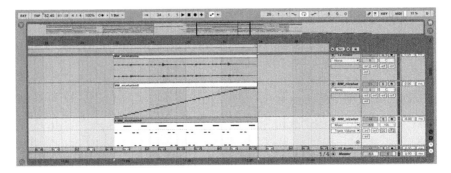

**Figure 2.5** The hi hats for this recording were sampled from MM_nicehatsm8. wav (from Splice/Medasin). Top section of sample = original sample; Middle section of sample = original sample sliced at transients; Bottom section of sample = the hi-hat pattern used on "Rinza" (played on a Novation Launchkey Mini keyboard controller).

instruments as though they occur together. According to the Gestalt *law of common fate*, sounds that occur within approximately 50ms of one another tend to be grouped together by listeners, depending on the signal's complexity (Blauert 1983: 201–271; Shelvock 2016: 80).

Once the song's main bass, kick, and snare groove were completed, I added hi hats. Following the example of Pro Era's Kirk Knight, as seen on *Rhythm Roulette*, I proceeded to beat slice a hi-hat loop called *MM_nicehatsm8.wav* (Figure 2.5).[26]

## Supporting Texture

The pads used on "Rinzai" also come from Splice's database. These samples are named (sample a) *ET_atonal_10* and (sample b) *LT_Skyward_B*. I panned sample a to the left by 92% and sample b to the right by 78% (Figure 2.6). Light saturation was also applied, via Fab Filter's *Saturn* plugin, to sample b in order to blend it with the distorted sounds of the main sample (3.1) and hi hats (3.2). These tracks also both feature a technique known as side-chain compression, which is triggered by the kick drum. This causes an audible amplitude *pumping* effect in these tracks each time the kick drum occurs.

## Additional Sounds

Sonically, "Rinzai" features a somewhat minimal production style (audio example 2.4). Only two of the tracks in the song's mix provide this type of function. One track uses a sample called *D_OLIVER_vocal_fx_ahhhhhh-h_01_150_A#.wav*, and the other uses a sample called *D_OLIVER_wet_vocal_oh_05_150_Gm*. I adjusted the pitch in both samples by +1 semitone (*vocal_fx_ahhhhhhh)* and +10 semitones (*wet_vocal_oh*) (Figure 2.7).

## Intermittent Mixing and Arrangement

Once I created a sequence of eight bars, using the materials described previously, I began to consider the song's overall arrangement and mix characteristics in more detail. However, mixing and sound design activities also occurred in the preceding steps (3.1–3.4). The act of selecting sounds and programming synthesizers, for instance, is an important sound design step. In addition, mix tools, such as EQs, compressors, saturation devices, reverbs, delays, chorus effects, pitch-shifters, and bus processing were altered throughout the entirety of this project. However, I also spent

**Figure 2.6** These samples act as pads on "Rinzai". The samples used are ET_Atonal_10.wav (Splice); LT_Skyward_B.wav. Top section = left pad; Bottom section = right pad. Heavy bus processing is also used (please see the second to last panel on the right) in conjunction with panning (same panel) to increase a sense of physical space around these pads.

**Figure 2.7** These sounds are classified more broadly as effects, even though they both originally come from vocal tracks. The samples used are D_OLIVER_vocal_fx_ahhhhhhh_01_150_A#.wav (top two rows of sample) and D_OLIVER_wet_vocal_oh_05_150_Gm.wav (bottom two rows). The top track was increased in pitch by 10 cents.

nearly one week *mixing* "Rinzai", as a separate process, once the arrangement details were decided upon.

In order to arrange "Rinzai", I simply experimented with various permutations of the initial eight-bar sequence. For instance, I (i) moved the MIDI events that comprise the song's main theme (3.1) to the left or right by an eighth note, (ii) repeated and deleted MIDI events, and (iii) replaced various MIDI events with new ones, causing different slices (i.e., sample regions) to play while maintaining the same rhythm. I followed a similar process for drums and bass.

## Mastering

Once I was satisfied with the song's arrangement and mix characteristics, I sent it to Jay Hodgson at Jedi Mastering/MOTTOsound. I was not present for the session, so the audio mastering activities for this record remain effectively *blackboxed*. This situation is quite common in digital music production, where beat producers of all types typically send their music to a separate mastering studio as a form of technical and creative quality assurance (Shelvock 2017b: 127).

## CONCLUSIONS

While I have not provided a comprehensive account of the hip-hop beat production network, this article takes the first steps towards doing so. As demonstrated in the preceding case study, production practice is full of sophisticated creative competencies that have hardly begun to receive direct scholarly attention, and ANT may provide one avenue for investigating them. One such competency is sampling, which is a common topic of discussion for critical theorists, but these studies are disconnected from the ways musicians actually use samples. For example, licensee-free samples have been pervasive since the 1990s, yet no author acknowledges the ways that hip-hop and electronic musicians use this type of sample library. Typically, sampling is instead discussed as an act of social resistance, wherein artists defiantly repurpose records made available by larger systemic forces (i.e., the music industry). However, most of the sampling that occurs in music production simply does not infringe on copyright in this way. By writing this chapter, I hope to enable increased clarity in future investigations of hip hop's production competencies, such as sampling, synthesis, mixing, and DAW-based editing.

This article also demonstrates one way to use ANT to analyze existing music production competencies, while, at the same time, clarifying some of the missteps of other researchers who have failed to apply ANT in a way which resembles Latour's recommendations. Indeed, all music production competencies involve complex interactions between humans and technology, and ANT provides a compelling tool for investigating these skills. While I've noted numerous exceptions throughout this paper (many of whom have contributed to this book), until recently, most researchers who discuss studio practices fail to accurately convey them, and this includes even the most remedial recording techniques and processes. Unfortunately, as a result, most faculty in higher education struggle to understand a medium which they routinely analyze in lectures, conference papers, and publications. And unlike other fields which examine recordings, such as film studies, where researchers have created a robust body of literature which analyzes the creative competencies required to make audiovisual recordings, no comprehensive body of literature exists in music. Although, since the inception of MPS, we are now beginning to see this change. Broadly speaking, ANT research may help rectify this lack of accurate production literature by encouraging researchers to focus on the interaction of humans and technology *per se*, without incorrectly reducing the complexity of these interactions to mere social issues.

## NOTES

1. For instance, please see Doyle, P. (2006). *Fabricating Space in Popular Music Recording, 1900–1960*. Middletown, CT: Wesleyan University Press.
2. Numerous examples of this are available online. For example, please see Mcintyre and Morey (2012), Campelo and Howlett (2013), Bourbon and Zagorsky-Thomas (2017), and Capulet and Zagorski-Thomas (2017).
3. For instance, please see Lefford (2015).
4. Although many more exist, here are a few prominent examples: Doyle (2005), Harkins (2015), Katz (2010a, 2010b), Lefford (2015), Liu-Rosenbaum (2012), O'Maly (2015), Ojanen (2015), and Zagorski-Thomas (2014).
5. Of course, a number of important books and articles led to the formation of this field, such as Zak's *Poetics of Rock* (2001). Even though little detail is provided on actual production technique throughout the book, his work made scholars begin to seriously consider recording practices. Hodgson's *Understanding Records* (2010) is one of the first steps towards refining Zak's work

and is one of the first examples of a *musicology of record production* that uses a practice-based research model.

6. A more in-depth analysis on the ontology of record production is available in Hodgson (2014: 8, 110).

7. As a result, a number of researcher/recordists have recently formed initiatives to address the lack of participation in music production analyses by music producers and engineers. For instance, please see Hodgson, J., & Hepworth-Sawyer, R. (2017). *Perspectives on Music Production: Mixing Music*. Pp. Xii–xiii. London: Routledge; Toulson, R., & Paterson, J. L. (2014). *Innovation in Music*. Future Technology Press; Paterson, J. L., Toulson, R., Hodgson, J., & Hepworth-Sawyer, R. (2016). *Innovation in Music II*. Shoreham-by-Sea: Future Technology Press.

8. Characteristics such as the design of circuits, transformers, or tubes, for example, can be ignored, or *blackboxed*, when applying ANT to music production. Traditionally, to satisfy the systemic nature of ANT, the internal specifications of technology do not need to be discussed. However, these discussions are highly relevant to other areas of music production study, such as Shelvock (2014, 2017).

9. One of the first texts that highlights the need for a musicology of record production is Albin Zak's *Poetics of Rock* (2001). However, a subfield that directly considers music production technique does not emerge until 2017 (i.e., music production studies). Please see Hodgson, J., & Hepworth-Sawyer, R. (2017). "Introduction". *Perspectives on Music Production: Mixing Music*. Pp. Xii–xiii. London: Routledge.

10. Hodgson, J., & Hepworth-Sawyer, R. (2017). "Introduction". *Perspectives on Music Production: Mixing Music*. Pp. Xii–xiii. London: Routledge

11. Please see Shelvock (2011, 2012a, 2012b, 2014, 2015, 2017a, 2017b).

12. It would be impossible, for instance, for researchers to perform a schenkerian reduction if they do not possess basic musical literacy. Those who analyze music production should demonstrate the same degree of literacy with recording methods, even though, historically, this has not been the case.

13. When presented with this point of view, I am reminded of Neil DeGrasse Tyson's words when defending the place of experts in our society; he states, "[T]he universe is under no obligation to make sense to you" (Quoted from *The Conan O'Brien Show*: http://teamcoco.com/video/neil-degrasse-tyson-the-universe-is-under-no-obligation-to-make-sense-to-you, accessed September 19, 2017).

14. For example, please see *Marvin Gay vs Robin Thicke and Pharrell Williams*: www.rollingstone.com/music/news/robin-thicke-and-pharrell-lose-blurred-lines-lawsuit-20150310, accessed September 27, 2017.

15. For an example of this technique, please see Mr. Carmack on the popular show *Rhythm Roulette*: www.youtube.com/watch?v=Hou9KevLeGU, accessed September 25, 2017.

16. More often, producers now use samples from original 808 devices, or they synthesize their own similar sounds. At the same time, Roland TR-808s are now highly coveted vintage machines, and they are costly (approximately $4–5,000 USD).

17. These songs are by no means an exception. In fact, *Who Sampled*, a website for DJs and producers, demonstrates exactly which samples were used on G-Funk records. Please see www.whosampled.com/song-tag/G-Funk/samples/, accessed September 21, 2017. Until recently, most scholarly literature on the topic of sampling features similar misconceptions as seen in Krims (2000).

18. Interestingly, a number artists at Canada's *Self Serve Records* also use tape samples this way. According to producer Ray Muloin (a.k.a., Ray Mosaic, one of Illa J's collaborators), he records live musicians to tape before sampling occurs.

19. This sample was removed from Splice and Loopmasters while I was writing this chapter. However, a comparison of Kendrick's song to the sample is available at the following site: www.whosampled.com/sample/494321/Kendrick-Lamar-FEEL.-Loopmasters-COF-125-Am-LaidOut-Underwater/, accessed September 27, 2017.

20. For example, nearly every producer on the popular web series known as *Rhythm Roulette* uses these software resources. Please see www.youtube.com/playlist?list=PL_QcLOtFJOUgNxURr8B4lNtf_3e9fWZzl, accessed September 27, 2017.

21. A flam is a note that is preceded by a grace note. Some producers extensively use this sound, such as Knxwledge on NxWorries (*Yes Lawd*, 2016). Others, such as the majority of trap producers (i.e., London on the Track), prefer to employ flams less often.

22. A portion of this research is also repeated online in Bourbon and Zagorsky-Thomas (2017a).

23. Reproduced from https://theproaudiofiles.com/producers-rough-mix/, accessed September 27, 2017.

24. Some notable exceptions are Hodgson (2010), Shelvock (2012), Cousins and Hepworth-Sawyer (2013), Nardi (2014), and Hepworth-Sawyer and Hodgson (2019).

25. For a demonstration of this technique on Mass Appeal's *Rhythm Roulette*, please see www.youtube.com/playlist?list=PL_QcLOtFJOUgNxURr8B4lNtf_3e9fWZzl, accessed September 27, 2017.

26. Please see www.youtube.com/watch?v=OzOkEYNXs1o&t=187s, accessed September 27, 2017.

## BIBLIOGRAPHY

Bijker,W., Hughes, T., & Pinch, T. (2012). *The Social Construction of Technological Systems, Anniversary Edition*. Cambridge, MA: MIT Press.

Blauert, J. (1983). *Spatial Hearing – the Psychophysics of Human Sound Localization*. Cambridge, MA: MIT Press.

Bourbon, A., & Zagorski-Thomas (2017). The ecological approach to mixing audio: agency, activity and environment in the process of audio staging. *Journal on the Art of Record Production (Issue 11)*. [online]

Campelo, I., & Howlett, M. (2013). The "virtual" producer in the recording studio: media networks in long distance peripheral performances. *Journal on the Art of Record Production, 8*. Available at: http://arpjournal.com/the-"virtual"-producer-in-the-recording-studio-media-networks-in-long-distance-peripheral-performances/, accessed October 3, 2018. [online]

Capulet, E., & Zagorski-Thomas, S. (2017). Creating a rubato layer cake: performing and producing overdubs with expressive timing on a classical recording for 'solo' piano. *Journal on the Art of Record Production, 11*. [online]

Danielsen, A. (Ed.). (2010). *Musical Rhythm in the Age of Digital Reproduction*. Farnham: Ashgate Publishing.

Devine, A., & Hodgson, J. (2017). Mixing in and modern electronic music production. *Perspectives on Music Production: Mixing Music*. Eds. Hepworth-Sawyer, R. & Hodgson, J. New York: Routledge/Focal Press.

Doyle, P. (2005). *Echo and Reverb: Fabricating Space in Popular Music Recording, 1900-1960*. Middletown, CT: Wesleyan University Press.

Elsdon, P. (2014). Review: the musicology of record production. *IASPM*. Available at: www.iaspmjournal.net/index.php/IASPM_Journal/article/view/801, accessed October 3, 2018. [online]

Harkins, P. (2015). Following the instruments, designers, and users: the case of the fairlight cmi. *Art of Record Production Journal, 10*. [online]

Hepworth-Sawyer, R., & Hodgson, J. (2017). *Perspectives on Music Production: Mixing Music*. New York: Routledge/Focal Press.

Hepworth-Sawyer, R., & Hodgson, J. (2019). *Audio Mastering: The Artists*. New York: Routledge/Focal Press.

Hodgson, J. (2010). *Understanding Records*. London: Continuum.

———. (2014a). *Representing Sound*. Waterloo: WLA Press.

JAMES. (2018). JAMES (Official Webpage). [online]

Katz, M. (2010a). *Capturing Sound: How Technology Has Changed Music* (Vol. 1). Berkeley: University of California Press.

———. (2010b). *Groove Music: The Art and Culture of the Hip-Hop DJ*. Oxford: Oxford University Press.

Koszolko, M. K. (2015). Crowdsourcing, jamming and remixing: a qualitative study of contemporary music production practices in the cloud. *Journal on the Art of Record Production, 10*. [online]

Krims, A. (2000). *Rap Music and the Poetics of Identity*. Cambridge: Cambridge University Press. p. 74.

Latour, B. (1996). On actor-network theory: a few clarifications. *Soziale welt*, 369–381.

———. (1999). *Pandora's Hope: Essays on the Reality of Science Studies*. Cambridge, MA: Harvard University Press.

———. (2005). *Reassembling the Social: An Introduction to Actor-Network-Theory*. New York: Oxford University Press.

Lefford, N. (2015). The sound of coordinated efforts: music producers, boundary objects and trading zones. *Art of Record Production Conference: Record Production in the Internet Age 04/12/2014–06/12/2014* (Vol. 10). [online]

Liu-Rosenbaum, A. (2012). The meaning in the mix: tracing a sonic narrative in 'when the levee breaks'. *Journal on the Art of Record Production, 7*. [online]

Marrington, M. (2016). Paradigms of music software interface design and musical creativity. *Innovation in Music II*. Eds. Patterson, J., Toulson, R., Hepworth-Sawyer, R., & Hodgson, J. Shoreham-by-sea: Future Technology Press.

McIntyre, P., & Morey J. (2012). Examining the impact of multiple technological, legal, social and cultural factors on the creative practice of sampling record producers in Britain. *Journal on the Art of Record Production (Issue 07)*. [online]

Mooney, J. (2011). Frameworks and affordances: understanding the tools of music making. *Journal of Music, Technology, and Education, 3*, 141–154.

Nardi, C. (2014). Gateway of sound: Reassessing the role of audio mastering in the art of record production. *Dancecult: Journal of Electronic Dance Music Culture, 6*(1), 8–25. [online]

Ojanen, M. (2015). Mastering Kurenniemi's rules (2012): the role of the audio engineer in the mastering process. *Journal on the Art of Record Production (Issue 10)*. [online]

O'Maly, M. (2015). The definitive edition (digitally remastered). *Journal on the Art of Record Production*. [online]

Pensado, D. (2014). Noah '40' Shebib – Pensado's place #151. Available at: www.youtube.com/watch?v=ESUHhXgIaos&index=19&list=PLh0d_FQ4KrAV F5LM2sCC_nncdjk7jCK7G, accessed October 3, 2017. [online].

Shelvock, M. (2011). Interview with Josh Leo. *Journal on the Art of Record Production (Issue 05)*. [online].

———. (2012a). Interview with Steve Marcantonio. *Journal on the Art of Record Production (Issue 06) [online]*.

———. (2012b). Interview with Ben Fowler. *Journal on the Art of Record Production (Issue 06)*.

———. (2014). *The Progressive Heavy Metal Guitarist's Toolkit: Analog and Digital Strategies. Innovation in Music*. Shoreham-by-sea: Future Technology Press.

———. (2015). Gestalt theory and mixing audio. *Innovation in Music Conference*. Anglia-Ruskin University, Cambridge, UK. Conference paper.

———. (2016). Gestalt theory and mixing audio. *Innovation in Music II*. Eds. Patterson, J., Toulson, R., Hepworth-Sawyer, R., & Hodgson, J. Shoreham-by-sea: Future Technology Press.

———. (2017a). Groove and the grid: mixing contemporary hip hop. *Perspectives on Music Production: Mixing Music*. Eds. Hepworth-Sawyer R., & Hodgson, J. New York: Routledge/Focal Press.

———. (2017b). *Audio Mastering as a Musical Competency*. Dissertation. University of Western Ontario. Available at: http://ir.lib.uwo.ca/etd/.

Strachan, R. (2017). *Sonic Technologies: Popular Music, Digital Culture and the Creative Process*. New York: Bloomsbury Publishing USA.

Tatnall, A., & Gilding, B. (1999). Actor-network theory in information systems research. *10th Australasian Conference on Information Systems (ACIS)*. Wellington, New Zealand.

Zagorski-Thomas, S. (2010). The stadium in your bedroom: functional staging, authenticity and the audience-led aesthetic in record production. *Popular Music, 29*(2), 251–266.

———. (2014). *The Musicology of Record Production*. Cambridge: Cambridge University Press.

Zak, A. (2001). *The Poetics of Rock: Cutting Tracks, Making Records*. Berkeley: University of California Press.

# 3

# *Boom Bap* Ex Machina

## Hip-Hop Aesthetics and the Akai MPC

## Michail Exarchos (a.k.a. Stereo Mike)

## INTRODUCTION

Over the past three decades, the growing literature on hip-hop musicology has paid ample tribute to Akai's range of MPCs (originally, MIDI production centers – currently, music production controllers), acknowledging their pivotal influence on rap production practices (D'Errico 2015; George 1998; Harkins 2009, 2010; Kajikawa 2015b; Morey and McIntyre 2014; Ratcliffe 2014; Rodgers 2003; Rose 1994; Schloss 2014; Sewell 2013; Shelvock 2017; Swiboda 2014; Wang 2014; Williams 2010). The technology's combined functionality of sampling, drum machine and MIDI sequencing features has been embraced by rap practitioners ever since the release of the standalone MPC60 in 1988. The time line coincides with particular sonic priorities in Hip-Hop that can be grouped under the "boom bap" aesthetic – an onomatopoeic celebration of the prominence of sampled drum sounds programmed over sparse and heavily syncopated instrumentation. But what is the association between subgenre aesthetics and MPC functionality, and what parallels can be drawn between the evolution of the technology and stylistic deviations in the genre? This chapter examines how MPC technology impacts upon the stylization of Hip-Hop as a result of unique sonic, rhythmic and interface-related characteristics, which condition sampling, programming and mixing practices, determining in turn recognizable sonic signatures. Furthermore, the boom bap sound is traced from its origins in the mid- to late 1980s to its current use as an East Coast production reference, honoring a sample-based philosophy that is facilitated by the MPCs' physical interface and operating script. The findings from a number of representative case studies form a systematic typology of technical characteristics correlated to creative approaches and resulting production traits, informing speculation about the future of the MPC, its technological descendants, and the footprint of its aesthetic on emerging styles and technologies.

## A BRIEF HISTORY SAMPLE

> When a new technology gets loose in the music world, it takes a few years to become front-page news. Consider the electric guitar. . . . The sampler is making a serious bid to be the electric guitar of the '90s. . . . What has reached critical mass is the complex of music (and social) meanings attached to sampling.
>
> (Aikin 1997: 47)

It was October 1997 when *Keyboard* magazine made sampling front-page news, featuring DJ Shadow alongside the Akai MPC2000 on the front cover and leading with the story "Samplers Rule". The story of sampling, however, had begun over a decade and a half ago (at this point, the MPC2000 represented the fourth generation of MPC technology), and the story of Hip-Hop had commenced before digital sampling was even possible. Rap aficionados associate hip-hop music automatically with sampling, but to put things in perspective, Hip-Hop's original "instrument" was the turntable (Katz 2012: 43–69). Pioneering DJs in the Bronx had used a pair of turntables and a mixer to extend the instrumental sections of 1960s and 1970s soul and funk recordings, providing the instrumental foundation that MCs would eventually rap over (Chang 2007; Toop 2000). It took a number of years before any rapping was actually committed to record and the first successful rap release arrived in the form of *Rapper's Delight* by the Sugarhill Gang in 1979 (Howard 2004). *Rapper's Delight* and all proto-rap releases of the era utilized disco, soul and funk musicians for the production of the instrumental backing (Kulkarni 2015; Serrano 2015), and turntables did not feature prominently on records until *The Adventures of Grandmaster Flash on the Wheels of Steel* in 1981. Flash's turntable performance stands both as a historical *record* of the performative hip-hop tradition (turntablism), and as a production that contains multiple phonographic segments that are then cut, manipulated and juxtaposed further by the DJ. It is these sonic artifacts that early hip-hop producers attempted to replicate when the first affordable digital samplers hit the market, a notion that is audible in mid- to late 1980s sample-based releases. Ingenuity on the side of the producers and evolving design on the side of the manufacturers meant that samplers would soon transcend the function of merely replicating turntable performance, unpacking new creative possibilities and becoming hip-hop instruments par excellence (in the hands of studio DJs who had now transitioned to fully fledged *producers*).

## What Is Boom Bap?

The story of Boom Bap is closely associated with the development and practice of sampling, and as such, Boom Bap is often described as a production technique, a sound, a style, or a subgenre. The term was first uttered in 1984 by T La Rock in the final ad-libs of "It's Yours" (Mlynar 2013), but it was popularized by KRS-One with the release of the *Return of the Boom Bap* album in 1991. It stands for an onomatopoeic celebration of the sound of a loud kick drum ("boom") and hard-hitting snare ("bap") exposed over typically sparse, sample-based instrumental production. It could be argued that these two words (*boom* and *bap*) conjure rhythmic, timbral and balance implications, and as such, Boom Bap could be better described as an overarching aesthetic that signifies hip-hop eras, production preferences, sonic traits, subgenre variations, geographical connotations and even authenticity claims. Mike D'Errico (2015: 281) defines "boombap" as a "sound that was shaped by the interactions between emerging sampling technologies and traditional turntable practice" by producers who "used turntables alongside popular samplers such as the Akai MPC and E-Mu SP-1200", resulting in "gritty, lo-fi audio qualities . . . and innovative performance practices that continue to define the sound of 'underground', 'old-school' hip-hop".

But how do we make the transition from particular mechanistic affordances[1] (Clarke 2005: 36–38) to a complex set of sonic signatures claiming their very own raison d'être? The connection lies in the sampling affordances that enabled the separation, reinforcement and stylization of individual drum sounds within a hip-hop context to such an extent that practitioners "baptized" the phenomenon with its own onomatopoeia. The significance of this is that the sonic variables that characterize Boom Bap are interrelated to production techniques and workflow approaches conditioned by technical characteristics found in digital samplers in general and the MPC range in particular. Through case-study analysis, this chapter will demonstrate how this mapping occurs and its implications for current hip-hop production; but also how it may predict future practices within the genre.

## BOOM BAP OUT OF THE MACHINE

The isolation of the "boom" and the "bap" can be traced back to pioneering hip-hop producer Marley Marl, who "discovered the power of sampling drums by accident during a Captain Rock session" (Weingarten 2010: 22) and, in his own words, found that he "could take any drum sound from any old record, put it in [t]here and get that old drummer sound" (cited in George 1998: 92). Kajikawa (2015b: 164–165) informs us that Marl must have "first experimented with sampled drum breaks in or around 1984 when the first devices with adequate memory and function, such as [the] E-mu Emulator II and the Ensonique Mirage, began hitting the market". The significance of this discovery – and Marl's influence on a genealogy of producers associated with Boom Bap, such as DJ Premier, Pete Rock, Q-Tip, RZA, Prince Paul, DJ Shadow, J Dilla and Madlib – is that it empowered rap producers to transition from "surface manipulators" (users of drum loops or breaks referential to a turntable affordance) to drum "scientists": samplist-programmers who could come up with new patterns altogether, layer multiple drum sounds upon one another and create original rhythms out of minimal sonic segments from the past. Soon, the techniques advanced to dense layering of sampled *and* synthesized sources (the latter often courtesy of a Roland TR-808), complex rhythmic appropriation, chopping and juxtaposition.

It is in the trajectory of this evolving production technique – from Marl's drum-hit isolation to later producers' intricate juxtaposition – that the development of the boom bap aesthetic can be observed, highlighting the tension between "liveness" and rigidity; organic and synthetic sonics. Talking about the Roland TR-808 and E-mu SP-1200, Kulkarni (2015: 43) observes that "The two most emblematic pieces of hardware hip hop has ever used both, in their way, crystallise that delicious dilemma, that tightrope between looseness/'feel' and machine-like tightness that hip hop's sound so engagingly steps on". Naturally, with powerful sampling technology integrated alongside drum machine and sequencer functionality in standalone production centers, future producers would go on to approach all past phonographic material – not just funk and soul drum breaks – with increasing microscopic focus, separating instrumental phrases into "stabs", assigning them to MPC drum pads, and performing and programming re-imagined sequences into new cyclical arrangements (loop-based compositions). The "boom" and the "bap" would evolve to represent not only a drum-inspired onomatopoeia but an overarching "chopped", manipulated and syncopated *aesthetic* founded upon the interaction of past records with new mechanistic sequencing. Table 3.1 maps distinct characteristics of the boom bap sound against affordances – and limitations – found specifically on MPCs. The left column highlights characteristic stylizations that

**Table 3.1** A Mapping of Boom Bap Stylizations Against MPC Affordances and Limitations

|  | Boom bap characteristics | MPC affordances/limitations |
|---|---|---|
| ***Balance*** | Prominent kick drum | Internal mix functionality |
|  | Prominent snare drum | Internal mix functionality |
| ***Timbre*** | Emphasized low-end (kick drum) | Internal processing (effects) / Resolution limitation |
|  | Hard-hitting snare drum (presence) | Internal processing (effects) / Resolution limitation |
|  | Low fidelity | Resolution limitation |
|  | Vinyl (sample) sources | Phono inputs |

(*Continued*)

**Table 3.1** (Continued)

| | Boom bap characteristics | MPC affordances/limitations |
|---|---|---|
| ⬤ | | |
| ***Dynamic*** | Compressed instrumental production | Internal processing (compression) / Resolution limitation |
| | Interpolation/filtering | Controllers (interface) |
| ***Sonic "Glue"*** | Instrumental production "glue" | Resolution limitation / Converters (I/O) / Internal processing (compression) |
| | Shared ambience on sampled elements | Internal processing (effects) |
| ***Arrangement*** | Isolated drum "hits" | (Auto-)Slice functionality / Memory limitation |
| | (Short) Other instrumental "stabs" | (Auto-)Slice functionality / Memory limitation |
| | Layered kick drum (often with 808) | Program functionality / MIDI out |
| | Layered snare drum | Program functionality / MIDI out |
| | "Chopped" breaks (drums) and other phonographic samples | (Auto-)Slice functionality |
| | Sparse instrumentation | Memory limitation / Program functionality (mono) |
| | Turntable effects / performance | Phono inputs |
| | Four-measure repetition / chorus variation | Sequencer / song functionality |
| ***Rhythmic*** | (Highly) Swung programming | MPC swing / quantization algorithm |
| | Tight drum-instrumental syncopation | Program functionality / MIDI out / MPC swing / quantization algorithm |
| ***Motivic*** | Re-arranged phrases/ rhythms/motifs | (Auto-)Slice functionality / Drum pads (interface) |
| | Percussive programming of instrumental phrases | Drum pads (interface) / Program functionality (mono) |

define the boom bap aesthetic, while the right column indicates software and hardware functionality within the MPC environment that promotes these stylizations or makes them possible. It is worth noting that many of these affordances are not exclusive to MPC technology anymore, but their combined integration on a standalone piece of hardware as early as 1988 became instrumental in allowing the aesthetic to develop, whilst also conditioning future workflow preferences mirrored in later generations of the same hardware (and competitive designs too). As such, the MPC workflow allowed 1990s hip-hop producers to perfect the sample-based art form, and the boom bap sound became synonymous with Hip-Hop's Golden Era (circa 1988–1998), as well as the East Coast's rather dogmatic reliance on phonographic sources.

East Coast producers continued to rely on boom bap methods not only because of their preference for phonographic sources but also as a reaction to the more synthesized subgenres coming out of the West Coast or US South, and as a form of conscious sonic signposting toward the birthplace of the genre – New York. The boom bap sound was later taken into more experimental, instrumental frontiers by producers such as J Dilla, Madlib, Prefuse 73 and Flying Lotus (Hodgson 2011), while it currently enjoys a resurgence in the form of a plethora of releases classified as Boom Bap[2] and mainstream releases increasingly tapping into it to support more conscious lyrical content (for example, Jay-Z's 2017 single, "The Story of O.J.," produced by No I.D.). In an interview with *Rolling Stone* magazine (cited in Leight 2017), No I.D. sums up the rationale behind his return to a sample-based approach by saying, "I began to play the samples like I would play an instrument . . . I had stepped away from my strength sometimes because the business makes you think you can't do it . . . I can do it. And I can create new art".

## HIP-HOP'S WEAPON OF CHOICE

It is important, however, to consider the point at which MPCs enter the historical time line and the rationale behind them replacing E-mus as preferred weapons of – hip-hop – choice. Akai released the MPC60 in 1988, bringing a number of improvements to the notion of integrated sampling, drum machine and MIDI-programming functionality. Ratcliffe informs us that:

> MPC is the model designation for a range of a sampling drum machine/sequencers, originally designed by Roger Linn and released by Akai from 1988 onwards (for instance, the MPC 60, MPC 2000, and MPC 3000). These instruments are favoured for sample-based hip-hop and EDM due to both the design of the user interface (featuring drum pads for real-time programming) and idiomatic performance characteristics (such as the swing quantisation algorithm).
>
> (Ratcliffe 2014: 113)

Rap producers made the switch from E-mus to Akais for different reasons, but it could be summarized that the MPCs' unique swing quantization parameter, the higher bit-depth resolution, the touch-sensitive drum pads of the physical interface and the internal mixing functionality were among the main reasons (Anderton 1987; Linn 1989). Roger Linn (Scarth and Linn 2013) himself attributes the "natural, human-feeling grooves in [his] drum machines. . . (i)n order of importance" to the factors of "(s)wing", "(n)atural dynamic response on [the] drum pads", the "(p)ressure-sensitive note repeat" function, programming accuracy, strong factory sounds and a user-friendly interface. Hank Shocklee of the Bomb Squad (the production collective behind iconic Public Enemy albums such as *It Takes a Nation of Millions to Hold Us Back* and *Fear of a Black Planet*) asserts:

> [The 12000] allows you to do everything with a sample. You can cut it off, you can truncate it really tight, you can run a loop in it, you can cut off certain drum pads. The limitation is that it sounds white, because it's rigid. The Akai Linn [MPC-60] allows you to create more of a feel; that's what Teddy Riley uses to get his swing beats.
>
> (Cited in Dery 1990: 82–83, 96)

Schloss (2014: 201–202) adds that "the circuitry and programming of different models of samplers are believed to impart special characteristics to the music (perhaps the best known of these characteristics is the legendary 'MPC swing', a rhythmic idiosyncrasy first noted in the Akai MPC 60 sampler, circa 1988)". And Kajikawa (2015a: 305) reports that the Akai MPC60's "touch-sensitive trigger pads allowed producers to approach beatmaking with renewed tactile sensitivity". Figure 3.1 provides a schematic representation of a time line mapping

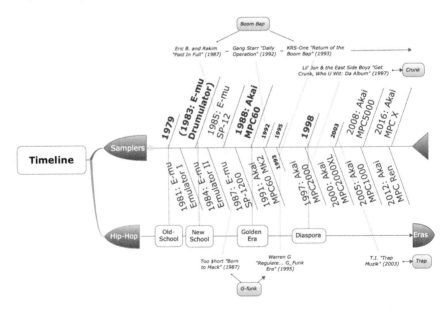

**Figure 3.1** Time line mapping hip-hop eras against E-mu and Akai products, with examples of seminal releases characteristic of rap subgenres

hip-hop eras against the releases of particular models of E-mu and Akai products, illustrated by examples of seminal releases signifying hip-hop subgenres.

## A TYPOLOGY

The technical characteristics of the MPC range can thus be grouped into variables relating to the operating script on the one hand and to physical attributes of the hardware interface on the other. These, in turn, influence sampling, programming and mixing tendencies in producers' workflows, with sonic, rhythmic and motivic implications for the musical outputs. It is beyond the scope of this chapter to detail every function, physical attribute, or parameter found on the MPC range, so the focus will remain instead on characteristics that create noteworthy affordances in producers' workflows. These will then be mapped to a number of predictable sonic signatures (potential aesthetic results), as can be observed through aural analysis of seminal works. Consequently, the observations here do not follow a one-way, technologically deterministic rationale but take into account the creative agency of producers illuminated through the discussion of key works, and informed by existing musicological literature and producer testimonials. A visual representation of the typology is provided in Figure 3.2.

Starting from the MPC's operating script, key characteristics highlighted in the preceding typology are (a) the quantization algorithm (and MPC's infamous "swing" parameter); (b) the MPC's onboard sequencer, its looping function and the included "song-mode" for the construction of longer phrases; (c) the note-repeat function (tied to the sequencer and quantization function); (d) the (auto-)slice option of the zone functionality (introduced in 1997 with the release of the MPC2000), which enables the separation of a longer audio sample into separate segments or "chops"; (e) the memory limitations of earlier designs (resulting in shorter sampling times); (f) the monophonic/muting functionality within programs; and (g) the internal routing functionality (including optional effects boards introduced after 1997 with the release of the MPC2000).

Key features of the physical interface highlighted in this typology are (a) the velocity-sensitive drum-style finger pads; (b) physical controllers such as sliders and rotary knobs found on the hardware; and (c) various aspects of the MPCs' input/output (I/O) functionality, which can be further subdivided to i) the type of sampling inputs available; ii) the (multiple) outputs functionality (and how this relates to internal routing and processing as mentioned earlier); and iii) the quality and resolution of the analogue-to-digital (AD) and digital-to-analogue (DA) conversion, including the bit-depth limitations of earlier designs (and their emulation thereof in later ones).

Next, I turn my attention to how these technical features have affected creative (ab)use in the context of hip-hop music making and what their influence has been on the boom bap sound. To answer this question, it

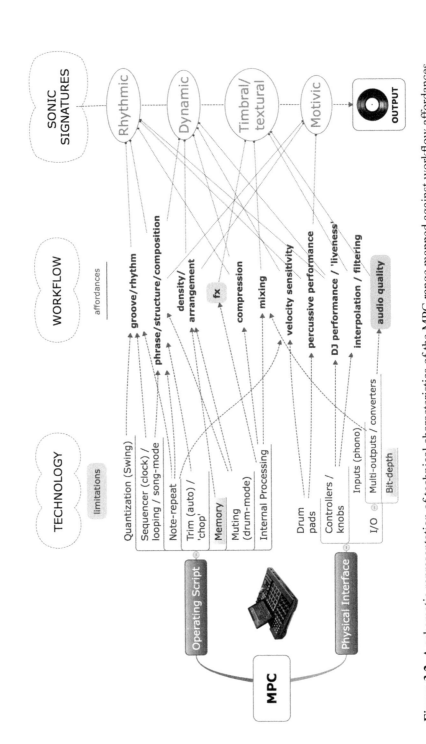

**Figure 3.2** A schematic representation of technical characteristics of the MPC range mapped against workflow affordances and sonic signature categorizations

is worth revisiting some seminal work in the genre and investigating the sonic signatures present in relation to the highlighted features. For a comparative analysis, two releases before and after 1988 are considered that have been classified as Boom Bap: Eric B. and Rakim's album *Paid in Full* (1987), largely produced by Eric B. but influenced by Marley Marl, featuring two of his remixes and, in fact, partly recorded at his home studio; and Gang Starr's *Daily Operation* (1992), featuring DJ Premier's archetypal – and highly swung – production footprint.

## Rhythmic Signatures

In a comparison of Marl's remixing of "My Melody" and "Eric B. Is President" to DJ Premier's programming on tracks such as "2 Deep" and "Take It Personal", there is a discernible difference evident in the swing quantization of various elements. We know for a fact that although DJ Premier learned his craft on an E-mu SP12, by the early 1990s, he had switched to an Akai MPC60 for his programming, triggering samples from an Akai S950 sampler (Tingen 2007). Based on previous research (George 1998; Kajikawa 2015b; Weingarten 2010), it is also safe to assume that what we are hearing on *Paid in Full* is E-mu technology, especially as this is the year prior to the release of the MPC60. Marl's work is indeed swung (as can be clearly heard in the rhythmic placement of the 16th kick-drum figure in the third measure of each four-measure loop on *Eric B. Is President*), but when compared to DJ Premier's programming (on elements such as the brass stabs of "2 Deep" and the swung kick-drum 16ths on "Take It Personal"), the latter is so highly swung it almost resembles a triplet feel. In fact, it is impossible to find any records prior to 1988 that are as highly swung as DJ Premier's work in the early 1990s, a fact that ties this characteristic musical figure to the MPC swing algorithm. But the sonic artifacts observed here cannot be attributed to technology alone. DJ Premier's love of jazz (as exemplified by his sampling choices famously featuring Charlie Parker on Gang Starr's 1989 debut release *No More Mr Nice Guy*) surely has a lot to do with his abuse of the swing affordance, and it is precisely this flux between absorbed influences (culture), technology and personal agency that results in stylistic evolutionary leaps.

Such was the effect of MPC quantization parameters on hip-hop outputs that in following years producers meticulously reaped the time intricacies of different generations of the hardware and imported them into digital audio workstations (DAWs), emulating the MPC "feel". Before these came prepackaged in contemporary DAWs, such as Ableton Live, the producing community would share them online (in the form of MIDI files and song templates), while producers would frequently extract them in person to ensure higher accuracy between sequenced elements on their version of the hardware and programmed elements running in parallel on their DAWs. Celebrated producer Just Blaze is one of the producers who made the switch from producing on hardware MPCs to Apple's Logic

DAW, and in an interview with uaudio.com, he provides valuable insight into his rationale:

> When I made the decision to move over to Logic, a couple of guys that I work with and I imported all of the actual sequencer grooves from three MPCs into Logic. . .
>
> We did this because, even though it's all ones and zeros and they're all computers, every processor and sequencer is a bit different. . . . The grooves that we imported over are the only thing that I always make sure that I have warmed up when I open a blank session.
>
> (Gallant and Fox 2016)

Just Blaze here points out another important quality of the MPC range (and hardware/sampling drum machines in general): the computerized "grid" is not actually set in stone, and timing idiosyncrasies in hardware sequencers do not stop with swung quantization templates. It has been shown that even when a straight feel is selected, the MPC impacts timing errors that deviate from a strict mathematical grid, and the complexity of the programmed material can have an incremental effect on timing accuracy (Perron 1994: 5).

## Dynamic Signatures

The combination of unique quantization templates and timing idiosyncrasies can therefore be seen as a useful binary of control versus randomization, which, in the right programming hands, allows for the creation of "groove" and "feel" signatures, balancing the quest for rhythmic tightness with hip-hop's "genetic" predisposition for a "live" heartbeat. But no rhythmic analysis of groove is complete without a mention of dynamics, and as any drummer (or drum programmer) would add, rhythmic feel is the result of not just timing but also accents and velocity variations. Here, the touch-sensitive drum pads of the physical interface enable further interaction between human and machine-like qualities. The drum pads on the MPC could be used in a fully touch-sensitive mode, registering any velocity that the performer/programmer exercised as a MIDI value, or they could be preset to threshold or stepped velocities, allowing for more controlled expression. Furthermore, the note-repeat function on MPCs would allow for automatic repetition of a sample assigned to a drum pad according to the preset quantization value, but the benefit of touch-sensitivity again presented a unique opportunity for mechanistic timing over expressive dynamics.

## Motivic Signatures

An important affordance that allowed further agency with the drum break – and closer interaction with individual hits – was the auto-slice option implemented as part of the zone functionality since the release of the MPC2000 (Avgousti 2009). Beat mavericks such as J Dilla and Madlib clearly abused this function on albums such as *Champion Sound* (2003), but – in fact – the automated process mirrored a practice long-exercised by boom bap producers. Whether laboriously "chopping" longer drum breaks (or motivic phrases) "by hand" or using automatic processes to separate them into shorter segments, sample-based producers took pride in meticulously subdividing sections to the shortest temporal denominator necessary – eighths, 16ths, or individual "hits" or "stabs" – to assign "cuts" onto drum pads for re-triggering. Although this process may now feel quite commonplace to the contemporary producer, the rhythmic, dynamic and motivic implications of this practice on the stylization of hip-hop production at the time were of massive importance. It allowed for increased rhythmic freedom, re-appropriation and syncopation, but it also had timbral and motivic implications as will be demonstrated next, especially when used in conjunction with the MPC's program-muting/monophonic functionality.

A typical boom bap practice would be to set a program's polyphony to mono, so that each segment triggered mutes the previous one already playing (for this to work samples would have to be set to "one-shot" triggering, which ensures they play out until interrupted by another event pertaining to the same program). Two positive side effects of the process were a highly rhythmical effect and the preservation of clarity in the harmonic progression of newly constructed patterns (by avoiding the juxtaposition of overlapping melodic or harmonic information present in sampled phrases). The monophonic triggering and muting would thus create tightly syncopated results due to the placement of the new "cut" (initiated by the percussive attack of the edit or a kick drum on the first beat) against rhythmical subdivisions already present in the previously playing segment. As a result, the original material would assume new rhythmic qualities due to its placement and truncation within the programming order sequenced on the MPC. A prime example of this rhythmic-motivic signature can be heard on Gang Starr's *Hard to Earn* album (1994). It could be argued that the resulting sensibility is quintessentially Hip-Hop: the *meta*-syncopation interacts favorably with the sampled material's inherent syncopation.

## Timbral/Textural Signatures

In the first volume of *Perspectives on Music Production*, Matt Shelvock (2017: 170) demonstrates how the notion of hip-hop production – as "beat-making", composition, or creation – is closely integrated with mixing practices and "that the lines between mixing and production are often blurred within this genre". As such, he highlights how hip-hop mix engineers may be closely involved with creative production decisions, and that clear stages between making and mixing a track may not be adhered to in hip-hop practice. Looking at this relationship from the perspective of beat-making, it is also true that the hip-hop producer assumes a plethora of mixing roles *while* creating a track. One of the defining features of the MPC has been the inclusion of mixing/processing functionality, which has empowered beat makers with sonic options traditionally reserved for the mixing stage. What's more, the MPC operating script caters for a flexible routing functionality that enables the insertion of effects at various stages (upon an individual sample, program, track or the master). The resulting "signal flow" outcomes create striking and unique sonic signatures, which would have required complex processes were they attempted within a DAW environment or via patch-bay routing in a hardware studio.

It is also important to discuss the limitations imposed by the internal effect processing capabilities of the DSP chips on early MPCs and note the aesthetic implications these have had on the musical outputs of respective eras. The early chips limited the amount of effects that could be internally patched in at the same time (as insert effects or in an auxiliary send configuration) to two, so for the sake of efficiency, practitioners would often share effects across a number of programs or tracks. For instance, a typical configuration would be to enable a master compressor for the whole internal MPC mix "buss", while also sharing a reverb effect across numerous elements. This limitation would often result in a notion of sonic "glue" or "blend", an illusion constructed out of shared spatial and dynamic characteristics.

Compression hereby deserves special mention, because its (ab)use by practitioners within the context of Hip-Hop and the MPC environment has resulted in very particular stylizations (these have been exponentially expressed in the more experimental outputs of Flying Lotus, Madlib, J-Dilla, and Prefuse 73 as discussed by D'Errico (2015) and Hodgson (2011)). Although compression is a dynamic effect common to most virtual or physical mixing environments, its inclusion within the MPC operating script, the possibility to insert master compression on the stereo mix-buss (as part of the internal processing matrix), and the combined effect of the compression algorithm with that of the reduced bit-depth resolution of earlier models (and its emulation on later ones) lends it a unique quality in this context.

Furthermore, the physical input and output connectors on the MPC range, the analogue-to-digital and digital-to-analogue converters, and the resolution capabilities of various models, contribute to particular mixing affordances, sonic signatures and lo-fi signifiers. The notion of being able to produce a complete musical piece on an MPC meant that it was possible for producers to bypass the use of a computer sequencer or DAW altogether and bring a complete instrumental idea into the studio for mixing. Akai facilitated multiple outputs on most models of the hardware (either as default or optional as an expansion) with the aim of allowing an instrumental production to be output directly onto a hardware studio mixer. The sonic character of the physical I/O on MPCs (including the vinyl preamp inputs on most models) would thus be imprinted onto both incoming – sampled – sources and outgoing multi-tracks, contributing to a particular timbral footprint. The sound of certain MPC models has become revered to such an extent that recent reincarnations of the technology feature emulations of earlier models (for example, "vintage mode" on the MPC Renaissance and MPCX provides emulations of the MPC60 and the

MPC3000). The reduced bit-rate (resolution) of earlier models is one of the notable variables impacting the audio quality of hip-hop outputs from respective eras. This brings us to the present, and it is important to investigate the current footprint of these stylizations and their relationship to contemporary musical outputs in Hip-Hop.

## EVOLUTION

> I give you bars, no microwave rap
> I can take it down South, but it's gon' be my version of trap . . .
> I don't hate the trap but give me that boom bap
> Yeah the 808 eating at the beats drill the 808
>
> (Statik Selektah 2017)

The preceding lyrics are from Statik Selektah's recent release "But You Don't Hear Me Tho" (2017), highlighting the aesthetic friction between Trap – the prevailing and highly synthesized contemporary rap style – and the retrospective boom bap sound. The track features many of the stylizations characteristic of Boom Bap, complete with prominent kick and snare drum sounds, highly swung programming, and chopped, soulful samples (albeit recorded freshly by soul-funk band Mtume with additional horns from Utril Rhaburn). Static Selektah is a contemporary hip-hop producer and DJ Premier protégé whose production duties are called upon when – according to XXLmag (Emmanuel 2017) – "classic hip-hop at its finest" is required. He represents a generation of producers and artists that Pitchfork (Ruiz 2017) describes – alongside rapper Your Old Droog – as "a nostalgic sonic wave currently being surfed by NYC contemporaries Roc Marciano, Action Bronson, and Joey Bada$$, rappers doing their best to embody the spirit of New York hip-hop without getting stuck in its past". Indeed, Action Bronson's 2015 album *Mr. Wonderful* – partly produced by hip-hop veteran The Alchemist – sits among a plethora of recent releases categorized as Boom Bap; meanwhile, the boom bap term enjoys a revival in the context of artists' lyrics and as a common stylistic descriptor in album reviews. Indicatively, Ruiz (2017) describes Your Old Droog's single "Bangladesh" (from the album *Packs*) as "an ill Bansuri loop over a simple boom-bap drum beat", and HipHopDX (Leask 2017) reviews his track "Help!" as "an unremittingly noisy blast of psychedelic boom-bap".

If Your Old Droog's *Packs* merges sparseness and a sample-based aesthetic with noisy psychedelia, then another recent release, Apollo Brown and Planet Asia's 2017 album *Anchovies*, takes the recipe down to the absolute rawest of materials; perhaps it is a sign of genre maturity when process is reduced to its leanest, and the overarching simplicity of *Anchovies* exposes Boom Bap's DNA in a minimal production approach that does away with obvious drum reinforcement, coming full circle to the turntablist tradition of chopped, flipped and rewound instrumentals (with rather extraneous amounts of vinyl noise). As an exception to the onomatopoeic boom bap dogma, it reveals the mechanics underneath the beat:

chopped instrumental samples programmed into re-imagined sequences and chord progressions, where swung quantization powers a highly rhythmic and hypnotic interaction between sourced phonographic material and the enforced temporal relationships of a sequencer. The customary boom bap beats are either minimal or simply implied, but the rhythmic placement of the chopped samples provides the very essence of the sample-based aesthetic. Perhaps producers require the distance of a couple of decades to identify the raw essence of a genre, and be ready to expose its fundamental mechanics and source materials with such transparency. At the heart of the art form lies a producer working with an instrumental formula that affords these sonic and temporal relationships to manifest: a formula originally inspired by the MPC.

## ENDTRODUCTION

Although the sample-based modus operandi that defines the boom bap sound is represented by a large number of contemporary releases, it has to be acknowledged that it is not the prevalent style of Hip-Hop in the mainstream. Conversely, Trap's reign over the genre for a considerable number of years requires further examination at a time when electronic music forms are subject to exponential "trans-morphing" into numerous subgenres (Sandywell and Beer 2005). Yet, it is not rare for forms that cross over into the mainstream to simultaneously undergo an aesthetic counter-reaction, a phenomenon expressed by underground purveyors tracing and practicing the mechanics and stylizations of older subgenre forms. This is encapsulated by subcultures becoming consciously retrospective and evolving their stylizations according to their own code of aesthetic conduct, one that is slower than the pace dictated by commercial pressures (Thornton 1995). The currently buzzing hip-hop underground certainly represents such a reaction, and the answer in artists' lyrics and producers' practices seems to point to a Golden Era boom bap recipe.

But a pragmatic challenge lies in the sourcing of raw materials (phonographic samples) necessary for the process to function. Boom Bap's dependence on the past is challenged by the legal context and finite pool of phonographic material available to producers in the decades prior to the birth of sample-based Hip-Hop. As a result, sample-based producers are forced to source alternative content should they continue to actualize boom bap practices. As noted in Statik Selektah's release, new content is sourced from live performers, while, for their most recent album, *And the Anonymous Nobody* (2016), De La Soul (2017) inform us that their "first album in 11 years was born of 300 hours of live material". J.U.S.T.I.C.E. League, in contrast, are a production collective "recreating every aspect of the original sample, down to the kind of room it was recorded in" (Law 2016) in order to power a plethora of contemporary rap hits (for artists such as Rick Ross, Gucci Mane, Drake and Lil Wayne), and Frank Dukes produces original music to function as sampling material (for artists such as Rihanna, Kanye West, Drake and Future), fusing vintage sonics into his productions to render the new content favorable for subsequent sample-based interaction (Whalen 2016). The Roots, of course, have chosen a predominantly live approach throughout their career but remain conscious of the aesthetic compromises resulting from not directly interacting with sampling technology (Marshall 2006). Ben Greenman (Thompson 2013: 101) comments on The Roots' debut (*Organix* 1993), "It was swag deficient, lacking the grit of sample, microchip, and identifiable urban narrative that, to this day, define the genre".

The latest incarnation of the MPC range (MPCX) seems to be acknowledging the methodological alternatives contemporary producers practice, retaining the interface, operating system and workflow affordances that powered Golden Era aesthetics, while maximizing the potential for recording new music directly into its interface and leveraging interaction with synthesized music forms (exemplified by direct inputs for live instruments, CV outputs for analogue synth control, pre-loaded Trap and EDM sound libraries, and a "controller" mode for working with a computer). If music-makers are adamant about pursuing and evolving the sample-based art form – going as far as reverse-engineering original sampling content – then the instrument that has been fueling sample-based divergences since 1988 may just be able to support this retrospective-futuristic oscillation that characterizes so much of meta-modern creative practice (Vermeulen and Van Den Akker 2010). Yet, the stylizations it has afforded, and the workflow tendencies it has conditioned, are now part and parcel of producers' global vocabularies practiced beyond the context of MPC technology and the confinements of the boom bap aesthetic: the sample-based, syncopated lo-fi "chop" may just have become the guitar riff of the microchip era.

## NOTES

1. Clarke expands Gibson's (1966: 285) coining of the term *affordance* from "a substitute for *values* . . . what things furnish. . . (w)hat they *afford* the observer. . . (which) depends on their properties" into a concept that takes into account the social interdependence between objective properties and the nature of human users/listeners; according to Clarke (2005: 38), "affordances are primarily understood as the *action* consequences of encountering perceptual information in the world."

2. Representative releases include *Mr. Wonderful* (Action Bronson 2015), *Bare Face Robbery* (Dirt Platoon 2015), *Underground With Commercial Appeal* (Fokis 2017), *That's Hip Hop* (Joell Ortiz 2016), *B4.DA.$$* (Joey Bada$$ 2015), *Super Hero* feat. MF Doom (Kool Keith 2016), *Which Way Iz West* (MC Eiht 2017), *Ode to Gang Starr* (Sam Brown 2017), *You Don't Hear Me Tho* feat. The LOX & Mtume (Statik Selektah 2017), *The Good Book, Vol. 2* (The Alchemist & Budgie 2017), *The Ghost of Living* (Vic Spencer & Big Ghost Ltd 2016), and *Packs* (Your Old Droog 2017).

## BIBLIOGRAPHY

Aikin, J. (1997) "Samplers Rule," *Keyboard* (October), p. 47.

Anderton, C. (1987) *SP-1200 Operation Manual*. Scotts Valley: E-MU Systems.

Avgousti, A. (2009) *Beatmaking on the MPC2000XL*. 3rd edn. MPC-Samples. com.

Chang, J. (2007) *Can't Stop Won't Stop: A History of the Hip-Hop Generation*. London: Ebury Press.

Clarke, E.F. (2005) *Ways of Listening: An Ecological Approach to the Perception of Musical Meaning*. New York: Oxford University Press.

De La Soul (2017) *Facebook*, 24 February. Available at: www.facebook.com/ wearedelasoul/ (Accessed: 25 February 2017).

D'Errico, M. (2015) "Off the Grid: Instrumental Hip-Hop and Experimentation After the Golden Age," in: Williams, J.A. (ed.) *The Cambridge Companion to Hip-Hop*. Cambridge: Cambridge University Press, pp. 280–291.

Dery, M. (1990) "Hank Shocklee: 'Bomb Squad' Leader Declares War on Music," *Keyboard*, (September), pp. 82–83, 96.

Emmanuel, C.M. (2017) *The Lox and Mtume Show Off on Statik Selektah's "But You Don't Hear Me Tho"*. Available at: www.xxlmag.com/rap-music/new-mu sic/2017/09/the-lox-and-mtume-show-off-on-statik-selektahs-but-you-dont-hear-me-tho/ (Accessed: 19 September 2017).

Gallant, M. and Fox, D. (2016) *10 Questions With Producer Just Blaze*. Available at: www.uaudio.com/blog/creating-hip-hop-with-just-blaze (Accessed: 19 September 2017).

George, N. (1998) *Hip Hop America*. New York: Viking Penguin.

Gibson, J.J. (1966) *The Senses Considered as Perceptual Systems*. Boston, MA: Houghton Mifflin.

Harkins, P. (2009) "Transmission Loss and Found: The Sampler as Compositional Tool," *Journal on the Art of Record Production* 4.

——— (2010) "Appropriation, Additive Approaches and Accidents: The Sampler as Compositional Tool and Recording Dislocation," *Journal of the International Association for the Study of Popular Music* 1(2), pp. 1–19.

Hodgson, J. (2011) "Lateral Dynamics Processing in Experimental Hip Hop: Flying Lotus, Madlib, Oh No, J-Dilla And Prefuse 73," *Journal on the Art of Record Production* 5.

Howard, D.N. (2004) *Sonic Alchemy: Visionary Music Producers and Their Maverick Recordings*. Milwaukee, WI: Hal Leonard Corporation.

Kajikawa, L. (2015a) "Bringin' '88 Back': Historicizing Rap Music's Greatest Year," in: Williams, J.A. (ed.) *The Cambridge Companion to Hip-Hop*. Cambridge: Cambridge University Press, pp. 280–291.

——— (2015b) *Sounding Race in Rap Songs*. Berkeley, CA: University of California Press.

Katz, M. (2012) *Groove Music: The Art and Culture of the Hip-Hop DJ*. New York: Oxford University Press.

Kulkarni, N. (2015) *The Periodic Table of Hip Hop*. London: Ebury Press.

Law, C. (2016) *Behind the Beat: J.U.S.T.I.C.E. League*. Available at: www.hot newhiphop.com/behind-the-beat-justice-league-news.23006.html (Accessed: 6 September 2017).

Leask, H. (2017) *Review: Your Old Droog Steps His Rap Game Up on "Packs."* Available at: http://hiphopdx.com/reviews/id.2917/title.review-your-old-droog-steps-his-rap-game-up-on-packs# (Accessed: 19 September 2017).

Leight, E. (2017) *'4:44' Producer No I.D. Talks Pushing Jay-Z, Creating '500 Ideas'*. Available at: www.rollingstone.com/music/features/444-producer-no-id-talks-pushing-jay-z-creating-500-ideas-w490602 (Accessed: 6 October 2017).

Linn, R. (1989) *MPC60 Midi Production Centre: Operator's Manual*. Yokohama: Akai Electric Co.

Marshall, W. (2006) "Giving Up Hip-Hop's Firstborn: A Quest for the Real After the Death of Sampling," *Callaloo* 29(3), pp. 868–892.

Mlynar, P. (2013) *In Search of Boom Bap*. Available at: http://daily.redbullmusicacademy.com/2013/11/in-search-of-boom-bap (accessed: 19 September 2017).

Morey, J. and McIntyre, P. (2014) "The Creative Studio Practice of Contemporary Dance Music Sampling Composers," *Dancecult: Journal of Electronic Music Culture* 6(1), pp. 41–60.

Perron, M. (1994) "Checking Tempo Stability of MIDI Sequencers," (from the Proceedings of the 97th AES Convention, San Francisco, 10–13 November 1994).

Ratcliffe, R. (2014) "A Proposed Typology of Sampled Material within Electronic Dance Music," *Dancecult: Journal of Electronic Dance Music Culture* 6(1), pp. 97–122.

Rodgers, T. (2003) "On the Process and Aesthetics of Sampling in Electronic Music Production," *Organised Sound* 8(3), pp. 313–320.

Rose, T. (1994) *Black Noise: Rap Music and Black Culture in Contemporary America*. Middletown: Wesleyan University Press.

Ruiz, M.I. (2017) *Featuring Heems, Danny Brown, and Wiki, the Young New York MC's First Official Album Feels Like a Graduation, With Bars and Beats That Feel Both Fresh and Lived-In*. Available at: https://pitchfork.com/reviews/albums/22988-packs/ (Accessed: 19 September 2017).

Sandywell, B. and Beer, D. (2005) "Stylistic Morphing: Notes on the Digitisation of Contemporary Music Culture," *Convergence* 11(4), pp. 106–121.

Scarth, G. and Linn, R. (2013) *Roger Linn on Swing, Groove & the Magic of the MPC's Timing*. Available at: www.attackmagazine.com/features/interview/roger-linn-swing-groove-magic-mpc-timing/ (Accessed: 10 May 2018).

Schloss, J.G. (2014) *Making Beats: The Art of Sampled-Based Hip-Hop*. 2nd edn. Middletown: Wesleyan University Press.

Serrano, S. (2015) *The Rap Year Book: The Most Important Rap Song From Every Year Since 1979, Discussed, Debated, and Deconstructed*. New York: Abrams Image.

Sewell, A. (2013) *A Typology of Sampling in Hip-Hop*. Unpublished PhD Thesis, Indiana University, Bloomington, IN.

Shelvock, M. (2017) "Groove and the Grid: Mixing Contemporary Hip Hop," in Hepworth-Sawyer, R. and Hodgson, J. (eds.) *Perspectives on Music Production: Mixing Music*. New York: Routledge, pp. 170–187.

Swiboda, M. (2014) "When Beats Meet Critique: Documenting Hip-Hop Sampling as Critical Practice," *Critical Studies in Improvisation* 10(1), pp. 1–11.

Thompson, A. (2013) *Mo' Meta Blues: The World According to Questlove*. New York: Grant Central Publishing.

Thornton, S. (1995) *Club Cultures: Music, Media and Subcultural Capital*. Cambridge: Polity Press.

Tingen, P. (2007) *DJ Premier: Hip-Hop Producer*. Available at: www.soundon sound.com/people/dj-premier (Accessed: 19 September 2017).

Toop, D. (2000) *Rap Attack 3: African Rap to Global Hip Hop*. 3rd edn. London: Serpent's Tail.

Vermeulen, T. and Van Den Akker, R. (2010) "Notes on Metamodernism," *Journal of Aesthetics & Culture* 2(1), pp. 56–77.

Wang, O. (2014) "Hear the Drum Machine Get Wicked," *Journal of Popular Music Studies* 26(2–3), pp. 220–225.

Weingarten, C.R. (2010) *It Takes a Nation of Millions to Hold Us Back*. New York: Bloomsbury Academic.

Whalen, E. (2016) *Frank Dukes Is Low-Key Producing Everyone Right Now*. Available at: www.thefader.com/2016/02/04/frank-dukes-producer-interview (Accessed: 6 September 2017).

Williams, J.A. (2010) *Musical Borrowing in Hip-Hop Music: Theoretical Frameworks and Case Studies*. Unpublished PhD thesis, University of Nottingham, Nottingham.

## DISCOGRAPHY

Action Bronson (2015) *Mr. Wonderful* [CD] US: Atlantic/Vice Records 549151-2.

The Alchemist & Budgie (2017) *The Good Book, Vol. 2* [Stream, Album] US: ALC.

Apollo Brown & Planet Asia (2017) *Anchovies* [Stream, Album] US: Mello Music Group.

De La Soul (2016) *And the Anonymous Nobody* [CD] US: AOI Records AOI001CDK.

Dirt Platoon (2015) *Bare Face Robbery* [Stream, Album] US: Effiscienz.

DJ Shadow (1996) *Endtroducing* [CD] UK: Mo Wax MW059CD.

Eric B. & Rakim (1987) *Paid in Full* [CD, Album] Europe: 4th & Broadway, Island Records 258 705.

Fokis (2017) *Underground With Commercial Appeal* [18 x File, FLAC, MP3, 320Kbps] US: Loyalty Digital Corp.

Gang Starr (1989) *No More Mr Nice Guy* [CD] US: Wild Pitch Records WPD 2001.

——— (1992) *Daily Operation* [CD] UK & Europe: Cooltempo, Cooltempo ccd1910, 3219102.

——— (1994) *Hard to Earn* [CD] US: Chrysalis 7243 8 28435 2 8.

Grandmaster Flash (1981) *The Adventures of Grandmaster Flash on the Wheels of Steel* [Vinyl] US: Sugar Hill Records SH 557.

James Brown (1970) *Funky Drummer* [Vinyl] US: King Records 45-6290.

Jaylib (2003) *Champion Sound* [Stream, Album] US: Stones Throw Records.

Jay-Z (2017) *4:44* [File, FLAC, Album] US: Roc Nation.

Joell Ortiz (2016) *That's Hip Hop* [CD] US: Deranged Music Inc. THHJ01.

Joey Bada$$ (2015) *B4.DA.$$* [17 x File, AAC, Album, 256] US: Pro Era, Cinematic Music Group.

Kool Keith feat. MF Doom (2016) *Super Hero* [File, AIFF, Single] US: Mello Music Group.

KRS-One (1993) *Return of the Boom Bap* [CD] US: Jive 01241-41517-2.

Lil' Jon & The East Side Boyz (1997) *Get Crunk, Who U Wit: Da Album* [CD] US: Mirror Image Entertainment MRR 4556-2.

MC Eiht (2017) *Which Way Iz West* [Stream, Album] US: Year Round/Blue Stamp Music.

N.W.A. (1988) Straight *Outta Compton* [Vinyl] US: Ruthless Records, Priority Records SL 57102.

Public Enemy (1988) *It Takes a Nation of Millions to Hold Us Back* [CD] UK & Europe: Def Jam Recordings 527 358-2.

Public Enemy (1990) *Fear of a Black Planet* [CD] Europe: Def Jam Recordings, CBS 466281 2.

The Roots (1993) *Organix* [CD] US: Remedy Recordings, Cargo Records CRGD 81100.

——— (1996) *Illadelph Halflife* [CD] UK & Europe: Geffen Records GED 24972.

Sam Brown (2017) *Ode to Gang Starr* [File, AAC, Single] US: Below System Records.

Statik Selektah feat. The LOX & Mtume (2017) *But You Don't Hear Me Tho* [File, AIFF, Single] US: Showoff Records/Duck Down Music Inc.

Sugarhill Gang (1979) *Rapper's Delight* [Vinyl] US: Sugar Hill Records SH 542.

T.I. (2003) *Trap Muzik* [CD] Europe: Atlantic, Grand Hustle 7567-83695-2.

T La Rock & Jazzy Jay (1984) *It's Yours* [Vinyl] US: Partytime Records, Def Jam Recordings PT 104.

Too $hort (1987) *Born to Mack* [CD] US: Jive 1004556.

Vic Spencer & Big Ghost Ltd (2016) *The Ghost of Living* [11 × File, FLAC, MP3, Album] US: Perpetual Rebel.

Warren G. (1994) *Regulate . . . G Funk Era.* [CD] US: Violator Records, Rush Associated Labels 314-523-335-2.

Your Old Droog (2017) *Packs* [CD, Album] US: Fat Beats FB5181.

# 4

## The DAW, Electronic Music Aesthetics, and Genre Transgression in Music Production

### The Case of Heavy Metal Music

## Mark Marrington

This chapter continues my ongoing research into the DAW and its role in re-configuring popular music practice (see Marrington, 2011, 2016, 2017a), with particular reference to genre characteristics in heavy metal music. My focus is on the manner in which those "rules" that determine the "compositive" elements of the metal genre (see Fabbri, 1982) are "transgressed" as a result of their interaction with new technological forms. These ideas are applied specifically in relation to DAW-based practices, highlighting the manner in which this medium might be regarded as an agent of genre deconstruction and destabilization. Key to my argument is the hypothesis that the DAW is essentially a *genre-specific* medium in the sense that it foregrounds specific notions of creative practice associated with the aesthetics of electronic music. To illustrate this, I trace the route by which the DAW and its inherent electronic music aesthetics have become gradually integrated into metal music practice, from metal musicians' uses of the DAW during the mid-1990s through to the activities of DAW-based metal practitioners during the mid- to late 2000s and the hybridized forms of "electronic" metal that emerged at this time, including djent, djent-step, and cyber-grind. The concluding part of the chapter discusses situations in which older metal bands have attempted to incorporate, from outside their idiom, the traits of DAW-derived genres, with a particular focus on the dubstep collaborations on Korn's 2012 album, *The Path of Totality*. The chapter's key assertion is that metal artists who have gravitated toward either the DAW, or DAW-based practitioners, are all engaging to some degree with the aesthetics of electronic music and, while this has enabled the domain of metal music practice to be re-imagined, it has not been without consequences for the genre's integrity and its standing among metal purists.

### MUSICAL GENRES: RULES AND TRANSGRESSION

Central to this discussion is the idea of genre, a term that requires some initial definition prior to considering the ways in which the DAW may be regarded as inculcating specific genre inclinations. For this purpose,

I draw on genre theory originating from the discipline of popular music studies and in particular Franco Fabbri's well-known essay "A Theory of Musical Genres: Two Applications" (1982), which defines genre as "a set of musical events (real or possible) whose course is governed by a definite set of socially accepted rules". Fabbri views genres as sets with their subsets (i.e., subgenres) – essentially individualized manifestations of structures within broader "musical systems" (or codes) – and his account outlines a number of "generic rules" that can be usefully brought to bear in considering the nature of a particular genre, respectively: formal and technical, semiotic, behavioral, social and ideological, economic and juridical. Of the rules of genre presented by Fabbri, this chapter's focus is on the formal and technical – namely, rules which operate on what he calls the "compositive level", essentially referring to the music's substance (codes governing "note systems"; "conception of musical time"; the nature of its melodic, harmonic, and rhythmic aspects, and so on) as well as rules relating to performance techniques and instrumental characteristics. Fabbri suggests that the "codification" of a genre is dependent to a certain degree on aspects of already existing musical systems defined by such rules, with those genres that are recognized as being new distinguishing (or "individualizing") themselves via some form of "transgression" of the rules of a particular system. Pertinent to this discussion, a particular factor identified by Fabbri as potentially transgressive in this regard is "the application of new techniques, made possible by technological development". It is thus the purpose of this chapter to consider the ways in which these kinds of genre elements, specifically in the context of metal practice, have been re-configured as the result of a particular technological development – the DAW – and the consequences of this for the genre's identity.

## ELECTRONIC MUSIC AESTHETICS AND THE DAW

Discussion of the DAW in the literature has ranged widely in its focus, along the way encompassing theories of affordance, metaphor, human-computer interaction, and the psychology of creativity (see for example, Duignan et al., 2004, 2005; Barlindhaug, 2007; Brown, 2007; Mooney, 2011; Zagorski-Thomas, 2014; Bell et al., 2015; Marrington, 2016, 2017a; Strachan, 2017; Bell, 2018). In effect, this body of work has come to constitute a detailed inventory of characteristics of the DAW that potentially inform creative practice, revealing what Duignan et al. (2005: 3) have referred to as "the underlying assumptions and structures that favor one form of musical structuring over all others".[1] My own research, which is indebted to much of this literature, initially took place within the educational context in reference to students' working practices within DAW environments (Marrington, 2011). In particular, I drew attention to the nature of the DAW as a mediating technology which re-configured musical thought as students transferred their ideas from a "traditional" technological context, for instance a guitar-oriented approach, to creating songs using chord shapes, to certain DAW-determined ways of working (for example, programming within the Arrange window). From this emerged the idea of DAW-specific literacies – in essence languages engendered by the DAW that condition the ways that users articulate their musical ideas and which they take on board when they immerse in these environments. This was then pursued in further articles (Marrington, 2016, 2017a) discussing the DAW in the context of songwriting practice in which I explored the idea that DAW-based music creation effectively inculcated a literacy derived from electronic music aesthetics, ultimately with consequences for the ways in which songs were conceived and produced.

The notion that the DAW engenders electronic music genre aesthetics requires a certain amount of unpacking. Essentially, however, what this refers to are creative practices associated with the various tools that the DAW has inherited from analogue and digital hardware and earlier forms of computer software. This includes, for example, the programming of musical ideas using sequencers (hardware and software) and drum machines, sound design processes typically associated with synthesizers, and the application of digital sampling technology to isolate audio objects as basic raw materials. Examples I have discussed in relation to this include Radiohead, whose move to early versions of Logic and Cubase in the early 2000s effectively converted the band into an electronica outfit (see the albums *Kid A* and *Amnesiac*), and more recently, James Blake (on his 2011 debut album *James Blake*), whose singer-songwriter leanings were shaped by dubstep-derived editing approaches and sound design. Recent work on the DAW, most notably Strachan (2017), has echoed this view that the DAW engenders electronic music aesthetics and has considered this directly in relation to concepts of genre. Strachan posits a number of what he calls "cyber-genres", including dubstep, hauntology, vaporwave, and the metal subgenre djent, asserting that these are "simultaneously resultant from, and reflective of, the contexts of digitization", and that the "widespread availability of computer-based music production technologies has been integral to these post-digital musics" (2017: 135).

It is arguable that of the various electronic music practices engendered by the DAW, the sampling aesthetic – characterized by the appropriation and re-combination of pre-existing audio materials – has become particularly dominant in DAW-based creative practice, with audio objects in essence, constituting a DAW's "primary fuel". This has to a degree been encouraged by the propensity of the typical DAW GUI to encourage the "atomization" of the musical material within its visual paradigm. Zagorski-Thomas (2014: 134), for example, has remarked on the "block diagram", which "would seem to encourage the user to think in terms of sound as an object rather than a stream", while Strachan (2017: 99) similarly draws attention to the "visual representation of sound as frozen in time" leading to "a conceptualization of musical material into distinct, temporally located blocks of musical information". A DAW's cut, copy, and paste functions, which are a staple of computer software environments, also contribute to this atomization, enabling relationships between the musical elements contained within these objects to be infinitely re-configured. This ultimately plays down the notion of there being a *necessary* relationship between musical ideas, which in the case of music that draws on a wide range of audio sources can lead to some striking juxtapositions. Another by-product of this mode of engagement with musical material, which I have discussed elsewhere (Marrington, 2016, 2017a), is the "loop paradigm", which describes the tendency to organize audio objects into repeating patterns which are layered one upon the other within the sequencer environment of the typical DAW.[2] The net result of this is, as Strachan (2017: 7) observes, that "computer-based music production has increasingly brought the physical and textural properties of sound to the fore within the creative process" and with the user working "directly with captured and generated sounds that are at a remove from processes and competencies of performance traditionally associated with musicianship". The examples of hybridized metal genres discussed later in this chapter in the work of Gautier Serre (Whourkr/Igorrr) and Remi Gallego (The Algorithm) epitomize this particular creative paradigm of the DAW.

In addition to the DAW's foregrounding of the sampling aesthetic and audio-object as a primary unit of musical design, the DAW has also inherited electronic music practices concerned with sound design and synthesis. For example, most DAWs are equipped with onboard tools that model synthesis engines of various types (including analogue, FM, and granular) and support numerous third-party plugins, many of which emulate analogue synthesizers of the past (such as Arturia's Minimoog and Native Instruments' B4). These have been harnessed by the current generation of DAW users in much the same way as their earlier electronic musician counterparts to generate music of comparable character, albeit with the greatly increased level of refinement that computer-based equivalents of these older tools bring to the editing process. Certain DAW-evolved cyber-genres owe much of their character to the application of such sound design tools in their production. Dubstep, for example, is defined by the attention to detail accorded to the sonic design of its bass and lead synth lines, which typically operate both as riffs and attention-grabbing sonic elements (see Pearson and Mullane, 2014). In essence, this is the characteristic "dirty" bass sound, which is heavily reliant on the unique qualities of particular third-party FM synthesizer plugins, such as Native Instruments' FM8 and Massive. This, in conjunction with careful MIDI programming, enables the achievement of what John von Soggerson describes as "increasingly twisted bass riffs with complex rhythmic modulations and subtle changes" (2012).

A final point, which also has significance in relation to genre, concerns the inherent "musical" leanings of certain DAWs with regard to the domains of musical practice they imply. This refers to the idea that most DAWs are, to varying extents, revisiting past technological paradigms of music making, often accompanied by skeuomorphic representations (see Bell et al., 2015; Marrington, 2016), and that these have the potential to inform the character of the music they are used to create. Propelleheads' Reason, for example, closely models the visual characteristics of older technologies – samplers and synthesizers and old-school sequencers – within a rack of gear, as well as their behavior. This arguably has the purpose of appealing to a user group whose musical outlook is sympathetic to the aesthetic characteristics of the music these kinds of technologies were originally used to create (see Barlindhaug, 2007). For example, it might be suggested that Reason's interface is designed to appeal to hip-hop producers and electronic musicians of various stylistic persuasions. Compare to this to a DAW such as ProTools, which, while possessing a number of comparable studio tools, by contrast, does not parade these in front of the user in this manner – instead the emphasis is on working in the multi-track recording environment and mixer window. Production tools – EQs, compressors, effects, and so on – are called upon when required rather than being pointed to by the software. ProTools remains today the standard for professional studio recording and production, appealing primarily to engineers whose focus is on recording musical performances, and the conservatism of its interface ultimately reflects this.

Interestingly these tensions between the DAW's various implied creative aesthetics can be discerned in the hobbyist literature that supported the rapidly expanding DAW user-base during the early- to mid-2000s, including (in the UK) *Computer Music*, *Future Music*, and *MusicTech*. On the one hand, these magazines focused on contemporary DAW-based artists and their innovative techniques, the vast majority of whom were working within the context of electronic music production. On the other, they ran features that looked backward in the "rearview mirror"[3] to musical genres that had evolved in studio contexts prior to the DAW. *Computer Music*, for instance, ran a series of short articles entitled "Sound like . . .", providing walkthroughs of the production styles of artists as stylistically varied as My Chemical Romance, the Police, and Orson, whose genre characteristics and production aesthetics were mimicked using the virtual studio tools provided by the DAW. In one issue, the magazine also ran a feature exclusively devoted to metal music (*Computer Music Magazine*, 2007), which took a similar approach, discussing the genre in terms of the production values associated with earlier classics, such as Metallica's *Metallica* ("black") album (1991). The perspective on the metal genre expressed here is a narrow one relative to contemporary developments and largely unaware of the transformations that had taken place since the 1990s, such as the move toward sequencing and sampling aesthetics in industrial and nu metal. What is particularly interesting about this article, however, is that although its primary focus appears to be traditional metal production techniques, the same magazine is also accompanied by a DVD containing "1815 24-Bit Metal Samples", which it claims provides "everything you could possibly need to create your own metal masterpiece. . . . From thrash to screamo, there's a little something for the metal maniac in everybody". The inclusion of this library of metal samples, coupled with such tongue-in-cheek remarks, is clearly designed to court the electronic musician – i.e., the magazine's core readership – and effectively places the audience for the publication some considerable distance from the traditional metal practitioner. There is an irony in this, however, in that a decade on from this article, it is ultimately the younger generation of DAW-based artists, fully immersed in the electronic music aesthetics of the DAW, who have contributed most to the re-situation of the metal genre.

## LOCATING THE METAL GENRE: RULES, BOUNDARIES, AND TRANSGRESSION

Having established some ideas regarding the DAW's relationship to the compositive elements of music and considered the sense in which DAWs can be deemed genre-specific in their leanings, I now wish to consider the nature of metal's music's code as evaluated by writers working in the academic context. The purpose here is to provide a context for this article's later discussions of the ways in which the metal genre has been modified through contact with the DAW and its inherent creative practices. Commentators have frequently discussed the metal genre in terms not unlike those adopted by Fabbri in his aforementioned article. For example, Weinstein (2000: 6) states that heavy metal is a musical genre with

> a code, or set of rules, that allows one to objectively determine whether a song, an album, a band, or a performance should be classified as belonging to the category "heavy metal". That code is not systematic, but it is sufficiently coherent to demarcate a core of music that is undeniably heavy metal. It also marks off a periphery at which heavy metal blends with other genres of rock music or develops offshoots of itself that violate parts of its code or develop new codes.

Specifically, for Weinstein, "the sonic, the visual, and the verbal dimensions all make crucial contributions to definition of the genre" (2000: 7), and the two earlier subgenres of heavy metal that she specifically discusses – namely lite metal and thrash metal – are treated "in terms of their similarities to and differences from the core of heavy metal. Each of these offshoots changes or even breaks the heavy metal code in some ways, but still retains enough of this code to be placed in the same 'family' with it" (2000: 7). The genre has also been discussed in terms of specific defining features that cohere with Fabbri's "formal and technical" attributes. Weinstein (2000: 22–27), for example, focuses on metal's "sonic dimension", drawing attention to the influence of the blues rock and acid rock traditions on the genre's musical language, the role of "loudness" and amplification, the importance of the lead guitar solo executed with "great manual dexterity" and "elaborate electronic technology that distorts and amplifies", the specific timbres of the drumkit, the wide expressive range and power of the vocalist, and so on. Walser (1993: xiv) also adopts an overtly music-focused approach to discussing the genre, which serves to support a cultural studies perspective that these musical details constitute "significant gestural and syntactical units" that come to constitute "a system for the social production of meaning – a discourse". He posits that "guitarists have been the primary composers and soloists of heavy metal music", in effect affirming the pivotal role of that particular technological form in shaping the genre's musical identity, as well as noting heavy metal's reliance on "effects that that can be created only with the help of very sophisticated technology" (1993: 16). Walser places particular emphasis on heavy metal's timbral characteristics, identifying the distorted powerchord as a key signifier of metal music's power, which is reliant on technology – i.e., the unique characteristics of amplifiers and guitar pickups – for its effect. Other key "rules" of the genre listed by Walser include volume, vocal timbre, modal (Aeolian/Dorian/Phrygian) and harmonic language (use of certain chords and progressions), rhythm (4/4 time, "monolithic inexorable pulse"), the downplaying of melody (relative to timbre), and the importance of the virtuosic guitar solo (which often borrows from classical music tropes).

Where "transgressions" of metal's genre rules are concerned, these have tended to be discussed in reference to the ways in which metal's musical language has been re-worked within particular subgenre contexts. Pieslak (2007), for example, has explored Meshuggah's "re-casting" of metal music in terms of the band's experimentation with rhythmic and metrical complexity in the progressive/math metal subgenre. The most extensive study of this type is Phillipov (2012), who has demonstrated, in reference to bands such as Cannibal Corpse and Carcass, the ways in which the death metal subgenre subverts musical conventions, such as harmony, melody, rhythm, and form, and redefines vocal performance techniques, including some reference to studio production strategies. Such writing, along with coverage of metal's technologically achieved sonic characteristics in Weinstein and Walser, has thus begun to hint at the role that technology plays in shaping/re-shaping metal's genre characteristics. More recently, academics working in the music production studies field have begun to explore the sonic aspects of metal in reference to the studio production process. Mark Mynett's work, for example, in his recent doctoral thesis (2013) and subsequent *Metal Music Manual* (2017), expands the discussion of the metal's timbral rules in reference to the concept of "heaviness", whose successful achievement in the production process is key to the articulation of the genre's musical elements. Similarly Williams (2015) has discussed the ways in which digital technologies (including DAW software) have effected significant changes in metal music's timbre. It is interesting to note, however, that both of these authors have expressed reservations regarding the boundaries of the metal genre relative to the role of new technological forms in the music's execution. Mynett, for example, stresses the metal genre's concern to foreground "idealized performative nuances" above mediation by technological forms:

> The majority of Contemporary Metal Music's sounds index sound producers in the form of performing musicians, rather than computer, or synthetic based sound production. Additionally, despite the tendency for high levels of technological mediation in CMM's recorded and mixed form, producers in the field invariably focus the results of this technological mediation toward the performative nuances, or idealised performative nuances, of performing musicians.
>
> (Mynett, 2013: 52)

In a similar vein, Williams's comments assert that the compositive elements of metal music have remained immune to the effects of new technological forms: "sampling, synthesis and digital editing techniques have given rise to entirely new genres of music, yet metal as a genre has remained ostensibly with much the same musical format" (2015: 39). While Mynett does acknowledge in his discussion the industrial metal subgenre, whose character owes much to a willingness to adopt new technologies, both statements imply the delimiting of the genre rules of metal to production contexts in which the music's integrity remains intact in the face of technological change. This is understandable given the focus of both writers on unique timbral aspects of metal production within a specified context, but it does tend to misrepresent metal's changing relationship with a range of new technologies over the last forty years.

## METAL, ELECTRONIC MUSIC AESTHETICS, AND THE ADOPTION OF THE DAW

Until very recently, the only writer on metal to give any serious attention to the role of digital technologies in re-conditioning metal aesthetics is Ian Christe (2004), in his detailed historical survey of metal's evolution. As Christe's account implies, while technological forms, such as samplers, sequencers, and drum machines (whose characteristics, as aforementioned, anticipate the DAW), have tended to be employed at the peripheries of metal music, they have nonetheless impacted significantly on the genre's musical character. Christe's particular focus is on the affinities between metal and electronic music aesthetics that began to become apparent in the 1990s, for example in regard to the influence of 1990s techno-derived genres, such as Gabber (2004: 334–345). In an attempt to expand this perspective, my own research (Marrington, 2017b) has traced the interchanges that have taken place since the 1980s between metal artists and musicians working in genres whose practices have been largely defined by new technological forms, including hip hop, electronica, and industrial music. In particular two subgenres of metal that evolved in the 1990s owe much of their identity to the fusion of conventional metal tropes (riff-based guitar and metal vocal styles) with the aesthetics of sampling and programming (sequencers and drum machines in particular) – namely, nu metal and industrial metal. Indeed, it was artists working within these subgenres who were the first to adopt the DAW as a creative tool within their production practice. The band Fear Factory, for example, whose industrial music leanings had previously led them to work with remix artists such as Frontline Assembly, as a means of imbuing their sound with a certain electronic character, began to adopt ProTools into their armory as soon as it began to become a studio fixture in the mid-1990s (Dean, 2013; Marrington, 2017b). For Fear Factory, this technology did not function simply as a virtual tape machine; rather, it provided the ultimate means for the realization of their "machine" aesthetic:

> We used technology even though there was a human playing it. After it was edited, and after the tones have been changed, we were trying

to go for that electronic sound. That was our purpose. We were trying
to create a machine. We thought "What would a machine sound like
if it was in music? What would it sound like if the guitars and kick
drums played the exact same thing over and over again"? It would
sound like a machine, so that's exactly what we were trying to create.

(Dean, 2013)

In practice, Fear Factory's machine aesthetic is characterized by the use of
the DAW to dissect and reiterate fragments of their own recorded performances (hence a manifestation of the DAW's sample loop paradigm) as
can be heard on albums such as *Fear Is the Mind Killer* (1993), *Demanufacture* (1995), and *Obsolete* (1998). Metal artists' gravitation toward the
sampling aesthetic is also apparent in the nu metal subgenre in the work
of artists as Slipknot and Deftones in the early 2000s. These bands were
notable for including in their line-up personnel chosen for their affinities
with forms of technology outside the traditional metal context, including
hardware-based samplers, synthesizers, turntables, drum machines, and
DAW software.[4] Thus even in this earlier period of the DAW's development, the rules of the metal genre had already begun to be modified as a
result of their interaction with new forms of technology and indeed in the
years leading up to the establishment of the DAW at the center of contemporary record production, there had already been a certain conditioning to
the technological aesthetics of that medium.

## BEDROOM-PRODUCED METAL:
## THE EMERGENCE OF DJENT

The first high-profile metal subgenre to emerge from a primarily DAW-
based production context was djent,[5] a form of progressive metal that was
evolved by individual metal practitioners (typically guitarists) working
within "bedroom production" environments. Its proliferation during the
mid-2000s was catalyzed by social media formats for the distribution of
independently produced music, most notably Soundclick, and by 2011, it
had become recognized as an identifiable metal subgenre.[6] Djent emerged
precisely during the period of the DAW's widespread acceptance into
home studio practice and neatly straddles the line between older genre
traditions of metal and the electronic music hybrids that began to appear
in the late 2000s. The subgenre's debt to the digital music production environment can be understood in two ways. Firstly, DAW-based tools contributed significantly to the subgenre's sonic characteristics, mainly as a
result of its practitioners' heavy reliance on sampler plugins to program
their drum parts – typically Native Instruments' Battery and the sampler
instruments provided within Propellerheads' Reason. The sounds triggered by these samplers were also "authentically" sourced from a library
recognized for its metal leanings – the Toontracks "Drumkit from Hell"
sample pack (Laing, 2011), which was notable for having been developed

in association with Tomas Haake, drummer in the Swedish progressive metal band Meshuggah (Lentz, 2012).[7] Djent's characteristic guitar sound was also shaped by digital amp modeling hardware, most commonly the affordable Line 6 POD units (Laing, 2011; Shelvock, 2013). Secondly, the DAW's influence on djent's character can be seen in its conditioning of the performance precision that became associated with its guitarists. This was in essence a by-product of spending many hours recording guitar parts against the sequencer grid using click tracks. To this effect, Misha Mansoor, guitarist with the influential djent band Periphery, has commented that:

> When you're playing that way you start to focus on parts of your technique that make all the difference in the world. . . . Things you'd never have noticed if you weren't sitting in front of a computer and hearing your playing back. It taught me how to play guitar.
>
> (Laing, 2011: 52)

The character of djent guitar performance thus owes something to the DAW's sequencer paradigm, which, in effect, led djent musicians to develop a machine-oriented virtuosity. Nonetheless, these various interactions of djent artists with the DAW were primarily the product of expediency – a means of achieving the effect of a band performance on a budget – rather than the conscious cultivation of an electronic music aesthetic. The first wave of Djent artists – including Periphery, Tesseract, and Animals as Leaders – regard themselves first and foremost as metal musicians, placing emphasis on the guitar performance aspects of the subgenre and foregrounding slick production values that reflect consummate engineering skills. Ironically, all these acts today perform live in a full-band context and refer to the DAW primarily as a vehicle for conventional recording and production operations.[8]

## BREAKING METAL'S BOUNDARIES: THE DAW AND GENRE HYBRIDIZATION

While djent can be regarded as a typically guitar-centered metal subgenre whose sonic characteristics were incidentally inflected by the DAW, other parallel developments during this period illustrate a more thorough integration of metal elements with electronic music aesthetics. These developments were typically spearheaded by artists who, while professing authentic metal leanings, also possessed a strong affinity with electronic music and the range of subgenres that were developing within the context of DAW-based practice, such as breakcore and dubstep. A number of these artists evolved simultaneously both as performing musicians, typically proficient on either guitars or keyboards, and experts in programming, sampling, and sound design skills. This naturally led to the creation of various forms of hybridized metal – which have typically been referred to using terms such as electrogrind or cybergrind – and, consequently, a certain amount genre instability.

Important early pioneers of this synthesis during the mid-2000s are Genghis Tron, a New York-based three-piece with prominent extreme metal guitar/vocal leanings but also a strong appreciation for contemporary EDM and even synth-pop styles.[9] The trio comprises an adept lead guitarist but neither a bass player nor a drummer, assigning these aspects to live-performed synthesizers and programmed drums using NI Battery run through Ableton Live (Surachai, 2008). Unlike djent artists, Genghis Tron do not use the latter as a harmonic/percussive backdrop to the guitar-led metal elements, rather synthesizers and programmed beats are there to represent defined electronic music styles, which appear as unique sections alongside the metal passages. The typical approach is to juxtapose the two genres within the course of a single track using hard edits, as can be heard on songs such as "Endless Teeth", "I Won't Come Back Alive", and "City on Hill" on the album *Board Up the House* (2008). As *New York Times* reviewer Ben Ratliff noted with insight, "Letting go of the old metal-band code of virtuosos playing in real-time helps these musicians: It lets the songs on *Board Up the House*, the band's second album, take sudden bizarre turns and allows them moods and textures that cross genres" (Ratliff, 2008).

This hybrid approach can be seen to have reached its extremes in the work of Gautier Serre, a French musician known both for his earlier Whourkr project and, at the time of writing, his profile as the artist Igorrr.[10] Serre's primary musical influences are extreme metal (as a performing guitarist) and computer-based electronic music, but, as his more recent music demonstrates, he also has affinities with classical music and traditional folk. Much of his earlier work in Whourkr (a two-man band featuring Mulk on vocals) is clearly situated within the death metal subgenre, characterized by fast "blast" beats,[11] growled vocals, and heavy guitar riffs. What makes Whourkr unique, however, is Serre's unquestionably "electronic" approach to programming and editing the duo's recorded performances in the DAW environment (in this case, Cubase). Typically, Serre's strategy, on albums such as *Concrete* (2008) and *4247 Snare Drums* (2012), is to cut up and reiterate fragments of the material at high speeds (using the stutter-editing technique), as well as subject it to various forms of processing, including audio transposition. In his later work as Igorrr, Serre has collaborated with musicians working in a wider range of genre contexts, including classical music and jazz, as a means of sourcing diverse recorded material for his composite multi-genre mashups. The result is not unlike Genghis Tron, although here the contrasting musical elements are presented one after the other with such rapidity that there is little chance to orient one's listening in genre terms. This can be heard, for example, in the track "Tendon" (on the album *Nostril*, 2010), which comprises a barrage of samples and passages of metal, opera, folk fiddle, and classical string ensemble, as well as speech, all interspersed with, or accompanied by, frenetic high-speed beats and samples. The latter, in particular, are a hallmark of the Igorrr style and reflect the influence of breakcore[12] in particular, which Serre suggests is the "base for mixing all the genres I love" (Ume, 2012). Like breakcore itself, Igorrr's work appears to be genre-*less*, and despite his claim to metal roots, there is little concern with maintaining the integrity of the metal elements he uses: "I don't know from where this idea is coming, I'm just making the thing I want to hear, I think there is no concept or something, I'm making the music I feel. I love Baroque Music, I love Metal and I love Breakcore, I'm just expressing everything at the same time" (Melo, 2012).

Another French artist who has been of equal importance in the resituation of the metal genre in relation to DAW-based electronic music practice is Rémi Gallego, a guitarist and DAW-based musician who is known as The Algorithm. Gallego, for whom "the idea of combining electronic music with metal just came naturally", bases his production practice within Ableton Live, which he also uses for onstage performance. In addition to being referred to variously as "a crossover between IDM and metal" and a pioneer of "heavy computer music", Gallego's work has notably received the label "djent-step" (Clarity, 2011; DemiGodRaven, 2012; NewFuryMedia, 2016), implying a fusion of the djent subgenre with dubstep. In reference to this, Gallego has stated that "Metal is like my first love, and still the music genre I prefer because it conveys the violence and the intensity of feelings very well. Dubstep is like an electronic version of metal, this is why combining the two of them can bring of the best of two worlds" (Clarity, 2011). In practice, Gallego's music is comparable to the work of Igorrr, comprising an eclectic and changeable collage of many elements, including djent-like metal riffs, blast beats, dubstep-influenced bass lines, and 1980s-style synthesizer arpeggio figures as well as samples drawn from video game music and electronic artists, such as Daft Punk (for a recent representative example, see the album *Brute Force*, 2016). Unlike Igorrr, however, there is a greater consistency in the preservation of stylistic coherence within individual tracks, with metal riffs and lead lines more seamlessly integrated with the electronic textures. Gallego has provided some useful insights into his creative process in an interview, including the following remarks on the influence of the DAW on his musical thinking:

> I compose songs directly on the DAW (Digital Audio Workstation), like an electronic artist, and it allows me to visualize directly what I'm writing. It's like a horizontal 2D retro game. Just imagine a character running through different universes and ambiances. In fact, this way I can actually "see" a song, it's not just random sounds. I think I'm trying to record each song just as if it were a story full of twists. That's why I love brutal changes, and abnormal progressions. I want each part to be unpredictable. Just like a great scenario!
>
> (Clarity, 2011)

Such comments, which recall ideas discussed earlier in relation to the visual affordances of the DAW, offer valuable clues as to the conceptual processes by which DAW-based artists, such as Gallego, Igorrr, and Geng-his Tron, have arrived at their particular juxtapositional approaches to the organization of their material.

## OLD DOGS, NEW TRICKS: KORN'S *THE PATH OF TOTALITY*

Finally, it is significant that DAW-based electronic musical aesthetics have also attracted "old school" metal artists whose idiom was evolved in the more traditional band format, most notably Korn on their 2012 album, *The Path of Totality*. This was a unique project in which the band consciously attempted to experiment with metal's genre rules through collaborations with North American dubstep artists, including Skrillex, Dutch production trio Noisia,[13] and Western Canadian "bass" musicians Downlink and Excision.[14] Lead singer Jonathan Davis's claim that in undertaking these collaborations, the band were "fighting the fight to break a new genre through" (*Billboard*, 2011), reflected the view that despite their nu-metal origins (see Udo, 2002; Pieslak, 2008), Korn were not confined within any metal-related genre boundaries: "There's a lot of closed-minded metal purists that would hate something because it's not true to metal or whatever, but Korn has never been a metal band, dude. We're not a metal band" (Giles, 2012).[15] As with Gallego's djent-step experiments, the project also appears to have been driven by Davis's notion that dubstep had a particular affinity with metal music: "North American dubstep is the new electronic heavy metal. It's the filthier, the better in that world, and with heavy metal, it's the heavier the better, so it's kind of the same thing" (Goodwyn, 2012). While some critics have been skeptical about this claim (for example Deviant, 2011), such a statement is not as far-fetched as it might seem. Excision, for exam-ple, in a 2011 Roach Coach Radio interview, revealed that his influences were a "strange" combination of metal and hip hop so "when dubstep came along it was kind of the culmination of those like heavy riffs that I liked in metal and the slower paced laid back beats of hip hop and it just kind of combined into one thing that I fell in love with and had to learn how to produce" (Scion AV, 2011; see also Davis, 2013). Downlink has also acknowledged his early rock and metal leanings prior to developing an interest in electronic music via the drum 'n' bass scene in the 1990s (Bass, 2012).[16]

The particular significance of Korn's association with these artists lies in the fact that the project allowed them to explore creative perspec-tives deriving from a range of DAW production contexts.[17] Indeed, in the sleeve notes to the album, Davis has included specific thanks to Ableton Live, Native Instruments, and Novation, acknowledging the impact of their software tools on the character of what was created. In practice, the interaction between the band and their collaborators is characterized by varying degrees of synthesis of metal and DAW-situated aesthetics.

In many cases, the band's approach was to build their guitar and vocal parts on sketches that had been worked out beforehand by their collaborators in the form of fragments of programmed beats, bass, and synth lines (*Billboard*, 2011). Here, Korn's strategy was to situate their metal-oriented songs in relation to these fragments, expanding the material where necessary to create more involved song structures. For example, Noisia's contribution on the track "Let's Go" was a 16-bar riff at 125 bpm, which was used as the foundation for the song's chorus, in combination with new verses written by Korn. Another track, "Sanctuary", a "pet project" with Downlink, involved Davis undertaking the programming of certain parts of the song to expand the material beyond the basic fragments he had been supplied with (Davis incidentally credits Downlink in the album's sleeve notes "for teaching me mad skills and dubstep production"). Sometimes there is greater integration of material with the band's idiom, showing that Korn were to some degree open to modifying their musical style in relation to ideas conceived in the electronic music context. This can be seen on the track, "Kill Mercy Within", generated from a 32-bar programmed "riff" written by Noisia, which Korn's guitarist Munky attempted to transfer to his instrument. Munky has remarked on the unusual character of this riff, which was "challenging to recreate" because it was "all over the neck" and unlike anything he had played before (*Billboard*, 2011).

The tracks written and coproduced by Skrillex ("Chaos Lives in Everything", "Narcissistic Cannibal", and "Get Up!"), whose work on the EP *Scary Monsters and Nice Sprites* (2010) had inspired Korn to undertake the project in the first place, constitute the most commercially effective song structures on the album and feel the most natural relative to Korn's idiom. This is unsurprising given that Skrillex had played in a post-hardcore band prior to moving into the EDM field – as Davis has commented, "he knows rock structures and it was very easy working with him" (*Billboard*, 2011). For the most part, however, the seams of the contrasting genres of metal and dubstep tend to be apparent, with Korn's sound benefitting from a kind of dubstep sonic colouration and bass heaviness, which ultimately does not compromise their essential musical aesthetic. In this way, the album recalls the remixing fad of the 1990s, when industrial artists such as Trent Reznor and Laibach re-worked a range of metal tracks with varying degrees of success. It is this lack of integration between the two genres that has invoked the most critical consternation, as in the following comments by one reviewer: "these aren't remixes, but two alternating components working side by side, and the end result is an awkward collision that fails to be heavy and yet is still too insistent to be mere background music" (Deviant, 2011). Compared to industrial music's heyday of the 1990s and early 2000s, there have been relatively few examples of collaborations of this nature between established metal practitioners and contemporary electronic musicians in recent years. Indeed, it is significant that since *The Path of Totality*, Korn themselves have neither repeated this experiment nor wholeheartedly adopted modern DAW-based electronic music practices into their idiom.

## IDEOLOGY, METAL AESTHETICS, AND GENRE SPECIFICITY

The comments quoted in response to Korn's album have been fairly typical of attitudes of the metal community toward metal artists who have attempted to court electronic music aesthetics, particularly those working in the industrial and nu metal subgenres. Vince Neilstein's comments in a recent *Metal Sucks* article discussing The Algorithm's 2016 *Brute Force* album highlight the persistence of these attitudes in reference to the more recent examples discussed in this chapter:

> Certain metalheads feel a need to shove a stick in the mud whenever bleeps and blips get anywhere within the vicinity of chugs and blasts. "Get that shit out of my metal!" And so I venture that The Algorithm will rub a lot of metalheads the wrong way within just a few seconds of pressing play.
>
> (Neilstein, 2016)

These kinds of observations draw attention to the ideological conflicts that exist within the wider community that surrounds a genre (e.g., critics, fans, record labels, and so on) and the exercise of what Fabbri calls its "competence" to validate the changes that may develop within it. Fabbri illustrates this as follows:

> Let us suppose that a new musical event is brought to the public attention. One part of the musical community, let's say the critics, can, thanks to their analytical competence codes consider it an admissible variant of a genre already known. But another part, let's say the audience, can consider a particular combination of rules to which the event conforms so unusual as to be significantly against the well-established ideology, so that the creation of a new genre is considered necessary.
>
> (Fabbri, 1982: 63)

In the present context, fans, critics, and practitioners alike have all demonstrated varying degrees of "analytical competence" in their bids to validate these metal-electronic music fusions. One strategy, as has been noted in regard to The Algorithm and Korn, is to try to make comparative relationships between metal music and the electronic musical aesthetics that are adopted in their work. This can be seen, for example, in Davis's aforementioned comments that electronic music genres such as dubstep have qualities that are somehow akin to metal in either feel or attitude, or in critic Vince Neilstein's concluding comments to his article on The Algorithm that, "It's incredibly complex music, and it's also decidedly heavy. No, it isn't for everyone, but those with an open mind about what heavy music can and should sound like will certainly find something to like" (2016).

Another approach is to sidestep the issue of the "creation of a new genre" and advocate the benefits that accrue to both parties when they explore the possibilities of unfamiliar genre rules in relation to one another. This is implied in Gautier Serre's (Igorrr) response to a question concerning metal fans' inability to react well to "artificial" metal-electronic fusions:

> Metal heads are not so much open specially for the electronic music. I feel pity for this, I like Metal as much as Electronic Music, it's just two really different perspectives of music. There are really amazing things on both sides, the primitive energy of the Metal and the mathematic precision of the Electronic Music, both are very different and worth the attention.
>
> (Melo, 2012)

Fans' commentaries have also offered insightful ways of articulating the relationship of metal/electronic music syntheses to the metal genre. In the following *Encyclopedia Metallum* review of Whourkr's *4247 Snares* album, for example, a fan employs terms such as "nihilistic", "antimusic", and "Dada" as a means of accounting for the band's avant-garde progressiveness while still acknowledging its extreme metal roots:

> The term "metal" might fail to describe what this is. *4247 Snare Drums* is nihilistic antimusic, an ugly, malformed creation that dares you to hate it and exists to mock the notion of arranging sounds in a pattern. The extreme metal scene's Dada movement crystallized in the form of a single album, *4247 Snare Drums* isn't for everyone. It might not even be for anyone.
>
> (Valfars Ghost, 2015)

The question of metal's evolving genre identity in the context of contemporary musical developments has also been addressed in the academic literature by Keith Kahn-Harris, who, in an essay entitled "Breaking Metal's Boundaries" (2014), has called for an open-minded view of metal's capacity for re-contextualization. Kahn-Harris's argument hinges on the idea that the essence of the metal genre – what he terms the "inner core" – is the distorted guitar riff, which remains consistent in all metal subgenres regardless of what else changes. His purpose in reducing metal's genre rules in this way, while clearly provocative, is designed to raise the question of how much modification the genre will bear before it begins to lose its identity as metal:

> Whatever metal is it almost always utilises distorted guitars that draw on a quite limited selection of riffs – this is metal's inner "core". To be sure, lite metal at one end, and grindcore at the other, differ substantially in song structure, instrumentation, timbre, vocals, tempo, lyrics and other aspects. Yet the repetition of distorted guitar riffs mobilising augmented fourths, and a limited set of other intervals, is common to both.
>
> (Kahn-Harris, 2014)

Kahn-Harris's remarks do in certain respects cohere with the attitude adopted by DAW-based metal musicians whose perspective has been conditioned by the sampling aesthetic, which in essence concerns the abstraction and re-situation of definitive metal gestures. Kahn-Harris goes even further than this, however, in an assertion within the same essay that further innovation within metal will be concerned with the introduction of musical elements "outside this riff-based format" as well as "questioning and transforming the core itself", begging the question – what actually should remain if the genre is to be regarded as metal at all?

## CONCLUSION

The foregoing commentary has highlighted a number of contexts of contemporary electronic music practice in which the compositive elements of the metal genre have been re-situated. Those artists who have pioneered the genre experiments discussed, namely Genghis Tron, Serre, and Gallego, on the one hand, exhibit a strong allegiance to traditional performance-oriented metal aesthetics. Hence, when aspects of metal's genre rules (guitar riffs, drum patterns, and so on) appear in their work, they are executed with conviction and a conscientious observance of appropriate genre conventions. However, at the same time, these metal elements have been subject to the aesthetics of computer-based electronic music production, in which context they possess a dual function as audio "objects" for re-situation alongside other, often strongly contrasting, genre tropes. This freedom to juxtapose materials is, to a certain degree, a by-product of DAW environments, whose cut and paste aesthetics are naturalized in the work of these artists, rendering the notion of genre-multiplicity unproblematic. What makes their metal re-imaginings ultimately convincing is their committed and proficient utilization of such DAW-based electronic music conventions. I have contrasted this with Korn, whose metal idiom was constructed outside the contemporary electronic music format and prior to the DAW's rise to prominence. Here the band's working approach prevented their musical identity from being reconstituted within the DAW environment, resulting in little more than a timbral re-imagining of their idiom, even with the involvement of DAW-based dubstep practitioners. Indeed, one might imagine the difference in outcome if, for example, Korn had instead provided their ideas in the form of samples to their respective collaborators. Finally, I have noted that if a sense of the metal genre is to be preserved in the light of the radical re-imaginings of its rules discussed in this chapter, it falls to both enlightened critics and the artists pursuing these electronic music fusions to demonstrate the analytical competence to connect these new developments to the metal genre concretely.

## NOTES

1. Much attention has been given, for example, to what Strachan (2017: 92) has termed the DAW's "visual affordances", namely the mode of representation

of musical material within the visual domain (or GUI) of the typical DAW and the implications of this for the conceptualization of musical material and its organization.

2. The naturalization of the loop paradigm / sampling aesthetic within the DAW is in effect confirmed by the default presence within most packages of onboard libraries of pre-created audio loops, which suggest an assumption on the part of developers that the processes involved in generating them (via traditional forms of musical performance and composition) are less important to the user than their value as raw material.

3. An expression used by Marshall McLuhan to account for the tendency to employ unfamiliar new technologies in terms of older practices.

4. Other bands of this period that demonstrate such technological inclinations are Static X and Pitchshifter.

5. Regarding pronunciation, the "d" is silent, thus, "gent", as in the shortened version of the word *gentleman.*

6. As evidenced by the press attention that it had begun to attract, see, for example, Thomson (2011).

7. Meshuggah also has an additional significance where the djent subgenre is concerned, as the word *djent* itself is commonly regarded as an onomatopoeic reference to the band's powerchord riffing style.

8. See, for example, Levine's (2015) interview discussing Periphery producer Nolly Getgood's recording approach.

9. Genghis Tron emerged in 2004 with their acclaimed EP, *Cloak of Love.*

10. The Igorrr project dates from c. 2005.

11. Derek Roddy (2008: 11) defines blast beats as "an alternating single-stroke roll broken up between the kick and snare, with your ride hand generally playing in unison with the kick drum". Blast beats are typically played at high tempos and have been associated with the death metal and grindcore scenes in particular.

12. Whelan (2008: 265–280), in an extensive discussion of breakcore aesthetics, has drawn attention to the notion of "breakcore as *method*" characterized by an "edit-tightening, plunderphonic aesthetic" and "incongruous sample juxta-position". Igorrr has cited the artist Venetian Snares as a particular influence on his approach.

13. A Dutch electronic music production trio based in Groningen, Netherlands, comprising Nik Roos, Martijn van Sonderen, and Thijs de Vlieger.

14. Based in Kelowna, British Columbia (as is Excision, aka Jeff Abel). Both are associated with the Rottun Recordings record label (established mid-2000s) and also members of the dubstep supergroup, Destroid. Other contributors on Korn's album include Datsik, Feed Me, and 12th Planet.

15. Davis has indicated that it was the EDM scene's genre fluidity and resistance to pigeonholing that appealed to him in particular, a situation which contrasted sharply with what he referred to as the "stale" (i.e., overcoded) rock genre (*Billboard*, 2011).

16. For further technically focused discussion of the musical connections between rock/metal and dubstep, see von Soggerson (2012).

17. Skrillex (aka Sonny Moore), the most high-profile collaborator on the album, was producing his tracks entirely "in the box" with Ableton Live (*Computer Music Specials*, 2011). Noisia, had evolved their "blend of Dubstep, Breaks and Drum 'n' Bass" (*Future Music*, 2013) using Cubase in conjunction with a

range of third-party plugins, including NI Massive and Sylenth1, Toontrack's Superior Drummer and iZotope's Trash for "creative distortion effects" (*Future Music*, 2013; *iZotope*, n.d.). Downlink, whose influence is felt on a number of tracks on *The Path of Totality*, was producing tracks in Logic using synthesizer plugins, such as NI Massive and Sylenth 1 (Bass, 2012), while his colleague Excision employed a combination of Cubase, Logic, Ableton, and FL (Excision, 2014) alongside NI Massive and Rob Papen's Albino and Predator synths for sound design (Excision, 2013).

## BIBLIOGRAPHY

Barlindhaug, G. (2007) Analog Sound in the Age of Digital Tools: The Story of the Failure of Digital Technology. In: R. Skare, N. Windfeld Lund & A. Vårheim eds. *A Document (Re)turn: Contributions From a Research Field in Transition*. Franfurt am Main, Peter Lang Publishing Group, pp. 73–93. Available from: <http://munin.uit.no/bitstream/handle/10037/971/paper.pdf?sequence=1>.

Bass, Vivien. (2012) Downlink interview for Bass Island. *Bass Island*. Available from: <https://www.youtube.com/watch?v=ntIJWK8OLu0> [Accessed 16 March 2018].

Bell, A. (2018) *Dawn of the DAW: The Studio as Musical Instrument*. Oxford, New York, Oxford University Press.

Bell, A., Hein, E. & Ratcliffe, J. (2015) Beyond Skeuomorphism: The Evolution of Music Production Software User Interface Metaphors. *Journal on the Art of Record Production*, *9*. Available from: <http://arpjournal.com/beyond-ske uomorphism-the-evolution-of-music-production-software-user-interface-met aphors-2/> [Accessed 17 July 2015].

Billboard (2011) Korn Talks 'Path of Totality': Video Track-By-Track. *Billboard*. Available from: <www.billboard.com/articles/news/42333/korn-talks-path-of-totality-video-track-by-track> [Accessed 21 October 2016].

Brown, A.R. (2007) *Computers in Music Education: Amplifying Musicality*. New York, Routledge.

Christe, I. (2004) *Sound of the Beast: The Complete Headbanging History of Heavy Metal*. London, Alison and Busby.

Clarity (2011) Exclusive: Interview With the Algorithm. *got-djent.com*. Available from: <http://got-djent.com/article/exclusive-interview-algorithm> [Accessed 15 October 2016].

Computer Music Magazine. (2007) The CM Guide to Metal. *Computer Music Magazine*, pp. 24–31.

Computer Music Specials. (2011) Interview: Skrillex on Ableton Live, Plug-Ins, Production and More. *MusicRadar*. Available from: <www.musicradar. com/news/tech/interview-skrillex-on-ableton-live-plug-ins-production-and-more-510973> [Accessed 16 March 2018].

Dean, J. (2013) Fear Factory: The Making of 'Demanufacture'. *Faster Louder*. Available from: <http://fasterlouder.junkee.com/fear-factory-the-making-of-demanu facture/847981> [Accessed 29 October 2016].

DemiGodRaven. (2012) No Clean Singing – the Algorithm. *No Clean Singing*. Available from: <www.nocleansinging.com/tag/the-algorithm/> [Accessed 28 November 2016].

Deviant. (2011) Review: Korn – the Path of Totality. *Sputnik Music*. Available from: <www.sputnikmusic.com/review/46804/Korn-The-Path-of-Totality/> [Accessed 16 March 2018].

Duignan, M., Noble, J., Barr, P. & Biddle, R. (2004) Metaphors for Electronic Music Production in Reason and Live. In: M. Masoodian, S. Jones & B. Rogers eds. *Computer Human Interaction*. Lecture Notes in Computer Science. Heidelberg, Springer Berlin Heidelberg, pp. 111–120. Available from: <http://link.springer.com/chapter/10.1007/978-3-540-27795-8_12> [Accessed 17 July 2015].

Duignan, M., Noble, J. & Biddle, R. (2005) A Taxonomy of Sequencer User-Interfaces. *Proceedings of the International Computer Music Conference*, pp. 725–728.

Excision. (2013) I Am Excision, dubstep producer/dj/robot dinosaur – Ask Me Anything: IAmA. *Reddit*. Available from: <www.reddit.com/r/IAmA/comments/1dwbbj/i_am_excision_dubstep_producerdjrobot_dinosaur/> [Accessed 16 March 2018].

Excision. (2014) @MuffinZ_ Depends on the Song, Cubase, Logic, Ableton, FL [Internet]. Available from: <https://twitter.com/Excision/status/506650111125225472?ref_src=twsrc%5Etfw> [Accessed 16 March 2018].

Fabbri, F. (1982) A Theory of Musical Genres: Two Applications. In: D. Horn & P. Tagg eds. *Popular Music Perspectives: Papers From the First International Conference on Popular Music Research*. Gothenburg and Exeter, International Association for the Study of Popular Music, pp. 52–81.

Future Music. (2013) Noisia in the Studio With Future Music. *MusicRadar*. Available from: <www.musicradar.com/news/tech/noisia-in-the-studio-with-future-music-581078> [Accessed 14 March 2018].

Giles, J. (2012) Korn's Jonathan Davis: 'We're Not a Metal Band'. *Loudwire*. Available from: <http://loudwire.com/korn-jonathan-davis-were-not-a-metal-band/> [Accessed 21 October 2016].

Goodwyn, T. (2012) Korn's Jonathan Davis: 'Dubstep Is the New Electronic Heavy Metal'. *NME*. Available from: <www.nme.com/news/music/korn-17-1258109> [Accessed 21 October 2016].

Kahn-Harris, K. (2014) Breaking Metal's Boundaries. *Souciant*. Available from: <http://souciant.com/2014/01/breaking-metals-boundaries/> [Accessed 16 November 2016].

Laing, R. (2011) What Is Djent? *Total Guitar*, (214), pp. 49–54.

Lentz, A. (2012) Tomas Haake: Meshuggah's Djentle Giant. *DRUM! Magazine*. Available from: <http://drummagazine.com/tomas-haake-meshuggahs-djentle-giant/> [Accessed 27 April 2018].

Levine, M. (2015) Tracking and Mixing With Nolly. *Audiofanzine*. Available from: <http://en.audiofanzine.com/sound-technique/editorial/articles/tracking-and-mixing-with-nolly.html> [Accessed 15 October 2016].

Marrington, M. (2011) Experiencing Musical Composition in the DAW: The Software Interface as Mediator of the Musical Idea. *Journal on the Art of Record Production*, 5. Available from: <http://arpjournal.com/experiencing-musical-composition-in-the-daw-the-software-interface-as-mediator-of-the-musical-idea-2/> [Accessed 17 July 2015].

Marrington, M. (2016) Paradigms of Music Software Interface Design and Musical Creativity. In: R. Hepworth-Sawyer, J. Paterson, J. Hodgson & R. Toulson eds. *Innovation in Music II*. Shoreham-by-Sea, Future Technology Press.

Marrington, M. (2017a) Composing With the Digital Audio Workstation. In: K. Williams & J. Williams eds. *The Singer-Songwriter Handbook*. New York, Bloomsbury Academic.

Marrington, M. (2017b) From DJ to Djent-Step: Technology and the Re-Coding of Metal Music Since the 1980s. *Metal Music Studies*, *3* (2), pp. 251–268.

Melo, F. (2012) Igorrr: Interview With Gautier Serre. *Groundcast*. Available from: <http://groundcast.com.br/igorrr-interview-with-gautier-serre/> [Accessed 25 March 2018].

Mooney, J. (2011) Frameworks and Affordances: Understanding the Tools of Music-Making. *Journal of Music, Technology and Education*, *3* (2), pp. 141–154.

Mynett, M. (2013) *Contemporary Metal Music Production*. Unpublished PhD thesis. University of Huddersfield.

Mynett, M. (2017) *Metal Music Manual: Producing, Engineering, Mixing and Mastering Contemporary Heavy Music*. New York: Taylor & Francis Group.

Neilstein, V. (2011) The Latest Metal Micro-Genre Bastardization: Dubstep + Djent = Djentstep. *MetalSucks*. Available from: <www.metalsucks.net/2011/02/16/the-latest-metal-micro-genre-bastardization-dubstep-djent-djentstep/> [Accessed 14 October 2016].

Neilstein, V. (2016) 'Pointers' on How to Be Both Metal and Electronic at the Same Time From the Algorithm. *MetalSucks*. Available from: www.metalsucks.net/2016/03/11/pointers-on-how-to-be-both-metal-and-electronic-at-the-same-time-from-the-algorithm/> [Accessed 29 March 2018].

NewFuryMedia (2016) Featured Interview: The Algorithm (Remi Gallego). *The New Fury*. Available from: <http://thenewfury.com/wordpress/featured-interview-algorithm/> [Accessed 15 October 2016].

Pearson, C. & Mullane, L.O. (2014) The Ultimate Guide to Producing Dirty Dubstep – Part One. *MusicTech*. Available from: <www.musictech.net/2014/10/dirty-dubstep-1/> [Accessed 16 March 2018].

Phillipov, M. (2012) *Death Metal and Music Criticism: Analysis at the Limits*. Lanham, Lexington Books.

Pieslak, J. (2007) Re-Casting Metal: Rhythm and Meter in the Music of Meshuggah. *Music Theory Spectrum*, 29 (2), pp. 219–245.

Pieslak, J. (2008) Sound, Text and Identity in Korn's 'Hey Daddy'. *Popular Music*, *27* (1), pp. 35–52.

Ratliff, B. (2008) Genghis Tron – Board Up the House. *New York Times*. Available from: <www.nytimes.com/2008/02/18/arts/music/18choic.html> [Accessed 25 March 2018].

Roddy, D. (2008) *The Evolution of Blast Beats*. Pembroke Pines, FL, World Music 4all Publications.

Scion AV (2011) *Excision – Interview With 12th Planet of Roach Coach Radio*. Available from: <www.youtube.com/watch?v=6jsXNCnn0-o> [Accessed 16 March 2018].

Shelvock, M. (2013) The Progressive Heavy Metal Guitarist's Signal Chain. In: Hepworth-Sawyer, J. Paterson, J. Hodgson & R. Toulson eds. *Innovation in Music*. Shoreham-by-Sea, Future Technology Press, pp. 126–138.

Strachan, R. (2017) *Sonic Technologies: Popular Music, Digital Culture and the Creative Process*. New York, Bloomsbury Publishing USA.

Surachai Workspace and Environment: Genghis Tron. (2008) *Trash_Audio*. Available from: <http://trashaudio.com/2008/03/workspace-and-environment-genghis-tron/> [Accessed 25 March 2018].

Thomson, J. (2011) Djent, the Metal Geek's Microgenre. *Guardian*. Available from: <www.theguardian.com/music/2011/mar/03/djent-metal-geeks> [Accessed 3 February 2017].

Udo, T. (2002) *Brave Nu World*. London, Sanctuary.

Ume (2012) MxCx Interview#1 'Igorrr'. *MxCx Blog*. Available from: <http://mxcx tokyo.blogspot.co.uk/2016/09/this-interview-was-recorded-on-november. html> [Accessed 25 March 2018].

Valfars Ghost (2015) What Insanity Sounds Like. *Encyclopaedia Metallum: The Metal Archives*. Available from: <www.metal-archives.com/reviews/Whourkr/4247_Snare_Drums/339803/> [Accessed 8 April 2018].

Vivien Bass (2012) *Downlink Interview for Bass Island*. Available from: <www.youtube.com/watch?v=ntIJWK8OLu0> [Accessed 16 March 2018].

von Soggerson, J. (2012) Electronic Music Theory: From Funk and Metal to Dubstep. *Dubspot Blog*. Available from: <http://blog.dubspot.com/electronic-music-theory-from-funk-and-metal-to-dubstep-excision-rage-against-the-ma chine/> [Accessed 16 March 2018].

Walser, R. (1993) *Running With the Devil: Power, Gender, and Madness in Heavy Metal Music*. Middletown, CT, Wesleyan University Press.

Weinstein, D. (2000) *Heavy Metal: The Music and Its Culture*. New York, Da Capo Press.

Whelan, A. (2008) *Breakcore: Identity and Interaction on Peer-to-Peer*. Unabridged edition. Newcastle, Cambridge Scholars Publishing.

Williams, D. (2015) Tracking Timbral Changes in Metal Productions From 1990 to 2013. *Metal Music Studies*, *1* (1), pp. 39–68.

Zagorski-Thomas, S. (2014) *The Musicology of Record Production*. Cambridge, Cambridge University Press.

## DISCOGRAPHY

Algorithm (2012) *Polymorphic Code*, Basick Records.

Algorithm (2016) *Brute Force*, FiXT.

Animals as Leaders (2009) *Animals as Leaders*, Prosthetic.

Downlink (2010) *Crippled Camel*, Rottun Recordings.

Excision (2013) *X Rated: Remixes*, mau5trap.

Fear Factory (1993) *Fear Is the Mindkiller*, Roadrunner.

Fear Factory (1995) *Demanufacture*, Roadrunner.

Fear Factory (1998) *Obsolete*, Roadrunner.

Genghis Tron (2008) *Board Up the House*, Relapse Records.

Igorrr (2010) *Nostril*, Ad Noiseam.

Igorrr (2012) *Hallelujah*, Ad Noiseam.

Igorrr (2017) *Savage Sinusoid*, Metal Blade Records.

Korn (2011) *The Path of Totality*, Roadrunner.

Noisia (2010) *Split the Atom*, Vision Recordings.

Periphery (2010) *Periphery*, Roadrunner.

Periphery (2011) *Icarus*, Roadrunner.

Skrillex (2010) *Scary Monsters and Nice Sprites EP*, Big Beat Records/Atlantic.

Slipknot (1999) *Slipknot*, Roadrunner.

Tesseract (2011) *One*, Century Media.

Various (2011) *Metal Hammer Presents Djent*, Future Publishing [Online]. [Accessed 3 December 2016].

Various (2012) *Basick 2012 Free Sampler*, Basick Records [Online]. Available from: <http://music.basickrecords.com/album/basick-2012-free-sampler> [Accessed 19 October 2016].

Whourkr (2008) *Concrete*, Crucial Blast.

Whourkr (2012) *4247 Snare Drums*, Ad Noiseam.

# 5

# Antonio Carlos Jobim

## The Author as Producer[1]

## Marcio Pinho and Rodrigo Vicente

## INTRODUCTION

Bossa Nova is one of the greatest paradigms of Brazilian popular music. Many critics and researchers (Campos 2008; Castro 1990; Mammì 1992; Napolitano 2007; Naves 2010) have taken as a starting point an evolutionist view on Brazilian music, positing that the movement is the stylistic apex of a national popular music. It is commonly regarded as the harmonic and melodic peak of the country's musical history, whose sonic characteristics (resulting especially from the singing style and the arrangement choices) represent the point from which all Brazilian music is read, both before and after its inception. This vision extrapolates the national borders and invades the global imagination, converting Bossa Nova into a synonym of Brazilian identity. What we intend to discuss here is not the way of constructing these ideas, but the way that the technological aspects existent in studios at that time had consequences for the elaboration of the arrangements. More accurately, the possibilities and limitations of those aspects and the importance of Tom Jobim as a conscious agent of the recording conditions of that epoch, which culminate in the consolidation of a pattern, a style that will guide this new musical aesthetic.

Antônio Carlos Jobim was born in Rio de Janeiro in 1927. Besides always playing popular music, Jobim had an education in classical music, studying piano with the renowned teacher Nilza Branco. He also had classes with Hans-Joachim Koellreutter, a German composer who had an important role in Brazilian music, as he disseminated new musical ideas and conceptions throughout the country, placing an emphasis on the importance of the "musical creation" and the "new" (Zanetta and Brito 2014). Jobim started his career as a piano player in bars and nightclubs. In 1952, he started to work at the recording company Continental, where he transcribed popular songs to be registered (Jobim 1983). With the support of Radamés Gnatalli, a very important musician of that time who worked as an arranger, composer, piano player, and conductor in different recording companies and radio stations, Jobim began to work increasingly as an arranger and his compositions began to be recorded by different artists. In

1956, he started to work at the recording company Odeon, as an artistic director and later arranger, managing the recording process of numerous albums (Jobim 1996). In this way, he was always in touch with the studio technologies, especially at Odeon, where he was accountable for the recording quality. As a result of hours of practice inside the studio, Jobim became aware of the possibilities given by the available technology of that period, that is to say, what would and wouldn't work in the recording process.

Magnetic tape recorders arrived in Brazil at the end of the 1950s. It is difficult to know precisely which were the equipment and the techniques used in the studios at that time, but as we can perceive from different sources, Bossa Nova recordings were possibly made on four-channel recorders, without the use of the technique of *overdubbing* (Vicente 1996: 22–24). The advent of vinyl represented an enormous advance in sound quality. Besides the name of the so-called High Fidelity (Hi Fi) technology, the sound quality of the recordings would still drastically improve in the following years (Vicente 2014).

The other crucial figure in this discussion is João Gilberto. He was born in 1931, in the state of Bahia. Obsessed with this way of singing and playing guitar, Gilberto was one of the main developers of the interpretative style of the music that would later be called Bossa Nova. Gilberto had moved to Rio de Janeiro in 1950 and worked as a singer in vocal groups and as a guitar player accompanying other crooners (Castro 1990). In 1958, his first 78 rpm disk was released, which contained "Chega de Saudade" (Tom Jobim and Vinicius de Moraes) and "Bim Bom", composed by Gilberto himself. This work consolidated his partnership with Tom Jobim and Vinícius de Moraes, which began a couple of months before, in the LP *Canção do Amor Demais* (1958), with songs by Jobim and Vinícius, sung by Elizeth Cardoso, and with João Gilberto accompanying on the guitar. Nevertheless, the repercussion and recognition of his work would not begin decisively until 1959, with the release of the LP *Chega de Saudade*, arranged by Tom Jobim. This album has, among other songs, compositions that would later be considered Bossa Nova standards, such as "Desafinado" (Tom Jobim and Newton Mendonça) and "Brigas Nunca Mais" (Tom Jobim and Vinícius de Moraes). In the following years, Gilberto would release two further LPs in partnership with Jobim: *O Amor, o Sorriso e a Flor* (1960) and *João Gilberto* (1961), to conclude the "trilogy" that began with the LP of 1959 and finishing the latter, known as the "first Bossa Nova stage" (Idem). This period marks the consolidation of a new vocal interpretative style and a new way of rhythmically and harmonically accompanying by guitar, which influenced a whole generation of singers and guitarists.

Equally important to the following generations is the sonority heard on the recordings of the initial Bossa Nova period, which uniquely occurs as a result of the way in which Tom Jobim worked on his conception of their instrumentation and arrangement. Here, Jobim exceeded the common functions of the composer and arranger, developing his own particular way of writing. This took into consideration not only João Gilberto's *sui generis* style of vocal interpretation and guitar accompaniment, but also the technical resources and equipment available in the recording studio at the time. Essentially, Jobim undertook the functions of the professional, which would later be known as a musical producer, permanently reflecting on the known "extra-musical" issues. Nevertheless, these issues are inseparable from the strictly aesthetic creation and are decisive to the final results of the work in the studio since the popular song is, ultimately, a recording, a commodity. In this sense, it is possible to say that Jobim's reflections were not just about the musical subjects – the composition forms; harmonic, rhythmic, and melodic construction; and so on – but also about the techniques necessary in order to make his creation become possible, i.e., his interest in choices and writing processes which had more efficacy and aesthetic relevance within the recording industry of that time.

As became apparent, Jobim had a clear understanding of the whole process of record production, valuing equally the work of both the recording engineer and the musician. This evokes the connection we are proposing in this work between the activity of Jobim and the notion of "producer" discussed by Walter Benjamin in a paper entitled "The Author as Producer" (1998). This paper was delivered to an audience composed of writers and progressive German intellectuals and dealt with issues in a context different to our focus in this chapter. However, what Benjamin highlights is the importance of the audience's presence in modern communication media, especially the press and radio. This was necessary in order to broaden their field of action and open new possibilities of political intervention at a time of turbulence within German society during the 1930s, which had witnessed the rise of the Nazi Party to power. Benjamin believed that it was necessary to avoid an aversion to the mass media, while at the same time, assimilate its rationality and assume its structures, developing new writing "techniques", suitable to the imposed reality. To illustrate his arguments, Benjamin cites the case of classical music on the context of mass culture, from the testimony of the German composer Hanns Eisler, one of the Arnold Schoenberg disciples:

> In the development of music, both in production and in reproduction, we must learn to recognize an ever-increasing process of rationalization. . . . The gramophone record, the sound film, the nickelodeon can . . . market the world's best musical productions in canned form. The consequence of this process of rationalization is that musical reproduction is becoming limited to groups of specialists which are getting smaller, but also more highly qualified, all the time. The crisis of concerthall music is the crisis of a form of production made obsolete and overtaken by new technical inventions.
>
> (Benjamin 1998: 95–96)

Heading back to Brazil at the end of the 1950s, Jobim, we suggest, acted as an "author-producer" as he became interested in the popular music market and renovated the musical material and the phonographic production methods. Jobim assimilated the idiosyncratic rationality of the industry, seeking to overcome both the old patterns of sonority and the studio's technical limitations. The following musical analysis looks to illustrate this dimension of the work and the activity of Jobim on the first LPs of João Gilberto.

## THE BASS RANGE

Through analyzing the first LPs of João Gilberto in terms of technical features of the recording process and the possible issues inherent to this practice in the 1950s, it is possible to identify a series of aspects, via a music-aesthetic point of view, which enable new perspectives on his work. The first of them is related to the treatment of the bass range (lowest pitches) of the recordings. One of the more evident peculiarities is the absence of double basses[2] in Tom Jobim's arrangements, as commented by researchers of Brazilian music. Walter Garcia, for example, states that the bass lines made by João Gilberto on the acoustic guitar do, in a certain way, cover the function of the double bass. Thus he plays the quarter notes on beats one and two of the binary measure, this being the basis for the non-regular variations of the higher structures of the chords (Garcia 1999). This combination, along with the technical accuracy of Gilberto, leads to the characteristic "swing" of the genre. Other authors, some of whom are collaborators for the book *Balanço da Bossa e outras bossas*, believe that the diminished number of instruments on Bossa Nova performances is due to the aesthetics of the style (Brito *apud* Campos 2008), which is marked by an economy of elements, their functionality, the laidback style, and avoiding the "excess" or the "easy" resources important to the popular music "opero-tango-boleristic" of the mid 1950s (Medaglia *apud* Campos 2008: 75). According to Augusto de Campos, editor of the aforementioned book, this aesthetic is compatible with other artistic and literary manifestations of this period, as "concrete" poetry.

These statements assume a meaningful importance in Bossa Nova's history, as they are constitutive elements of the discursive tradition consolidated since the 1960s among artists and intellectuals. In a teleological concept of history, this vision established parameters of analysis for understanding Brazilian popular music "pre-" and "post-" Bossa Nova.

Going back to the "bass" issue, besides the opinions of some authors, there are other details which are also revealing. Ruy Castro states that the third LP of João Gilberto (1961) had some arrangements made by Walter Wanderley and some others by Tom Jobim. "Saudade da Bahia", "Bolinha de Papel", "Trenzinho", "Samba da Minha Terra", and "A Primeira Vez" were arranged by Wanderley, who tried to absorb to the maximum the countless "tips" of João Gilberto during the beginning of the LP's production (Castro 1990: 295–296). Tom Jobim would join the project just five months later, after accepting the invitation of the record label Odeon to finish the LP, which would be considered the last album of the first Bossa Nova stage. This happened because Gilberto interrupted the recordings as he was dissatisfied with the previous results (Idem). From then on, only these unpublished compositions were recorded: "Insensatez", "O Bar-quinho", "O Amor e a Paz", "Você e Eu", and "Coisa Mais Linda".

It is interesting to note that an attentive and careful listening reveals that the songs in which Walter Wanderley acted as arranger have a double bass, while Tom Jobim did not utilize the instrument on the first two albums – *Chega de Saudade* (1959) and *O Amor, o Sorriso e a Flor* (1960). It is not possible to know exactly the reason for their different attitudes. However, knowing the singular function of the bass lines of the guitar – as described by Walter Garcia – and adding it to the limitations imposed by the studios of that time, it is possible to suppose that Tom Jobim chose not to use the guitar and the double bass at the same time. The simultaneity between the two instruments could have resulted in an undefined sonority or, in other words, a certain saturation of the bass range. As an example of this, whilst listening to "Saudade da Bahia" and "Bolinha de Papel", both arranged by Walter Wanderley, it is hard to distinguish between the double bass and the guitar in that range.

At first, it is easy to assume that the bass line of the guitar and double bass do the exact same lines (Garcia 1999: 25–26).[3] But, as it is possible to realize with more attentive listening, it is recurring on those occasions when the guitar does not change the bass of the chords, whereas the double bass plays a variation between the root and the fifth of the chords (which is typical of the samba style). The following transcriptions are two examples of how this happens (cf. examples 5.1 and 5.2):

**Example 5.1** Excerpt of "Saudade da Bahia" (Dorival Caymmi), localized around 0'24" on João Gilberto's performance

**Example 5.2** Excerpt of "Bolinha de Papel" (Geraldo Pereira), localized around 0'16" on João Gilberto's performance

These different notes played at the same time in the bass range raise the difficulty of clearly discerning the different sounds at these moments. This kind of problem occurs throughout the recording history in the first half of the 20th century, as Eduardo Vicente (1996) illuminated in his research about the Brazilian recording industry:

> The techniques of disc recording also imposed restrictions to the utilization of low range instruments, which were determined by the groove geometry. During the recording technique's early phases, for example, the groups of strings had to be substituted by groups of wind instruments. . . . But at even later stages, the disc recording of low percussion instruments that emitted high volume of sound were still not viable. . .
>
> (Vicente 1996: 20, translation by the authors)

The hypothesis that Jobim recognized in the potential technical issues in letting both instruments play together can be reinforced by listening to his previous work, especially in some arrangements recorded on the afore-mentioned album *Canção do Amor Demais* (1958), which contain pieces sung by Elizeth Cardoso and composed by Jobim and his partner, Vinícius de Moraes.

In the example of "Outra Vez" (Tom Jobim), where João Gilberto is accompanying on guitar, it is possible to note meaningful similarities with the version that was later recorded by Gilberto in the LP *O Amor o Sorriso e a Flor* (1960). The guitar accompaniment is almost the same, as well as the instrumentation, while the main difference is the employment of double bass and *surdo*,[4] which appear only in the second exposition of the theme. Before this, the percussion was typically composed of a snare drum using brushes and a wood block – whose timbre resembles a snare rim – i.e., similarly to the first three Bossa Nova albums.

Looking further into the dynamics of this song, the arrangement has a continuous *crescendo*, and that is a possible reason for the insertion of double bass and *surdo* on the second half of the track, which also has a brief intervention of strings. It is important to highlight that Jobim seems to have been careful on the simultaneous use of "bass" range instruments alongside the acoustic guitar, as their lines and rhythms often coincided. This is not the case in isolated exceptions where the double bass alternates between the fundamental and the fifth of the chord whereas the guitar does not.

The care taken with the bass range is evident in "Eu Não Existo Sem Você" (Tom Jobim and Vinícius de Moraes): the song begins with a guitar playing an introduction using a chord melody style[5] straight after accom-panying the voice. Before the end of the A part, the guitar stops and an orchestral accompaniment begins (strings, woodwinds, brass, percussion, and harp), which includes the double bass. This instrumentation occurs throughout almost the whole recording, apart from the final verses, when the orchestra stops and the guitar comes back as the only rhythm-harmonic support. Thus, guitar and double bass do not play at the same time.

A less sudden transition between these instruments can be heard on "Caminho de Pedra" (Tom Jobim and Vinícius de Moraes), whose introduction has the acoustic guitar playing the fundamental notes of the chords on the first beat of a four-beat measure. The double bass begins to play along with strings and percussion on the four measures that come before the A part, in melodic (one octave lower) and rhythmic synchronicity with the bass line played by the guitar. The first exposition of the initial part is characterized by this combination, which does not reveal serious conflicts as in the situations where alternations occur between the fundamental and the fifth on the double bass. However, at the beginning of the second part, the acoustic guitar stops, leaving the double bass free to cover notes on beats 3 and 4 of the measure.

The waltz "Luciana" (Tom Jobim and Vinícius de Moraes) is maybe the better example of rhythmic and melodic simultaneity between the bass line of the acoustic guitar and double bass. Both play on the first beat of the measure, in general with one octave interval between them. In this case, except for the first eight measures of the A part – where the guitar plays without the double bass – both play together throughout the song, which appears to be possible because of a great adjustment between them, possibly the arranger's demand. Furthermore, it is important to emphasize that the triple meter favors the rhythmic simultaneity on the first beat – the stressed one.

## Other Notes Are Bound to Follow, But. . .

Tom Jobim's awareness of the technical implications of the recording process does not simply revolve around a singular treatment of the arrangement's bass range. Through our intention of incorporating the physical conditions to the theoretical reflection within the work of art, it is possible to map the creative process of the arranger and composer from the analysis of other elements of the recordings.

Before discussing this topic, it is worthwhile to point out what Jobim himself has stated (in a 1967 interview) regarding the broad historical context that shaped the first Bossa Nova LPs:

> The harmony in a broad way has simplified itself. Simplified and enriched. Many notes were taken away from the chords: the fifths were taken away as they already sound on the bass harmonics, etc. due to the recording precariousness of our studios. We wanted the public to hear the voices that we wished. This made us reduce the number of chord notes. It was a result of a long study, combining sounds of orchestra with the piano sound, the voice sound, and the guitar sound, in an attempt for it to be possible to hear something that could not be heard until then. It wasn't worth that amorphous mass of 100 violins. So, the total economy happened: a little flute, 4 violins playing generally in unison, in an attempt to reach the listener with an idea. . . . A series of things like that were introduced in instrumentation techniques, which till then used crowded chords . . . full (the more notes, the better). We started to empty the chords, "to empty" in a good

sense, that is, to make the essence appear, the core, the pith. . . . The studio destroyed all the equilibrium that existed before on live music, for example, the relation between the number of violins to a viola or the number of violas to a cello. These relations were all destroyed. Nowadays, you can have a bass flute with 20 brass instruments accompanying it. So, it created a different dynamic. *The classical dynamic of writing to an orchestra has absolutely changed: the sound engineer becomes more important than the musician, as the dynamic depends on the volume that each microphone has.*

(emphasis added) (Coelho and Caetano 2011,<br>translation by the authors)

In a few lines, Tom Jobim reveals an attitude that, without doubt, not all musicians with a classical background or who were recognized on the artist milieu of Rio de Janeiro (the main country's cultural center of that time) would have possessed in the 1950s – that is, an attitude of recognizing and highlighting the work of sound engineers. This is not an exaggeration, because, as he had himself commented, his knowledge of the details was intrinsic to the work inside the studio as a consequence of a "long study", i.e., Jobim seems to have been self-aware of the physical circumstances of record production. After all, as mentioned, Jobim had a long experience as an arranger and orchestra conductor in the recording companies Continental and Odeon.

When Jobim alludes to the necessity to "empty the chords", for example, "to make the essence appear", one of the problems he aims to avoid is the superimposition of the volume of some instruments in detriment to others. The arrangements were economical in relation to the number of instruments besides percussion, piano, and guitar but included just one or two woodwinds (a concert flute and, in some cases, tenor saxophone), one brass (trombone), a group of strings (violins in general, but also violas and cellos). This supports the idea that each instrument could be individually better heard. However, even in this situation, when performing chords, the risk of losing the sound of a more fragile instrument – such as the flute – still existed. Undoubtedly, this also depended on the way each element was placed on the arrangement and not just on how it was treated in the studio or on the instrumentation chosen.

There are very few cases when all instruments play together on early Bossa Nova recordings, and this denotes a certain cautiousness of the arranger. This occurs at very specific moments, as in final chords with *fermata*. And these are the cases that really illustrate the arranger's strategies (cf. example 5.3).

The song finishes in the first degree of the minor key (G#m), and the chord's triad (G#, B, D#) is played with all the available extensions on the Aeolian mode, i.e., from the tonic on: major ninth (A#), perfect fourteenth (C#), and minor thirteenth (E). This makes the chord sound dissonant, especially to the standards of the mainstream Brazilian popular music of that period. Besides that, there is no doubling apart from violins and violas, which play in unison as do cello and piano on a single pitch (E). This is in terms of the usual proceedings, where low-volume instruments play at the same time as high-volume instruments, such as the trombone.

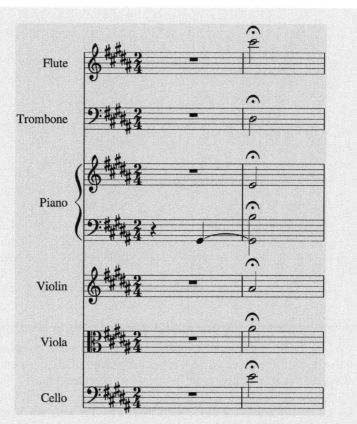

**Example 5.3** Final chord of "O Amor em Paz" (Tom Jobim and Vinícius de Moraes), as recorded on the LP *João Gilberto* (1961)

Another striking detail is the spacing between the chord notes, placed apart by intervals higher than a perfect fourth. This contributes to the hearing of each note among the whole chord, or, as Jobim said, "to make the essence appear". The musical texture becomes wide, bigger than three octaves, revealing an interesting fact about the flute, which is placed more than an octave higher than all other instruments. If this was not the case, then it would hardly be perceived.

Summarizing, even with a reduced instrumentation, these writing procedures enable an effect of orchestral density, very suitable to musical endings.

"Insensatez" (Tom Jobim and Vinícius de Moraes), which was composed in a minor key (Bm), also had a singular relation between the instruments, which occurs in a kind of interlude, a bridge between first and second theme expositions (cf. example 5.4).

As it is possible to see in the example, while the guitar plays the chord Bm7, tonic of the key, other notes are added in an ascending direction by each instrument or group of instruments (violins). Saxophone and flute insert structural notes of the rest chord, in this case, the root (B) and the

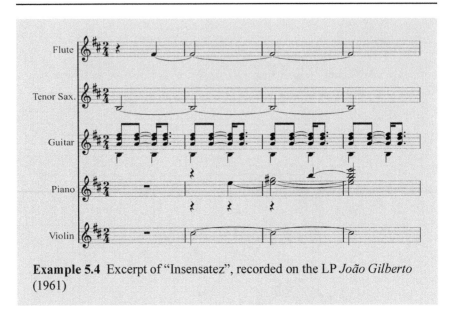

**Example 5.4** Excerpt of "Insensatez", recorded on the LP *João Gilberto* (1961)

perfect fifth (F#), while violins and then the piano intervene with extensions: major ninth (C#), perfect fourteenth (E), and major thirteenth (G), respectively, with doubles of the root and the perfect fourteenth. This way, despite the instrumentation being more reduced than in "O Amor em Paz", it is possible to see once again the spacing between chord notes and the effect of orchestral density by virtue of the wide tessitura of the arrangement – larger than three octaves (the guitar is one octave lower than the real sound).

In reference to the gradation of the orchestral color and the instrument's intensity, it is possible to note the arranger's attention to detail in the way that he masterminds the excerpt: saxophone is on mid-low range, playing *mezzo piano*. However, the note (B) is doubled by the guitar by one octave lower, avoiding the disappearance on the sound texture; the flute is on its low range, consequently, with low volume. To avoid overshadowing the note, Jobim placed it in *forte* dynamic and left it alone in its range, separated from the closer instruments by a perfect fifth (the saxophone is below the flute and the violins are above), pretty long intervals for this situation. The violins do not have this problem in this case, as they are grouped (albeit in a small number) playing the same note (C#) in the mid-high range and their naturally keen timber. Finally, the piano distinguishes itself in the context as it is introducing a new melodic and percussive color, i.e., distinct from the winds and the bowed strings, and, furthermore, its notes are on the higher pitch of the excerpt.

## CONCLUSION

Analyzing some of the songs from the Bossa Nova initial phase (1958–1961), it is possible to see that the instruments were linked in an organic way, as the guitar is a constitutive part of the arrangements, not an autonomous element.

The synthesis made by Tom Jobim and João Gilberto reveals a high degree of planning, evincing the parts that constitute the whole. This occurs because of the economy of elements and the interpretative style, which characterize their performances, marked by the subtle interrelationship between the instruments and by the dynamic variation of the sound texture. In this sense, it is possible to agree with Gilberto when stating that he "changes the whole music structure, without modifying anything" (Souza and Andreato 1979: 49–56 *apud* Garcia 1999: 127–128, translation by the authors), as, with Jobim, the originality of his reinterpretations are in the *sui generis* interpretative style, which gives new forms to the compositions.

As has been demonstrated through this analysis, some choices of Tom Jobim were oriented not only by aesthetic parameters but also by the technical ones, as some solutions found by the arranger can be seen as more effective considering the material conditions of recording production of that time. In other words, the study of the recordings along with the data survey of the structure of Brazilian studios in the 1950s revealed that Jobim articulated the styles of interpretation and writing with the knowledge acquired through the work within recording companies of that time, i.e., the practical and professional experience in the market of popular music.[6] This allows us to approach the work of Jobim with the Benjaminian concept of author as producer.

Lastly, the results presented here show how technology directly influenced the aesthetics of Bossa Nova. It is a practical example of some ideas of Friedrich Kittler, who worked with the consequences of the method of data production, distribution, and reproduction, thus the consequence of sound manipulation, highlighting the cultural aspect of music. His ideas asserted to a certain extent how the material conditions of a medium determine the contents transmitted through it (Kittler 1990). To conclude, it is interesting to note that, even with the development of the technology in the recording studios, the reference point for the Bossa Nova sonority remains that established on the first three LPs.

## NOTES

1.  We would like to thank the professors José Roberto Zan (Institute of Arts – University of Campinas), Walter Garcia (Institute of Brazilian Studies – University of São Paulo), and Eduardo Vicente (School of Communications and Arts – University of São Paulo) for the valuable contributions, without which this work would not have been possible.
2.  The double basses were the more usual instrument used to play the bass lines of that kind of music at that time.
3.  The author asserts that this occurs in the album *Getz/Gilberto* and in another three that were released later on.
4.  *Surdo* is a kind of floor tom, a deep percussion instrument, very common in *samba*, where it generally plays both beats of the binary compass, as does the bass.

5.  Chord melody is a technique where the guitarist plays both harmony and melody at the same time.
6.  It is also interesting to highlight João Gilberto's consciousness of the importance of recording equipment and methods to the construction of a particular sonority. There is a famous episode that happened in the beginning of the recordings of his first 78 rpm, in which Gilberto demands two microphones for himself: one for his voice and one just for his guitar, causing confusion with the recording technicians, as it was not a usual practice at that time (Castro 1990: 181).

## REFERENCES

Benjamin, W. (1998). "The author as producer". In *Understanding Brecht*. London: Verso.

Campos, A. (Org.). (2008). *Balanço da bossa e outras bossas*. São Paulo: Ed. Perspectiva.

Castro, R. (1990). *Chega de Saudade: A história e as histórias da Bossa Nova*. São Paulo: Companhia das Letras.

Coelho, F. O. & Caetano, D. (Orgs). (2011). *Tom Jobim – Coleção Encontros*. Rio de Janeiro: Beco do Azougue.

Garcia, W. (1999). *Bim Bom: a contradição sem conflitos de João Gilberto*. São Paulo: Paz e Terra.

Jobim, A. C. (1983). *A vida de Tom Jobim: depoimento*. Rio de Janeiro: Rio Cultura, Faculdades Integradas Estácio de Sá.

Jobim, H. (1996). *Antonio Carlos Jobim: um homem iluminado*. Rio de Janeiro: Nova Fronteira.

Kittler, F. (1990). *Discourse Networks: 1800/1900*. Stanford: Stanford University Press.

Mammì, L. (1992). "João Gilberto e o projeto utópico da bossa nova". In: *Novos Estudos*, Cebrap, n° 34.

Napolitano, M. (2007). *A síncope das ideias: a questão da tradição na música popular brasileira*. São Paulo: Editora Fundação Perseu Abramo.

Naves, S. C. (2010). *Canção Popular no Brasil: A canção crítica*. Rio de Janeiro: Civilização Brasileira.

Souza, T. & Andreato, E. "Entrevista: João Gilberto". In *Rostos e Gostos da Música Popular Brasileira*. Porto Alegre: L&PM, 1979.

Vicente, E. (1996). *A Música Popular e as Novas Tecnologias de Produção Musical: uma análise do impacto das novas tecnologias digitais no campo da produção da canção popular de massas*. 159f. Dissertação (Mestrado em Sociologia), Universidade Estadual de Campinas (UNICAMP), Campinas-SP.

Vicente, R. A. (2014). *Música em Surdina: sonoridade e escutas nos anos 1950*. 252f. Tese (Doutorado em Música), Universidade Estadual de Campinas (UNICAMP), Campinas-SP.

Zanetta, C. C. & Brito, T. A. (2014) "Hans-Joachim Koellreutter em movimento: ideias de música e educação". In *XXIV Congresso da Associação Nacional de Pesquisa e Pós-Graduação em Música, São Paulo*.

# 6

## Compost Listening

### Vaporwave and the Dissection of Record Production

## Josh Ottum

### INTRODUCTION

Click, click, click, click, ring, shake. Click, click, click, click, ring, pop. The opening percussive moments of "Earth Minutes" sound out a thin, digital tick-tock figure as synthetic "ahhs" bend in harmony. The rhythmic figures are misaligned, sounding thrown together at random. A triumphant, fake timpani resounds from dominant to tonic key at full force. As the "ahhs" ring out, the percussive motif insists on slight inaccuracy. It sounds like a musical theater overture played by defiant robots.

With three contrasting sections of calm, less calm, and frantic, "Earth Minutes" makes its way through a series of synthetic sonic spaces. What blends the sounds together is the hiss of blown-out digital keyboard presets mixed out of balance. The drums are suddenly too loud. The bass drops out inexplicably. The form rides the edge of falling apart completely. But the timbre is consistent: it is plastic, catchy, and repulsive. The sounds, by their "nature", interrogate the very idea of natural sounds. By attuning to Raymond Williams's (1972) three perspectives on nature as an essential quality, inherent force, and material aspects, we open the door to investigate how to make sense of this deeply familiar, hyper-synthetic sonic space.

This chapter examines a 2011 album by experimental electronic musician James Ferraro entitled *Far Side Virtual* (*FSV*). Consisting of 16 instrumental songs made up of preset digital sounds, rhythmic loops, and sonic logos, the composer set about to write a "rubbery plastic symphony for global warming, dedicated to the Great Pacific Garbage Patch" (Gibb, 2011). With its emphasis on uneven audio mixes, grating synthetic timbres, and humorous engagements with robotic voices, Ferraro has set up a metaphorical sonic imaginary: an oscillating world of trashy tones not unlike the aquatic clutter that converges in the North Pacific Gyre.

By positioning Ferraro's work alongside other environmentally engaged creative activity, I aim to illuminate the gap that invites unexpected and productive intersection between art and environmental issues. I argue that *FSV*, and the vaporwave genre from which it emerges, exploits the aesthetics of functional music as a way to sound out environmental imaginaries consistent with the complexities of life lived in the Anthropocene. By calling attention to the ubiquitous sonic infrastructure that permeates consumptive interactions with technology, *FSV* challenges the listener to examine not only these seemingly insatiable appetites but the ways in which such behaviors manifest themselves as we engage with environmental issues. In short, this chapter considers how particular record production practices can make sense of place, as it is rapidly redefined in an era wherein the human species has made irreparable impacts on geologic conditions and processes.

Central to this investigation is the unique process of sense-making that occurs through sound. Steven Feld's analysis of how sound functions for the Kaluli people of Papua New Guinea poses questions important for this study. In asking how "the perceptual engagements we call sensing [are] critical to conceptual constructions of place", Feld foregrounds the potential of sound to mold relationships with physical environments. Building on Feld's idea of acoustemology, or "acoustic ways of knowing", I focus on how "sonic presence and awareness" shapes how "people make sense of experiences" (Feld, 1996). At the same time, I consider how Feld's concept operates in dialogue with the realities of senses of place which reflect the 21st century, deterritorialized planet. This perspective is necessarily urban and distinctly virtual in its communicative processes. The idea of "being somewhere", has drastically changed with telecommunications allowing parties to interact across massive distances. Ecological disasters reflect this dispersal of time and space as well, proliferating across great portions of the planetary environments. To this end, ecocritic Ursula Heise's idea of "eco-cosmopolitanism" provides a productive way to engage with dispersed practices of planetary sense-making. This idea of "environmental world citizenship", in Heise's view, is sufficient to address the "challenges that deterritorialization poses for the environmental imagination" (Heise, 2008). This chapter places Feld and Heise in conversation in order to sound out the particular environmental imaginaries constructed by Ferraro's mutated production approach to functional music and the equally distorted realities of decentralized ecological disasters.

## DECENTERING POP

In an August 2009 edition of *The Wire*, a magazine devoted to experimental music and sound, writer David Keenan coined the term "hypnagogic pop" (Keenan, 2009). In Baudrillardian fashion, Keenan simply calls this newly christened, post-millennial form of outsider music "pop music refracted through the memory of a memory" (Keenan, 2009: 26). Artists such as Ariel Pink, Macintosh Plus, Oneohtrix Point Never, and James Ferraro have all approached '80s culture, the decade in which most of them were born, with a mixture of awe and irreverence. Like noise music cultures, hypnogogic pop "fetishizes the outmoded media of its infancy, releasing albums on cassette, celebrating the video era and obsessing over the reality-scrambling potential of photo-copied art" (Keenan, 2009: 26).

These varied attempts to conjure up *indistinct* impressions of an era of utopic ideals, Reaganomics, and the promises of digital technology speak to the genre's self-aware capacity to express ambivalence. By projecting and interrogating the ecstatic shortcomings of digital claims of authenticity, hypnagogic pop revels in its "drive to restore the circumstances of early youthful epiphanies while re-framing them as present realities, possible futures" (Keenan, 2009: 26) These "youthful epiphanies" are shot through with an updated brand of nostalgia, described by music critic Simon Reynolds as more of an addiction than a progressive political posture. As he notes in *Retromania*, "nostalgia is now thoroughly entwined with the consumer-entertainment complex: we feel the pangs for the products of yesteryear, the novelties and distractions that filled up our youth. . . . The intersection between mass culture and personal memory is the zone that spawned retro" (Reynolds, 2011). Perhaps this is nothing new.

Sometime around 2009, the term *vaporwave* emerged as a generic term for a style of electronic music that turned to the early '90s corporate Internet aesthetics for guidance. The term is thought to have roots, albeit shallow roots, in the term *vaporware*, which refers to a software project announced and marketed by a technology company that never comes to fruition. Vaporware can also refer to what musicologist Adam Harper calls "the deliberate fabrication of future products, with no intention to eventually release them, so as to hold customers' attention" (Harper, 2012). By slowing down micro-loops of '80s and '90s library music, embracing digital preset timbres of consumer synths, and recontextualizing sound logos and ringtones, vaporwave artists accentuate the ubiquitous sounds of the background, effectively turning the spotlight on the spotlight itself.

At the end of 2013, Ferraro had well over 100 releases, upwards of 30 under his own name and nearly 30 aliases with their own string of releases. After releasing multiple recordings in the mid-2000s as one half of the noise duo The Skaters, Ferraro released his first official solo record *Multitopia* (2007). The record uses clips of reality TV shows and tabloid shows to weave a post-9/11 tapestry of "extreme, baroque-style consumerism" (Ferraro, 2011). From the beginning of his solo career, Ferraro has employed hyperreal rhetoric in song titles such as "Wired Tribe/Digital Gods", "Roaches Watch TV", "Condo Pets", and so on, and in interviews, he often connects his work to theoretical work by Foucault and Baudrillard dealing with power relations and simulation. Much like the ephemeral quality of his continually transforming aliases, Ferraro's biography remains mysterious. Born in Rochester, New York, in the 1980s, Ferraro remains transient, calling both Los Angeles and New York home.

While Ferraro tours and releases physical albums, there remains a fleeting quality of copies without originals that spills forth from his work. Small-batch DIY releases on primitive formats such as CD-Rs and cassettes amplify the ephemeral halo around Ferraro's oeuvre as these limited releases fetch inflated prices on a regular basis. Furthermore, there is even evidence of Ferraro enacting the musical equivalent to *vaporware*, releasing *James Ferraro and Zac Davis Are Thieves*, a recording currently fetching upwards of $100 and bearing no evidence of its actual existence.

This performative act, wherein concepts take precedence over execution, promising a release of music that doesn't exist, underscores the virtual quality of the vaporwave experience. Ontological questions abound: Is this music you even need to hear to know its sound? Is there an original contribution from the artist? Should the music be available for purchase? Indeed, a central paradox of Ferraro's work lies in a drive toward a critical aesthetic autonomy, while simultaneously generating a glut of commodified content only to be fetishized as rare objects for consumption. This move obliterates easy high/low divides, bringing into question resolute boundaries between commercial and underground aesthetics. Ferraro's nods to the Great Pacific Garbage Patch (GPGP) can then be extended beyond a metaphor for widespread habits of consumption. The very identity of the GPGP (and Ferraro's *FSV*) hinges on attempts to amplify the imprecise.

Fredric Jameson's ideas of the role of mass culture in the postmodern moment speak to this tendency to relinquish the precision in order "to transform the transparent flow of language as much as possible into material images and objects we can consume" (Jameson, 1991: 133). Contrapuntal sonic logos and unidentifiable microplastics breaking down in the sea reflect decentralized conceptions of sonic expression and ecological disaster. This imprecision can be interpreted as the resultant anonymity of a networked world, or as Arjun Appadurai puts it, the consequences of how "locality as a property . . . of social life comes under siege in modern societies" (Appadurai, 1996: 179).

Sociologist Anthony Giddens further articulates the tensions inherent in the increased pressures of modernization. Through a "disembedding" of social systems, Giddens points toward a "'lifting out' of social relations from local contexts of interaction and their restructuring across indefinite spans of time-space" (Giddens, 1990: 21). While environmentalists can chant the familiar response to "think globally, act locally", what happens when the very idea of locality becomes indiscernible from globalized activities. The seemingly complementary visions eschewed by thinking about the globe in the abstract and religiously composting last night's dinner come under critique in Heise's project of redefining place in a globalized world.

Heise's idea of eco-cosmopolitanism arises from a view that attends to the realities of cultural formation outside of "naturally" arising, physically embedded circumstances. Her argument rests on the problem that "ecologically oriented thinking has yet to come to terms with one of the central insights of current theories of globalization: namely, that the increasing connectedness of societies around the globe entails the emergence of new forms of culture that are no longer anchored in place . . ." (Heise). The challenge, then, to the realities of decentering "is to envision how ecologically based advocacy on behalf of the nonhuman world as well as on behalf of greater socio-environmental justice might be formulated in terms that are premised . . . on ties to territories and systems that are understood to encompass the planet as a whole". Heise demonstrates her investment in eco-cosmopolitanism through exemplary works of literature and art which "deploy allegory in larger formal frameworks of dynamic and interactive collage or montage". I situate Ferraro's project, and the formal dimensions of vaporwave, in conversation with Heise's redefinition of place and Feld's acoustemology, or means of "sounding as a condition of and for knowing", in order to hear how functional sounds can be used to sound out the realities of expanded, decentered environmental disasters, such as anthropogenic climate change and the GPGP (Feld).

After experimental music magazine *The Wire* named *FSV* the 2011 "Album of the Year", Ferraro's project solidified the validity of the hypnagogic subgenres everywhere, making room for conceptually ambitious tendencies to be clothed in amateurish production techniques. It should be noted that by naming these "ambitious tendencies", I am referring to Ferraro's relentless posture which welcomes the "unexpected and imperceptible introduction of commodity structure into the very form and content of the work of art itself" (Jameson, 1991: 132). Just as Jameson expands on Adorno and Horkheimer's insights into the culture industry (41–72), we find an alignment with generic practices intent on sonically articulating a whole-hearted, virtual engagement with the commodity structure. There is no better example of these qualities in Ferraro's discography than on *Far Side Virtual*.

## TUNING IN TO THE FAR SIDE

A digital piano figure enters as the flutter of a synthetic, watery ringtone lingers in the background. Thin ersatz woodblocks hammer out a repetitive call and response pattern as they drown in gated digital reverb. The mood is light and anxious. A second ringtone-like piano figure enters supplying a crucial major third that cues the listener to stay light and bury the anxiety. But unease is what guides James Ferraro's "Linden Dollars" to its circular endpoint. Just after the second piano appears, a 32nd note tambourine is introduced, submerging and surfacing with each piano cue. Functioning as an audible representation of the daily engagement with endlessly clickable links, "Linden Dollars" just won't let up. More digital voices "ooo" and "ahh" and occupy the background as the piece concludes abruptly at 1:34. After an inexplicable 16 seconds of silence, a lower digital voice enters with a final trite ringtone figure. With no time to process, the album takes off into "Global Lunch": a scenario which pairs synthesized speech with loud digital triangle figures, innumerable Skype sign-on sounds, and a horrifically catchy sitar riff responded to by a digital voice saying "duh" on repeat. Welcome to *Far Side Virtual*, where "Dubai Dream Tone" is on full volume, relentlessly taking the listener on "Adventures in Green Foot Printing", exploding with major chords hopefully leading toward a true understanding of "Google Poeisis".

**Figure 6.1** *Far Side Virtual*

As voices from the virtual world of *Second Life* and text-to-speech voices cue the listeners to order sushi and that they have reached their destination, *FSV* is imbued with a distinctly mobile listening experience. It is not a stretch to imagine moving through virtual and literal spaces, panned and edited like a commercial, filled with goods to purchase, consume, and throw away. The sounds and melodies equally reflect these consumable qualities, mixed and pieced together as if to demonstrate the full and limited capabilities of a freshly purchased Yamaha DX-7 synthesizer in the year 1983. General MIDI (musical instrument digital interface) triggered sounds are often considered ground zero in the digital music landscape, functioning as a place to start and hopefully move away from as quickly as possible. Electronic musicians habitually have worked to shape and tweak preset timbres to create unrecognizable sounds, fresh to the ears. As sampling techniques have grown by leaps and bounds, what were once hailed as "authentic" sonic replications are often met with ironic cynicism or outright disgust to the distinguishing listener. Yet these dated mid-80s and 90s digital sounds are exactly what Ferraro privileges on *FSV*; by illuminating falsified representations of "the real thing", the listener begins to question the very idea of real. In context, these sounds beckon the user in and out of virtual and digital experiences.

Writing about plastic, Barthes expands on its ability to blur the lines between natural and artificial through its sound, stating that "what best reveals it for what it is, is the sound it gives, at once hollow and flat; its noise is its undoing, as are its colors, for it seems capable of only retaining the most chemical-looking ones" (Barthes, 1977). The hyperreal relentlessness of *FSV* does just this. It calls attention to the aural characteristics and relationship of plastic physical objects, such as the smartphone, a bottle, a lighter, and a computer keyboard, into razor-sharp focus. *FSV* subverts the expected dream-inducing, symmetrical melodic fragments of Reich-ian minimalism by pushing them into a realm of digital maximalism. We are being told each day: "Let's go! Create content! Look . . . another new thing! Buy it! Throw it away! Click that link! Login!" If this rapid series of mobile commands sums up time/space of everyday life in a capitalist society, why on earth choose an amped-up musical reflection of this virtual/real threshold? Isn't the noise of everyday life enough? In Ferraro's world, the difficult choice to surrender to these sounds promotes a conscious engagement with the ambivalences of our current environmental situation and the processes of framing and consuming it. If Ferraro is weaving a metaphorical sonic tapestry of dispersed disasters inherent to the Anthropocene, we must consider how sound and environment gained attention alongside the increasing concerns for the environment as they emerged in the late-60s.

## BEYOND GOOD AND BAD

Eliding pristine sonic production and *any* analogue sounds at all on *FSV*, Ferraro pushes forward the project of interrogating Schaferian notions of moralistic binary divisions within acoustic ecology. The project of acoustic ecology was started in the late-60s at Simon Frasier University in Vancouver as part of the World Soundscape Project. Led by R. Murray Schafer and his colleagues, the core of the project was to restore a sense of harmony to the soundscape by "finding solutions for an ecologically balanced soundscape where the relationship between the human community and its sonic environment is balanced" (Westerkamp, 1991). Guiding this original mission is a sense of a proper aural imaginary, that things need to be fixed so we can return to a "natural" state of calm. This "natural" sound seems to imply a world without the dissonances of synthesized speech, ringtones, and sound logos. Yet such devices are integral to the process of capturing and preserving a pure, hi-def representation of soundscapes. Like the promise of the Hollywood compositional devices infecting the ear with nostalgia, this Schaferian call to fix a world out of tune refuses to recognize what can be gleaned from dissecting the very idea of garbage, excavating tossed out timbres, and zooming in on the nuances of microscopic changes in the biosphere.

While Timothy Morton's ideas of dark ecology and Francisco Lopez's dark nature recordings aim to expose the full spectrum of ecological connection, there remains an issue of removing technological mediation from the constructed "scape". (Morton, 2010: 16). Whether one is documenting the sounds of a rainforest free from mechanical noise or a day at the landfill, the process is transduced, translated, and disseminated through technological means. The character of microphones, preamps, and digital audio workstations embodies specific ideas of how sound should sound. As Bennett Hogg notes, "the production practices and public use of music mediated by technology can be revealed as constitutive participants in the constellation of power relations . . ." (Hogg, 2005: 212). By attending to the complexities of the always-already mediated nature of sound, Ferraro's work is liberating in its hyper-artificiality. Schafer's "aesthetic moralism" has been critiqued at length by scholars who point out the flat-footed divisions between noise as bad and silence as good (Thompson, 2017). In Ferraro's domain of sonic simulacrum, *FSV* fires a shot over the bow with the following question: If these aren't the sounds of a globalized world, what are? Still, the questions persist as to the project of balancing out, or de-polluting, the global soundscape. Are seemingly clumsy audio mixes, harsh digital presets, and an overload of sonic logos helpful in articulating the sound of a planet rife with deterritorialized eco-catastrophes? A consideration of how visual culture conceives of the changing scapes of the planet through maximal, detached, dystopic aesthetics sheds light on the question.

## DISSECTING THE EARWORM

Althusser's idea of interpellation resonates through Ferraro's approach to record production as it gives shape to the process of subject formation in a capitalist system. Through a kind of hailing, or interpellation, the individual recognizes herself as a subject within an ideological matrix (Althusser, 1972). As Althusser defines ideology as "the imaginary relationship of individuals to their real conditions of existence", subjectivity is brought about in the circulatory, simultaneous experience of interpellation (Althusser, 1972: 162). The aesthetics of functional music reflects this infectious process of neoliberal capitalist subject formation through its ubiquitous and attenuated role in society. By implicating the listener through an endless suggestion of sonic logos, the braiding process of imagined and real conditions is smoothed out through the encouragement of continuous temporary states of feeling, one passing to the next.

*FSV* lends a refreshing flavor to the familiar sounds and narratives of environmental crises. By implicating the listener in this sonic examination of lines between the artificial and natural, Ferraro reflects and reacts to the fluid definitions of "nature" by creating an auditory time/space filled with swarming sounds. These sounds are inherently mobile, as they are born, live, and die in the digital space. There is nothing *natural* about the situation, except that it seems to be the real and virtual sound of our *natural* everyday life. Just as paradoxically natural is the act of discarding waste. We do it every day, but where does it go?

It is a stretch to imagine pieces of household plastics floating in remote parts of the Pacific Ocean. The garbage dump down the street or wastebaskets in the kitchen are familiar sites for waste disposal. Just as the trash begins to smell, it is buried or taken out, taken out to the next, bigger wastebasket. Pondering the transfiguration of garbage with its inherent mobility and anonymous character connects us to what commercial composers call the "sound logo". A sound logo identifies a product, aurally cuing the listener to attach meaning to the sound, hopefully developing a Pavlovian response. Commercial composers Eric Siday and Raymond Scott pioneered the idea of the sound logo, often using early analogue synthesizers to create aural components that would enable instant association of a sound with a brand. As Daniel Jackson, CEO of Sonicbrand Ltd., puts it, "The aim of sonic branding, in relation to music is not to pollute the art form but to more accurately express the emotions of individual brands through fabulous music" (Jackson, 2003: 44). Sound logos have arguably weathered the transformational storms of the media far better than their lyrical counterparts of the jingle. Familiar sound logos, such as signing into Skype, shutting down Windows, or a GPS voice are imbued with space-producing qualities. The rising "pop" of the Skype sign-on is moving the user out of the thick sludge of other applications into a clear, fluid world of communicative possibilities. The back-and-forth "question and answer" sounds mimicked by iPhone's iMessage maintain this fluid state, underscoring the already liquefied bubbles of text. By surfing our way through the technoscape, we are encouraged to stay mobile, to be distracted, and to continue exploring. *FSV* turns this mobile ideology up to 11. The sounds are in hi-def but too loud; the timbres are in focus, but instead of transporting the listener to the next place, they cut into the ear, resulting in a near catatonic immobility. The jarring, unbalanced mix of *FSV* pushes the listener toward a meditative state of anxiety: you want to turn it off, but you can't; you want to hate it, but you don't.

Garbage, clutter, and trash are concepts that have been constructed to carry with them undesirable sensorial qualities, but they remain highly transformational. A thing becomes trash once it is relocated, and as degenerative processes take hold, the trash changes shape at both internal and external levels. As microplastic in the GPGP has become small enough to seamlessly take root in the ecosystem, the biosphere begins to change. Throw-away, utilitarian sounds and music reflect similar transformational potential as they converge, reorganizing our relational habits. By allowing these micro-melodies to reproduce at magnified levels, the sounds effectively exhibit the potentialities of the ohrvurm, or earworm, as "an infectious musical agent" (Goodman, 2010: 147). As Steve Goodman notes in *Sonic Warfare*, "a catchy tune is no longer sufficient; it merely provides the DNA for a whole viral assemblage" (148). Fully immersed in the generative stench of capitalist compost, the earworms of *FSV* mate with each other, cobbling together a fertile leitmotif out of a variety of seemingly forgettable motifs. If the assemblage of recycled auditory viruses mirrors the ideological posture of Muzak and sonic branding practices, we must open the scope of what passes for garbage and what passes for art.

The final track on *FSV*, "Solar Panel Smile", opens with the sound of Windows XP shutting down. Ferraro then mutates the four-note motif, deconstructing it, highlighting the nuances of the descending melody. A hazy, bubbly ringtone shimmers on loop in the background as the XP shutdown continues to unravel. As digital chordal pad aimlessly descends and ascends, a thin, abrasive string motif appears. Just as the repetition of the figure begins to annoy (or become catchy), an incredibly loud digital bell slams the right side of the mix. The pain of the bell opens up to an endless drum fill ripped from Apple's consumer digital audio workstation GarageBand. The fill continues to support a series of micro(plastic) melodic fragments. But the fragments never open into a full melody. These melodies are a bridge to somewhere else, just like the endless drum fill should unfold into the persistent predictability of a spacious pop beat, just like the Windows XP motif was designed for a singular purpose.

Ferraro puts the listener smack in the middle of these transitory functional sounds and beckons us to linger, paying attention to the complexities of often ignored processes. As the journey of "Solar Panel Smile" unfolds, the ephemeral role of the tritone is embraced with ambivalence. This is the same musical interval that often functions in sound logos for movie production companies; these motifs exude the promise of fulfillment that imbues the spectacle of popular cinematic productions. In Ferraro's hands, constant repetition of the motif reveals its overuse as a device to achieve the "natural" sound of hope. Such ambitions are self-consciously thwarted by the arrival of a short digital crash cymbal that brings "Solar Panel Smile" to a grinding halt. In this way, Ferraro is adept at conjuring up musical situations that reflect, through rarefaction, our own relationships with the 21st century challenges of attaching comprehensible narratives to deterritorialized ecological crises. This paradoxical narrative is one of crisis and decision, but it moves slowly, without a center, through the sea of sound logos, background music, and virtual scenarios, meditating on how such discourses originated. This dispersal of audio logos metaphorically reflects a consummate disaster of our times.

## COMPOST LISTENING

Much like hip hop and many other forms of electronic music, *FSV* maintains an affinity for recycling and recontextualizing sonic material. Yet, it does so in a self-consciously naive way, steering clear of reusing forgotten content in the name of sharpening the cutting edge. Instead of dependable practices of musical recycling, Ferraro opts for the compost bin, listening to the re- and degeneration of discarded objects. Upon the release of *FSV*, Ferraro noted that the true form of the record would be to exist in the mode of ringtones, sounding off anywhere at any time (Hoffman, 2011). Additionally, Ferraro expressed ambitions to have the pieces be performed by an orchestra using "ringtones instead of tubular bells, Starbucks cups instead of cymbals" (Gibb, 2011). With such grand goals, Ferraro remains acutely aware of the ephemeral quality of the project, noting that "consumer transience was in fact always a part of the concept of the record. The process of disregard or transience elevated to assisted readymades. The last note of my record will be the sound of the uninterested listener disposing the album into the trash bin and emptying out their desktop's trash bin" (Gibb).

In this chapter, I have considered how making sound can make sense of place, as it is rapidly redefined in an era wherein the human species has made irreparable impacts on geologic conditions and processes. By examining the formal properties of an exemplary work of vaporwave record production practice, I have aimed to complicate the possibilities of a sonic ecology that allows for rapidly changing discourses that reflect the complexities of constructing environmental knowledge in the current moment. An analysis of *FSV* reveals and revels in the large chasm between blockbuster and academic approaches to creative work that takes environmental issues as its subject matter.

By asserting a direct connection to the natural reactions of the biosphere to constructed "natural" processes and the effects of plasticizing the ocean, Ferraro shines a light on the natural sounds of the modern environment. Propagating earworms are framed for aural exhibition, and synthetic voices assume the role of the lead singer. Here in this deterritorialized space, on the far side, the spaces of sonic ecology unexpectedly widen to confront the synthetic as natural, unflinching at the dispersed fragments of plastic sound that simultaneously mobilize and paralyze.

## BIBLIOGRAPHY

Adorno, Theodor, and Max Horkheimer. "The Culture Industry: Enlightenment as Mass Deception." In *Dialectic of Enlightenment*, translated by John Cumming, 94–136. New York: Herder and Herder, 1969.

———. *Essays on Music.* Edited by Richard Leppert, Translated by Susan H. Gillespie. Berkeley, CA: University of California Press, 2002.

———. *Introduction to the Sociology of Music.* New York: Continuum, 1988.

Allen, Aaron S. "Ecomusicology: Music, Culture, Nature . . . and Change in Environmental Studies?" *Journal of Environmental Studies and Sciences* 2, no. 2 (2012): 192–201.

Althusser, Louis. "Ideology and Ideological State Apparatuses." In *Lenin and Philosophy and Other Essays.* New York: Monthly Review Press, 1972.

Altman, Rick, ed. *Sound Theory-Sound Practice.* New York: Routledge, 1992.

Anderson, Mark. "New Ecological Paradigm (NEP) Scale." In *The Berkshire Encyclopedia of Sustainability: Measurements, Indicators, and Research Methods for Sustainability*, 260–262. Great Barrington: Berkshire Publishing, 2012.

Anderson, Tim J. *Making Easy Listening: Material Culture and Postwar American Recording.* Minneapolis, MN: University of Minnesota Press, 2006.

Andrady, Anthony L. "Microplastics in the Marine Environment." *Marine Pollution Bulletin* 62, no. 8 (2011), accessed 1 December 2015, DOI: 10.1016/j.marpolbul.2011.05.030.

Appadurai, Arjun. *Modernity at Large: Cultural Dimensions of Globalization*. Vol. 1. Minneapolis, MN: University of Minnesota Press, 1996.

Attali, Jacques. *Noise: The Political Economy of Music*. Translated by Brian Massumi. Minneapolis, MN: University of Minnesota Press, 1985.

Barthes, Roland. *Image Music Text*. New York: Hill and Wang, 1977.

———. *Mythologies: Roland Barthes*. New York: Hill and Wang, 1972.

———. *S/Z: An Essay*. New York: Hill and Wang, 1977.

Baudrillard, Jean. *The Gulf War Did Not Take Place*. Bloomington, IN: Indiana University Press, 1995.

———. *Selected Writings*. Stanford, CA: Stanford University Press, 1988.

———. *Simulacra and Simulation*. Translated by Sheila Faria Glaser. Ann Arbor, MI: University of Michigan Press, 1994.

Birosik, Patti Jean. *The New Age Music Guide: Profiles and Recordings of 500 Top New Age Musicians*. New York: MacMillan Publishing Company, 1989.

Blake, William. *Complete Poetry and Prose*. Edited by David Erdman. New York: Anchor, 1997.

Born, Georgina. "Music, Modernism and Signification." In *Thinking Art: Beyond Traditional Aesthetics*, edited by Andrew Benjamin and Peter Osborne. London: ICA, 1991.

———. *Rationalizing Culture: IRCAM, Boulez, and the Institutionalization of the Musical Avant-Garde*. Berkeley, CA: University of California Press, 1995.

Boym, Svetlana. *The Future of Nostalgia*. New York: Basic Books, 2001.

———. "Nostalgia and Its Discontents." *The Hedgehog Review* 9, no. 2 (2007): 7–18.

Brend, Mark. *The Sound of Tomorrow: How Electronic Music Was Smuggled Into the Mainstream*. New York: Bloomsbury, 2012.

Brown, Steven P., and Douglas Stayman. "Antecedent and Consequences of Attitude Toward the Ad: A Meta-Analysis." *Journal of Consumer Research* 19 (1992): 34–51.

Brown, Steven P., and Ulrich Volgsten. *Music and Manipulation: On the Social Uses and Social Control of Music*. New York: Berghahn Books, 2006.

Bruner, Gordon C. "Music, Mood, and Marketing." *The Journal of Marketing* (1990): 94–104.

Buck-Morss, Susan. *The Origin of Negative Dialectics*. New York: The Free Press, 1977.

Buell, Frederick. "A Short History of Oil Cultures: Or, the Marriage of Catastrophe and Exuberance." *Journal of American Studies* 46 (2012): 273–293.

Buell, Lawrence. *The Environmental Imagination: Thoreau, Nature Writing, and the Formation of American Culture*. Cambridge, MA: Harvard University Press, 1995.

———. *Writing for an Endangered World: Literature, Culture, and Environment in the U.S. and Beyond*. Cambridge, MA: Harvard University Press, 2001.

Carpenter, E. J., and K. L. Smith Jr. "Plastics on the Sargasso Sea Surface." *Science* 175 (1972): 1240–1241.

Carter, Paul. "Ambiguous Traces, Mishearing, and Auditory Space." In *Hearing Cultures: Essays on Sound, Listening, and Modernity*, edited by Veit Erlmann. New York: Berg, 2004.

Chion, Michel. *Audio-Vision: Sound on Screen*. Translated by Claudia Gorbman. New York: Columbia University Press, 1994.

———. *Film, a Sound Art*. Translated by Claudia Gorbman. New York: Columbia University Press, 2009.

Cook, Nicholas. *Analysing Musical Multimedia*. Oxford: Clarendon Press, 1998.

———. "Music and Meaning in the Commercials." *Popular Music* 13, no. 1 (1994): 27–40.

Cox, Christoph. "Beyond Representation and Signification: Toward a Sonic Materialism." *Journal of Visual Culture* 10, no. 2 (2011): 145–161.

Cox, Christoph, and Daniel Warner, eds. *Audio Culture: Readings in Modern Music*. New York: Continuum, 2004.

Cronon, William. "The Trouble With Wilderness." In *Uncommon Ground: Rethinking the Human Place in Nature*, edited by William Cronon, 69–90. New York: W.W. Norton and Company, 1996.

Damluji, Mona. "Petrofilms and the Image World of Middle Eastern Oil." In *Subterranean Estates: Life Worlds of Oil and Gas*, edited by M. Watts, A. Mason, and H. Appel. Ithaca, NY: Cornell University Press.

Davis-Floyd, Robbie. "Storying Corporate Futures: The Shell Scenarios." In *Corporate Futures* edited by George Marcus. Chicago: University of Chicago Press, 1998.

Debord, Guy. *Society of Spectacle*. Detroit: Black and Red, 2000.

Deleuze, Gilles. *Cinema 1: The Movement-Image*. Translated by Hugh Tomlinson and Barbara Habberjam. Minneapolis, MN: University of Minnesota Press, 1986.

Deleuze, Gilles, and Felix Guattari. *A Thousand Plateaus: Capitalism and Schizophrenia*. Translated by Brian Massumi, 310–350. Minneapolis, MN: University of Minnesota Press, 1987.

Dunlap, Riley, and Kent Van Liere. "The 'New Environmental Paradigm': A Proposed Measuring Instrument and Preliminary Results." *Journal of Environmental Education* 9 (1978): 10–19.

Dyson, Frances. *Sounding New Media: Immersion and Embodiment in the Arts and Culture*. Berkeley, CA: University of California Press, 2009.

Allen, Aaron S. "Ecomusicology." *Grove Dictionary of American Music*, Second Edition. New York: Oxford University Press, 2013.

Eremo, Judie, ed. *New Age Musicians*. Cupertino: GPI Publications, 1989.

Eriksen, Marcus, Laurent C. M. Lebreton, Henry S. Carson, Martin Thiel, Charles J. Moore, Jose C. Borerro, Francois Galgani, Peter G. Ryan, and Julia Reisser. "Plastic Pollution in the World's Oceans: More Than 5 Trillion Plastic Pieces Weighing Over 250,000 Tons Afloat at Sea." *PloS One* 9, no. 12 (2014): e111913.

Feld, Steven "Waterfalls of song: An acoustemology of place resounding in Bosavi, Papua New Guinea." In *Senses of Place*, edited by S. Feld and K. H. Basso, 91-135. Santa Fe, NM: School of American Research Press, 1996.

Fink, Robert. "Orchestral Corporate." *Echo: A Music-Centered Journal* 2, no. 1 (2000).

———. *Repeating Ourselves: American Minimal Music as Cultural Practice*. Berkeley, CA: University of California Press, 2005.

Fleming, James Rodger, Vladimir Janković, and Deborah R. Coen. *Intimate Universality: Local and Global Themes in the History of Weather and Climate*, Vol. 1. Sagamore Beach: Watson Publishing International, 2006.

Forman, Janis. *Storytelling in Business: The Authentic and Fluent Organization*. Stanford, CA: Stanford University Press, 2013.

Forsyth, Tim. "Critical Realism and Political Ecology." In *After Postmodernism: An Introduction to Critical Realism*, edited by Jose Lopez and Garry Potter, 146–154. New York: The Athlone Press, 2001.

Gibb, Rory. "Adventures on the Far Side: An Interview with James Ferraro." *The Quietus*, December 15, 2011, accessed November 13, 2014. Available at: http://thequietus.com/articles/07586-james-ferraro-far-side-virtual-interview

Giddens, Anthony. *Consequences of Modernity*. Stanford: Stanford University Press, 1990.

Glotfelty, Cheryll. "Introduction: Literary Studies in an Age of Environmental Crisis." In *The Ecocriticism Reader: Landmarks in Literary Ecology*, edited by Cheryll Glotfelty and Harold Fromm, xv–xxxvii. Athens, GA: University of Georgia Press, 1996.

Goggin, Peter. "Introduction." In *Rhetorics, Literacies, and Narratives of Sustainability*, edited by Peter N. Goggin, 2–12. New York: Routledge, 2009.

Goldmark, Daniel, Lawrence Kramer, and Richard D Leppert. *Beyond the Soundtrack: Representing Music in Cinema*, edited by Daniel Goldmark, Lawrence Kramer, and Richard D. Leppert. Berkeley, CA: University of California Press, 2007.

Goldstein, Miriam, Marci Rosenberg, and Lanna Cheng. "Increased Oceanic Microplastic Debris Enhances Oviposition in an Endemic Pelagic Insect." In *Biology Letters* 8, no. 5 (2012), accessed 11 November 2013, DOI: 10.1098/rsbl.2012.0298.

Goodman, Steve. *Sonic Warfare: Sound, Affect, and the Ecology of Fear*. Cambridge, MA: The MIT Press, 2010.

Gopinath, Sumanth. *The Ringtone Dialectic: Economy and Cultural Form*. Cambridge, MA: The MIT Press, 2013.

Gorbman, Claudia. *Unheard Melodies: Narrative Film Music*. Bloomington, IN: Indiana University Press, 1987.

Harper, Adam. "Vaporwave and the Pop Art of the Virtual Plaza." *Dummy Mag* (2012). Available at: www.dummymag.com/features/adam-harper-vaporwave

Heckman, Don. "Suzanne Ciani Has Designs on Music." *Los Angeles Times*, 11 September 1987.

Heise, Ursula K. *Sense of Place and Sense of Planet*. New York: Oxford University Press, 2008.

Hoffman, Kelley. "James Ferraro's Versace Dreams." *Elle Magazine* (2011). Available at: www.elle.com/culture/celebrities/news/a7617/james-ferraros-versace-dreams-31526/

Hogg, Bennett. "Who's Listening?" In *Music, Power, and Politics*, edited by Annie J. Randall. New York: Routledge, 2005.

Huron, David. "Music in Advertising: An Analytic Paradigm." *Musical Quarterly* 73, no. 4 (1989): 557–576.

Hutcheon, Linda. "Irony, Nostalgia, and the Postmodern." In *Methods for the Study of Literature as Cultural Memory*, edited by Raymond Vervliet and Annemarie Estor, 189–207. Atlanta: Rodopi, 2000.

Ingram, David. *The Jukebox in the Garden: Ecocriticism and American Popular Music Since 1960*. New York: Rodopi, 2010.

Jackson, Daniel M. *Sonic Branding: An Introduction*. Hampshire: Palgrave MacMillan, 2003.

Jameson, Frederic. "The Cultural Logic of Late Capitalism." In *Postmodernism, or the Cultural Logic of Late Capitalism*, 1–54. Durham, NC: Duke University Press, 1991.

Jhally, Sut. *The Codes of Advertising: Fetishism and the Political Economy of Meaning in the Consumer Society*. New York: Routledge, 1990.

Johnson, Randal. "Introduction." In *The Field of Cultural Production*, edited by Peirre Bourdieu, 1–25. New York: Columbia University Press, 1993.

Kahn, Douglas. *Earth Sound Earth Signal: Energies and Earth Magnitude in the Arts*. Berkeley, CA: University of California Press, 2013.

Keenan, David. "Childhood's End: Hypnagogic Pop." *The Wire* (August 2009): 26-31.

Kramer, Lawrence. *Music as Cultural Practice, 1800–1900*. Berkeley, CA: University of California Press, 1990.

Kupiers, Joyce. "Reality by Design: Advertising Image, Music and Sound Design in the Production of Culture." Order No. 3352224, Duke University, 2009.

LaBelle, Brandon. *Sound Culture and Everyday Life*, 163–200. New York: Continuum, 2010.

Ladino, Jennifer. *Reclaiming Nostalgia: Longing for Nature in American Literature*. Charlottesville and London: University of Virginia Press, 2012.

Lanza, Joseph. "Adventures in Mood Radio." In *Radiotext(e)*, Vol. VI: 1, edited by Neil Strauss, 97–105. New York: Semiotext(e), 1993.

———. *Elevator Music: A Surreal History of Muzak, Easy Listening, and Other Moodsong*. Ann Arbor, MI: University of Michigan Press, 2004.

———. "Rhapsody in Spew: Romantic Underscores in *The Ren & Stimpy Show*." In *The Cartoon Music Book*, edited by Daniel Goldmark and Yuval Taylor, 269–274. Chicago: A Cappella Books, 2002.

Moore, Captain Charles, and Cassandra Phillips. *Plastic Ocean: How a Sea Captain's Chance Discovery Launched a Determined Quest to Save the Oceans*. New York: Avery, 2011.

Morton, Timothy. *The Ecological Thought*. Cambridge, MA: Harvard University Press, 2010.

———. *Hyperobjects: Philosophy and Ecology After the End of the World*. Minneapolis, MN: University of Minnesota Press, 2013.

———. "Zero Landscapes in the Time of Hyperobjects." *Graz Architectural Magazine* 7 (2011): 78–87.

Nisbet, James. *Ecologies, Environments, and Energy Systems in Art of the 1960s and 1970s*. Cambridge, MA: The MIT Press, 2014.

Pinkus, Karen. *Alchemical Mercury: A Theory of Ambivalence*. Stanford, CA: Stanford University Press, 2010.

————. "Ambiguity, Ambience, Ambivalence, and the Environment." *Common Knowledge* 19, no. 1 (2012): 88–95.

Pirages, Dennis, and Paul Ehrlich. *Ark II: Social Response to Environmental Imperatives*. New York: Viking, 1974.

Powers, Devon. "Strange Powers: The Branded Sensorium and the Intrigue of Musical Sound." In *Blowing Up the Brand: Critical Perspectives on Promotional Culture*, edited by Melissa Aronczyk and Devon Powers, 285–306. New York: Peter Lang, 2010.

Price, Jennifer. "Looking for Nature at the Mall: A Field Guide to the Nature Company." In *Uncommon Ground: Rethinking the Human Place in Nature*, edited by William Cronon, 186–203. New York: W.W. Norton and Company, 1996.

Reynolds, Simon. *Retromania: Pop Culture's Addiction to Its Own Past*. New York: Macmillan, 2011.

Schafer, R. Murray. *The Soundscape: Our Sonic Environment and the Tuning of the World*. Rochester: Destiny Books, 1977.

Sterne, Jonathan. *The Audible Past: Cultural Origins of Sound Reproduction*. Durham, NC and London: Duke University Press, 2003.

————. "Sounds Like the Mall of America: Programmed Music and the Architectonics of Commercial Space." *Ethnomusicology* 41, no. 1 (Winter 1997): 22–50.

Tacchi, Jo. "Nostalgia and Radio Sound." In *The Auditory Culture Reader*, edited by Michael Bull and Les Back, 281–295. New York: Berg, 2003.

Tagg, Phillip. *Kojak-50 Seconds of Television Music: Towards the Analysis of Affect in Popular Music*. Dissertation, 1979.

————. "Nature as a Musical Mood Category." *IASPM Norden's working papers series* (1982).

Taylor, Timothy. "The Avant-Garde in the Family Room: American Advertising and the Domestication of Electronic Music in the 1960s and 1970s." In *The Oxford Sound Studies Handbook*, edited by Karin Bijsterveld and Trevor Pinch. New York: Oxford University Press, 2013.

————. *Beyond Exoticism: Western Music and the World*. Durham, NC and London: Duke University Press, 2007.

————. *The Sounds of Capitalism: Advertising, Music and The Conquest of Culture*. Chicago: The University of Chicago Press, 2012.

Thompson, Marie. *Beyond Unwanted Sound: Noise, Affect, and Aesthetic Moralism*. London: Bloomsbury Academic, 2017.

Thompson, Emily. *The Soundscape of Modernity: Architectural Acoustics and the Culture of Listening in America, 1900–1933*. Cambridge, MA: MIT Press, 2002.

Unattributed. "Music of the Future?" *Mademoiselle* (December 1959): 94–97.

Valentino, Thomas J. *Major Records Mood Music for Television, Films, Radio, Drama, Newsreels, Theatre*. New York: Thomas J. Valentino, Inc., 1968.

Vanel, Herve. *Triple Entendre: Furniture Music, Muzak, Muzak-Plus*. Champaign, IL: University of Illinois Press, 2013.

Washburne, Christopher, and Maiken Derno. *Bad Music: The Music We Love to Hate*. Edited by Christopher J Washburne and Maiken Derno. New York: Routledge, 2004.

Westerkamp, Hildegard. *The Soundscape Newsletter*, no. 1 (1991).

Williams, Raymond. "Ideas of Nature." In *Ecology: The Shaping Enquiry*, edited by Jonathan Benthall, 146–164. London: Longman, 1972.

Ziser, Michael G. "Home Again: Peak Oil, Climate Change, and the Aesthetics of Transition." In *Environmental Criticism for the 21st Century*, edited by Stephanie LeMenager, Teresa Shewry, and Ken Hiltner, 181–195. New York: Routledge, 2011.

———. "The Oil Desert." In *American Studies, Ecocriticism, and Citizenship: Thinking and Acting in the Local and Global Commons*, edited by Joni Adamson and Kimberly N. Ruffin, 76–86. New York: Routledge, 2013.

## DISCOGRAPHY

Asher, James. *Commerce*. 1981 by Bruton Music. BRL9. LP.

Bastow, Geoff. *Tomorrow's World*. 1978 by Bruton Music. BRI21. LP.

Eno, Brian. *Ambient One: Music for Airports*. 1978 by Polydor AMB001. LP.

Ferraro, James. *Far Side Virtual*. 2011 by Hippos in Tanks. HIT 013. CD.

———. *KFC City 3099: Pt. 1 Toxic Spill*. 2009 by New Age Tapes. CD.

———. *Multopia*. 2007 by New Age Tapes. CD.

# 7

## Classical Production

### Situating Historically Informed Performances in Small Rooms

## Robert Toft

One of the challenges facing performers who wish to record works from the past in an historically informed manner centers on finding acoustic spaces similar to those in which the music would have been heard originally. Unfortunately, many musicians today regard large churches as ideal locations for recording much of the music composed in the Renaissance and Baroque eras, even though these rooms can be far too reverberant for a significant portion of the repertoire, especially solo songs accompanied by quiet instruments (such as the lute, guitar, or harpsichord). Curiously, artists often spend a great deal of time learning how to bridge the gap between the written descriptions of performing practices found in old treatises and actual performance yet choose less-than-optimal locations to record vocal music that seems to have been conceived for intimate spaces (for drawings of the modest rooms in which Giulio Caccini probably sang his own compositions in the late 16th and early 17th centuries, see Markham 2012: 200–203).

In fact, recent research has shown that musicians in the 16th to 18th centuries regularly performed in private chambers or small music rooms (Howard and Moretti 2012: 106–107, 111–114, 185–189, 200–203, 248–249, 320), and because many of the *camere per musica* measured no more than seven by eleven meters (23 x 36 feet), with a ceiling height of five to six meters (16 to 20 feet), their volume (approximately 385 cubic meters or 13,250 cubic feet) produced a reverberation time of less than a second, which means that these spaces tended more to clarity and intimacy than reverberance (Orlowski 2012: 157–158). Moreover, since listeners would have been seated close to the performers in these rooms, direct sound would predominate, and early reflections from the walls and floor would further contribute to the sonic impression of clarity and intimacy (Orlowski 2012: 158).

Most recordists, however, do not have access to intimate historic spaces for recording purposes, but they can employ modern studio technology to replicate the aural sense of these *camere*, not only through convolution reverbs based on impulse responses taken from rooms in 16th- to 18th-century buildings but also with artificial reverberation of the reflection-simulation type designed to mimic the characteristics of generic spaces. This chapter discusses the procedures a group of recordists followed, first to create an historically informed interpretation of a 17th-century solo song and then to place that performance in an acoustic that would make listeners feel as though they were sitting in the same small room as the performers. Specifically, the chapter focuses on a track from Studio Rhetorica's recording *Secret Fires of Love*, which features tenor Daniel Thomson and Baroque guitarist Terry McKenna under my musical direction (Talbot Productions, TP 1701, 2017). The recording of "Si dolce è'l tormento", a song composed by Claudio Monteverdi and published in Carlo Milanuzzi's *Quarto scherzo delle ariose vaghezze* (Venice, 1624), was produced by me and recorded and mixed by Robert Nation (Kyle Ashbourne, assistant engineer) at EMAC Recording Studios in London, Canada (the track is available on my website, www.talbotrecords.net, or in full-track preview on CD Baby). I am most grateful to Robert for generously sharing his philosophies/procedures of recording and mixing with me, as the production and post-production sections of this chapter could not have been written without his input, for they blend Robert's explanations of his practices with my contributions as producer.

The first part of the chapter considers the pre-production phase of the project, during which Daniel, Terry, and I finalized the interpretive strategies that would be employed, and the following sections focus on the ways studio production practices at EMAC helped us transfer our historically informed conception of "Si dolce è'l tormento" to disk.

## PRE-PRODUCTION

Older principles of interpretation differ considerably from those currently used by classical musicians, and in order for people interested in historical performance to recover the old methods, they must reconstruct the practices from surviving sources of information. Fortunately, a great deal of material comes down to us, and this has allowed us not only to root the interpretive strategies employed in *Secret Fires of Love* in period documents but also to take a fresh approach to Renaissance and Baroque songs. Recent research has shown that from the 16th to the early 19th centuries, singers modeled their art directly on oration and treated the texts before them freely to transform inexpressive notation into passionate musical declamation (see in particular Toft 2013, 2014).

Daniel adopts the persona of a storyteller, and like singers of the past, he uses techniques of rhetorical delivery to recreate the natural style of performance listeners from the era probably would have heard. This requires him to alter the written scores substantially, and his dramatic singing combines rhetoric and music in ways that sympathetically resonate with performance traditions from the Renaissance and Baroque eras. In "Si dolce è'l tormento", Daniel sings prosodically, emphasizing important words and giving the appropriate weight to accented and unaccented syllables;

employs a highly articulated manner of phrasing; alters tempo frequently through rhythmic *rubato* and the quickening and slowing of the overall time; restores *messa di voce*, the swelling and diminishing of individual notes, as well as phrases, to its rightful place as the "soul of music" (Corri 1810: i. 14); contrasts the tonal qualities of chest and head voice as part of his expression; and applies *portamento*.

Among these principles, highly articulated phrasing, alterations of tempo, and variations in the tonal quality of the voice represent the most noticeable departures from modern practice. Singers of the past inserted grammatical and rhetorical pauses to compartmentalize thoughts and emotions into easily discernible units (that is, stops at punctuation marks and in places where the sense of the sentence called for them), and this frequent pausing gave listeners time to reflect on what they had just heard so they could readily grasp the changing sentence structure. In 1587, Francis Clement explained the rationale behind the addition of unnotated pauses: "the breath is relieved, the meaning conceived . . . the eare delited, and all the senses satisfied" (pp. 24–25; for further information on pausing, see Toft 2013: 20–45; Toft 2014: 84–98). Moreover, writers from Nicola Vicentino (1555: fol. 94v) to Giambattista Mancini (1774: 150) observe that singers best convey the true sense and meaning of words in a natural way if they derive the pacing of their delivery from the emotions in each text segment. Or to use Vicentino's words, tempo fluidity has "a great effect on the soul/effetto assai nell'animo" (1555: fol. 94v).

Similarly, the use of appropriate vocal timbres to carry the text's emotions to the ears of listeners requires singers not only to differentiate their registers (so that the lowest and highest parts of the range contrast with the middle portion) but also to link timbre and emotion (smooth and sweet, thin and choked, harsh and rough) – "the greater the passion is, the less musical will be the voice that expresses it" (Anfossi c.1840: 69). In earlier eras, a versatile tonal palette prevented the monotony of what David Ffrangcon-Davies dismissed in 1905 as the "school of sensuously pretty voice-production". Indeed, as Ffrangcon-Davies suggests, the then new monochromatic approach to timbre meant that if audiences had heard a singer in one role, they had heard that singer in every role (pp. 14–16).

Armed with a collection of historic principles, our first task in recreating period style was to study the lyrics of "Si dolce è'l tormento" to find all the places a singer might wish to insert grammatical or rhetorical pauses (see Figure 7.1).

This exercise involved following principles described in treatises on rhetoric and oration to add pauses of varying lengths at points of punctuation (grammatical pauses) and to separate subjects from verbs and verbs from objects, as well as conjunctions, relative pronouns, participles, and prepositional phrases from the text that precedes them (rhetorical pauses). The compartmentalization of ideas and emotions organizes and paces the content of the poem so that listeners can easily grasp the story, and since some ideas require a slower or quicker delivery than others, compartmentalization also provides appropriate places for singers to change the speed of delivery to match the emotional character of the phrases.

| | | |
|---|---|---|
| 1. | Si dolce * è'l tormento * | So sweet * is the torment * |
| | ch'in seno mi stà * | which resides in my breast * |
| | ch'io vivo contento * | that I live in contentment * |
| | per cruda beltà. * | because of the cruel beauty. * |
| | Nel ciel di bellezza * | In the heaven of beauty, * |
| | s'accreschi fierezza * | if arrogance increases * |
| | et manchi pietà * | and pity decreases, * |
| | che sempre qual scoglio * | then always like a rock * |
| | all'onda d'Orgoglio * | in a wave of disdain * |
| | mia fede * sarà. * | my faith * shall exist. * |
| 2. | La speme fallace * | False hope * |
| | rivolgam'il piè * | overwhelms my foundation, * |
| | diletto, * ne pace | neither delight * nor peace |
| | non scendano a me * | comes to me, * |
| | e l'empia * ch'adoro * | and the wicked one * that I adore * |
| | mi nieghi ristoro * | denies me the comfort * |
| | di buona mercé: * | of kind mercy: * |
| | tra doglia infinita * | amidst infinite pain, * |
| | tra speme tradita * | amidst hope betrayed * |
| | vivra * la mia fe. * | my faith * shall have life. * |
| 3. | Per foco, * e per gelo * | Because of the fire, * and because of the ice * |
| | riposo non hò * | I have no rest * |
| | nel porto del Cielo * | [but] in the refuge of heaven * |
| | riposo haverò. * | I shall have rest. * |
| | Se colpo mortale * | If a deadly blow * |
| | con rigido strale * | with a sturdy arrow * |
| | il cor m'impiagò, * | wounds my heart, * |
| | cangiando mia sorte * | changing my fate * |
| | col dardo di morte * | with the dart of death * |
| | il cor * sanerò. * | shall heal * my heart. * |
| 4. | Se fiamma d'Amore * | If the flame of love * |
| | già mai non * sentì. * | never before * was felt. * |
| | Quel riggido core * | That merciless heart * |
| | ch'il cor mi rapì. * | who stole my heart. * |
| | Se nega pietate * | If you deny pity, * |
| | la cruda beltate * | the cruel beauty * |
| | che l'alma invaghì * | who ravished my soul, * |
| | ben fia * che dolente * | it shall be right * that in sorrow, * |
| | pentita, * e languente * | repentant * and languishing, * |
| | sospirimi * un dì. * | you will sigh for me * one day. * |

**Figure 7.1** Text of "Si dolce è'l tormento" with grammatical and rhetorical pauses marked; asterisks represent pauses of various size

But apart from organizing and pacing ideas and emotions, singers must decide which word or words within a phrase should be emphasized, and treatises usually combine the discussion of emphasis with that of accent. Accent denotes the stress placed on a single syllable to distinguish it from the others in a word (this is known as speaking or singing prosodically), whereas emphasis refers to the force of voice laid on an entire word or group of words to bring the associated ideas to the attention of listeners. Proper accentuation, then, adhered to the normal pronunciation of words in ordinary speech, while emphatic delivery varied according to the meaning performers wished to convey (for more information on accent and emphasis, see Toft 2013: 73–79, 2014: 98–108).

Emphatic words, then, receive the greatest force within a sentence, and these important words are situated in an overall hierarchy of emphasis in which speakers reserve the strongest sound of voice for the most significant word or idea, firmly and distinctly pronouncing substantives (nouns), adjectives, and verbs, while relegating unimportant words (the, a, to, on, in, of, by, from, for, and, but, and so on) to relative obscurity (Walker 1781: ii. 15, 25). This hierarchy allows performers not only to arrange words into their proper class of importance but also to achieve a distribution of emphases that would prevent sentences from being delivered monotonously with uniform energy (Herries 1773: 218). Thus, the application of accent and emphasis creates light and shade and helps speakers (and singers) clearly project the meaning of long and complex ideas. Inappropriate shading would force listeners to decipher a sentence's meaning from an ambiguous or confusing delivery (Murray 1795: 153).

After we completed our analysis of the song's text using the principles just discussed, Daniel proceeded to create a dramatic spoken reading of the poem. Initially, this meant deciding which pauses would be employed and what ideas would be exhibited prominently (that is, emphasized), as well as what variations in the speed of delivery would suit the changing emotions of the text. As part of this process, Daniel also took note of where *messa di voce* and *portamento* occurred as he spoke, for in these places both the swelling and diminishing of the voice and the sliding between pitches would sound the most natural in singing (*messa di voce* and *portamento* are discussed more fully in Toft 2013: 45–69). Once we were satisfied with the spoken narrative, we transferred Daniel's interpretation to the song, altering Monteverdi's melodic lines to accommodate the dramatic reading.

By rooting our performance in historical documents, we were able to model our understanding of the relationship between performer and score directly on principles from the past. Indeed, singers in earlier times viewed scores quite differently from their modern counterparts. They realized that because composers wrote out their songs skeletally, performers could not read the notation literally, and to transform inexpressively written compositions into passionate declamation, vocalists treated texts freely and personalized songs through both minor and major modifications. In other words, singers saw their role more as one of recreation than of simple interpretation, and since the final shaping of the music was their responsibility, the songs listeners heard often differed substantially from what appeared in print (Toft 2013: 4–6 discusses the relationship between notation and performance).

Composers of the past did not notate subtleties of rhythm, phrasing, dynamics, pauses, accents, emphases, tempo changes, or ornamentation. Clearly, they had no desire (or need) to capture on paper the elements of performance that moved listeners in the ways writers from the time described. In the middle of the 16th century, Nicola Vicentino commented that "sometimes [singers] use a certain method of proceeding in compositions that cannot be written down"/"qualche volta si usa un certo ordine di procedere, nelle compositioni, che non si può scrivere" (1555: fol. 94v), and along these lines, Andreas Ornithoparchus, writing in 1517, praised singers in the Church of Prague for making "the Notes sometimes longer, sometime[s] shorter, than they should" (p. 89 in John Dowland's translation). Around 1781, Domenico Corri characterized the relationship between performance and notation candidly: "either an air, or recitative, sung exactly as it is commonly noted, would be a very inexpressive, nay, a very uncouth performance" (vol. 1, p. 2). Charles Avison had already made this notion explicit in 1753 (p. 124): "the Composer will always be subject to a Necessity of leaving great Latitude to the Performer; who, nevertheless, may be greatly assisted therein, by his Perception of the Powers of Expression", and a hundred years later, voice teachers like Manuel García (1857: 56) continued to suggest the same thing – performers should alter pieces to enhance their effect or make them suitable to the power and character of an individual singer's vocal capability.

In 1555, Nicola Vicentino suggested why performers valued flexibility of tempo (fol. 94v):

> The experience of the orator teaches this [the value of changing tempo (mutare misura) within a song], for one sees how he proceeds in an oration – for now he speaks loudly and now softly, and more slowly and more quickly, and with this greatly moves his auditors; and this way of changing the tempo has a great effect on the soul./La esperienza, dell'Oratore l'insegna, che si vede il modo che tiene nell'Oratione, che hora dice forte, & hora piano, & più tardo, & più presto, e con questo muove assai gl'oditori, & questo modo di muovere la misura, fà effetto assai nell'animo.

Hence, vocalists sang *piano e forte* and *presto e tardo* not only to conform to the ideas of the composer but also to impress on listeners the emotions of the words and harmony, and Vincenzo Giustiniani (c.1628: 108) characterized the approach singers from Mantua took in the latter part of the 16th century:

> [B]y moderating and increasing their voices, forte or piano, diminishing or swelling, according to what suited the piece, now with dragging, now stopping, accompanied by a gentle broken sigh, now continuing with long passages, well joined or separated [that

is, legato or detached], now groups, now leaps, now with long trills, now with short, and again with sweet running passages sung softly, to which one unexpectedly heard an echo answer/col moderare e crescere la voce forte o piano, assottigliandola o ingrossandola, che secondo che veniva a' tagli, ora con strascinarla, ora smezzarla,

**Figure 7.2** Claudio Monteverdi, "Si dolce è'l tormento" (Milanuzzi 1624); letters above each system indicate the chords the guitarist should play.

con l'accompagnamento d'un soave interrotto sospiro, ora tirando passaggi lunghi, seguiti bene, spiccati, ora grupi, ora a salti, ora con trilli lunghi, ora con breve, et or con passaggi soavi e cantati piano, dalli quali tal volta all'improvviso si sentiva echi rispondere.

Primarily, the period-specific alterations we made to Monteverdi's text and melodic lines involved adding pauses and adjusting the rhythmic values of the notes so that the delivery of the syllables and words came as close to speaking as possible. But Daniel also varied tempo along the lines Vicentino and Giustiniani had suggested and employed light and shade (accent and emphasis) in an historic way, for if he were to sing the melodies exactly as Monteverdi had notated them, he would, in Domenico Corri's 18th-century view, be guilty of an "inexpressive" and "uncouth" performance (a modern edition of Monteverdi's skeletal notation for "Si dolce è'l tormento" appears in Figure 7.2). By placing his persona as a storyteller in this older guise, Daniel has provided what one writer, John Addison (c.1850: 29), called the "finish" to the song in a way that approximates early 17th-century style.

## PRODUCTION

The main goal in producing "Si dolce è'l tormento", as well as the other songs included in *Secret Fires of Love*, was to enhance period interpretation through modern studio practices, especially isolated sound sources recorded by closely placed microphones. We chose to blend the worlds of recording and "live" performance so that we could capture a dramatic reading of the song, while achieving sonic clarity. In other words, we did not consider the two activities to be mutually exclusive, and because we felt that making records and archiving a "live" event differed fundamentally, we decided to use punch-ins to perfect excellent takes rather than completely re-record those sections that contained minor imperfections.

From these perspectives, the project benefited from having one person assume the roles of music director and producer, for a single conception of the song could then emerge from the various sonic possibilities available to the artists and engineers. Indeed, decisions made throughout the process, from those that led to an historically relevant interpretation of the printed score to those that guided the design of the soundscape in which the performance was presented, came from imagining how one world might inform the other. Knowledge of both historical performance and recording practices focused the energies of everyone involved in the project on an idealized conception, and the various elements of production, when combined with sympathetic strategies for editing and mixing, helped shape the recording along historical lines.

Robert tracked in ProTools HD at 24 bits / 96 kHz to provide an excellent signal-to-noise ratio, as well as an increased dynamic range, and microphone selection and placement figured prominently at the beginning of the sessions. Since Baroque guitars can be somewhat "noisy" to record, the question arose of how we might best capture the sound of the instrument. The mics chosen would need to produce as "natural" a stereo sound as possible when positioned closely, so microphones with too much boost on the top end would not be suitable. In fact, a stereo pair of omnidirectional microphones with a linear frequency response would probably be ideal for this application, as omnis would allow the characteristic resonance of the instrument to be portrayed realistically, without proximity effect. For the voice, a microphone that could provide a consistent frequency response across Daniel's range, while keeping sibilance to a minimum, would be preferable, and both ribbon and omnidirectional microphones were obvious possibilities.

A "shoot-out" using mics from EMAC's and the author's collections resulted in the following choices:

*Voice* – Royer R-122 active ribbon mic (a Schoeps small-diaphragm condenser with an MK 2 capsule, omnidirectional with a flat frequency response, was also quite attractive)
*Baroque guitar* – stereo pairs of DPA's omnidirectional 4006A condenser and Royer's R-122 active ribbon

Daniel sang in an isolation booth with his microphone, shielded by a pop screen, placed approximately 12 inches (30 cm) in front of his mouth. Although ribbon mics exhibit a fairly strong proximity effect, the Royer R-122 not only provided the consistency of frequency response we desired but also eliminated most of the problems associated with sibilance. Terry performed in the main tracking room, and because the dynamic range of Baroque guitars is not large, the first pair of mics for his guitar (DPA 4006A) were placed next to the instrument in an A-B arrangement, one pointing at the lower bout and the other at the upper bout. A second pair of stereo mics, R-122s in Blumlein configuration, were positioned about 24 inches (61 cm) from the guitar along its center line to create the sense of space a two millisecond delay would produce.

Once appropriate levels had been set, the tracking procedure consisted of several initial takes of the whole song. The entire team then listened to these recordings to determine if one of them could become a master take that would be refined through punch-ins. At this point, Terry suggested he try some new accompaniment ideas that had occurred to him as he listened (his part was improvised from the chord symbols shown in the preceding score), and since the take that resulted from this extemporaneous performance, complete with an impromptu introduction, had the level of spontaneity we were seeking and did not need any punch-ins, it became the master track which would be edited and mixed.

## POST-PRODUCTION

The editing process not only helped us achieve our ideal historically informed conception of "Si dolce è'l tormento" but also allowed us to reduce any noise in the recording that might distract listeners. Because audio recordings lack the visual connection of "live" performance, noise that mars the sensory surface of a disk can be much more disruptive, especially string noise and extraneous breathing in guitar tracks or plosives in vocal tracks. Kyle Ashbourne, the assistant engineer, and I carefully listened to the master take and used the spectral repair feature of iZotope's *Rx Advanced* to reduce string noise to an amount a listener would hear when seated 10–12 feet (three to four meters) away from the performers. In other words, we wished to present the guitar from the perspective of the listener rather than the player.

Similarly, plosives that resulted from the singer's closely placed ribbon microphone were reduced using the high-pass filter in Universal Audio's *Massenburg DesignWorks MDWEQ5*. Moreover, Pro Tools' clip gain helped us achieve the prosodic manner of vocal delivery prized in the past, for we lowered the level of those syllables that close micing had heightened. In addition, whenever the Royer R-122 exaggerated a *messa di voce* with too much energy at its peak, automation of the dynamics in ProTools helped bring those phrases into line with the way they would be heard a short distance from a singer (here we were thinking of how the inverse square law affects the propagation of the sound waves to soften the swells naturally).

The mixing sessions focused on three main elements – reverb, compression, and EQ. As we began to consider possible models for creating a suitable ambience, we decided to listen to a number of lute and Baroque guitar performances that had been recorded in large churches, the typical locations for such recordings. These rooms were, of course, far too reverberant for our purposes, and we realized that without appropriate models to emulate, we needed to imagine a performance space that did not yet exist on a recording. Robert then set about designing an artificial ambience that would approximate the small rooms in which the music was often performed originally.

He chose the same two algorithmic reverbs for the voice and the Baroque guitar, so that Daniel and Terry would sound like they were performing together in a room. The first was the Large Hall B in Universal Audio's *Lexicon 224 Digital Reverb*, with a decay time of 2.0 s in the vocal and 1.7 s in the guitar, and the second was the Small Plate 2 in Eventide's *UltraReverb*, set to a short decay time of 837 ms to "tighten" the ambience surrounding both performers. Robert used the two reverbs quite subtly, mixing them in at a low level, and because Daniel has a large dynamic range, Robert compressed the input to the *Lexicon 224* to bring the level of the loudest passages down by 2–3 dB, so that Daniel's voice would not over-trigger the reverb. Compression, then, helped make the vocal ambience less obvious. Baroque guitars, in contrast, have quite a small dynamic range, and since they could never over-trigger the *Lexicon 224*, Terry's input did not need to be compressed.

Beyond the reflection simulation Robert crafted to place the performers in an appropriately sized room, he included two parallel effects busses on the vocal track and adjusted the frequency balance of Daniel's voice. A compression back bus, combining Universal Audio's EL7 FATSO Sr with Sonnox's Oxford Inflator, was mixed in at a very low level (-24.0 dB) to increase the intimacy of the quietest passages, and on another bus, Nugen Audio's Stereoizer created a bit of extra space in the mix. Specifically, delays of 12 ms on one side and 19 ms on the other added the sense of early reflections, which, together with the reverbs, compensated for recording into a baffle.

To lessen the proximity effect inherent in the Royer R-122, the Massenburg DesignWorks MDWEQ5 parametrically removed some of the energy around 310 Hz, while the high-pass filter of the same plugin disposed of any rumble below 40 Hz. Close micing of a singer can also produce some mild harshness at the loudest moments, and a cut of 1.4 dB in Brainworx's dynamic equalizer dynEQ V2 was used to alleviate this tension around 2357 Hz. Brainworx (n.d.: 4) describes a dynamic EQ as "a filter that is not limited to being set to a specific gain level, but which changes its gain settings dynamically – following the dynamics of a certain trigger signal". Robert also applied the dynEQ V2 plugin to the guitar track to lessen the effect of some low "thumps" around 119 Hz (a cut of 2.5 dB).

On the master bus, because downstream codecs can increase the peak level of the signal somewhat, a true-peak limiter (Nugen Audio's ISL 2) was set at -0.7 to leave room for file conversion, and the general loudness characteristics of the track were analyzed through Nugen Audio's MasterCheck Pro.

*******

In our recording of "Si dolce è'l tormento", as well as in the other tracks on *Secret Fires of Love*, we clearly embraced the notion that "the most important reverberation exists within the recording, not the playback space" (Case 2007: 263), and since we did not want listeners to experience the music from a distance, as if they were in a large church or concert hall, we decided to create an artificial ambience that would situate them about 10 to 12 feet (three to four meters) from the artists. Hence, a blend of close micing and the digital processes described earlier, all in the service of an historically informed performance, allowed everyone involved in the project (artists, producer, and engineers) to realize on disk what we imagined someone in the 17th century might have heard in the small rooms in which the music was frequently performed.

## REFERENCES

Addison, J. (c.1850) *Singing, Practically Treated in a Series of Instructions*. London: D'Almaine.

Anfossi, M. (c.1840) *Trattato teorico-pratico sull'arte del canto . . . A Theoretical and Practical Treatise on the Art of Singing*. London: By the Author.

Avison, C. (1753) *An Essay on Musical Expression*. London: C. Davis. Reprint, New York: Broude, 1967.

Brainworx (n.d.) *bx_dynEQ V2 Manual*. Leichlingen: Brainworx.

Case, A. (2007) *Sound FX: Unlocking the Creative Potential of Recording Studio Effects*. Boston, MA: Focal.

Clement, F. (1587) *The Petie Schole*. London: Thomas Vautrollier. Reprint, Leeds: Scolar, 1967.

Corri, D. (c.1781) *A Select Collection of the Most Admired Songs, Duetts, &c.*, 3 vols. Edinburgh: John Corri. Reprint, Richard Maunder. *Domenico Corri's Treatises on Singing*, vol. 1. New York: Garland, 1993.

——— (1810) *The Singer's Preceptor*. London: Longman, Hurst, Ress, and Orme. Reprint, Richard Maunder, *Domenico Corri's Treatises on Singing*, vol. 3. New York: Garland, 1995.

Ffrangcon-Davies, D. (1905) *The Singing of the Future*. London: John Lane.

García (the Younger), M. (1857) *New Treatise on the Art of Singing*. London: Cramer, Beale, and Chappell.

Giustiniani, V. (c.1628) *Discorso sopra la musica*. Lucca: Archivio di Stato, MS O. 49. Edited Angelo Solerti, *Le origini del melodramma*. Torino: Fratelli Bocca, 1903. Reprint, Hildesheim: Georg Olms, 1969.

Herries, J. (1773) *The Elements of Speech*. London: E. and C. Dilly. Reprint, Menston: Scolar, 1968.

Howard, D. and L. Moretti, eds. (2012) *The Music Room in Early Modern France and Italy: Sound, Space, and Object*. Oxford: Oxford University Press for the British Academy.

Mancini, G. (1774) *Pensieri, e riflessioni pratiche sopra il canto figurato*. Vienna: Stamparia di Ghelen.

Markham, M. (2012) "Caccini's Stages: Identity and Performance Space in the Late Cinquecento Court." In Howard and Moretti 2012: 195–210.

Milanuzzi, C. (1624) *Quarto scherzo delle ariose vaghezze*. Venice: Vincenti.

Murray, L. (1795) *English Grammar*. York: Wilson, Spence, and Mawman. Reprint, Menston: Scolar, 1968.

Orlowski, R. (2012) "Assessing the Acoustic Performance of Small Music Rooms: A Short Introduction." In Howard and Moretti 2012: 157–159.

Ornithoparchus, A. (1517) *Musice Active Micrologus*. Leipzig: Valentin Schumann. Translated John Dowland, *Andreas Ornithoparcus His Micrologus, or Introduction*. London: Thomas Adams, 1609. Reprint of both editions in one volume, New York: Dover, 1973.

Studio Rhetorica. (2017) *Secret Fires of Love*. London, ON: Talbot Productions, TP 1701.

Toft, R. (2013) *Bel Canto: A Performer's Guide*. Oxford: Oxford University Press.

——— (2014) *With Passionate Voice: Re-Creative Singing in Sixteenth-Century England and Italy*. Oxford: Oxford University Press.

Vicentino, N. (1555) *L'antica musica ridotta alla moderna prattica*. Rome: Antonio Barre. Reprint, Kassel: Bärenreiter, 1959. Translated Maria Rika Maniates, *Ancient Music Adapted to Modern Practice*. New Haven, CT: Yale University Press, 1996.

Walker, J. (1781) *Elements of Elocution*. London: T. Cadell.

Part Two

# Music Production and Technology

# 8

# The Politics of Digitizing Analogue Recording Technologies

Pat O'Grady

## INTRODUCTION

In this chapter, I examine the ways in which ideas about vintage analogue sound are reproduced in digital music production contexts. Over the past decade, a renewed interest in analogue technologies has occurred in the field of pop music production. The audio recording company Universal Audio has capitalized on this trend and offers a range of digital plug-ins that reproduce a number of elements of vintage analogue technologies. These technologies include hardware produced by the company itself as well as other audio companies. Historically, these technologies were associated with large studio recording spaces and practices from the 1960s to 1980s. I examine the strategies deployed by UAD in order to reproduce these technologies within digital contexts. These strategies include references to scientific research, skeuomorphism, associations with analogue discourse, and "artist endorsement". In addition to the sound UAD plug-ins produce, I argue that these strategies also play a crucial role in the reproduction of these technologies.

## THE POLITICS OF REPRODUCING "VINTAGENESS"

The relationships between UAD and pop music production can be understood as part of a political discourse. Stuart Hall draws on Michel Foucault's work to argue that a discourse is "a group of statements which provide a language for talking about – a way of representing the knowledge about – a particular topic at a particular historical moment" (Hall 2013: 291). For Hall (2013), discourses produce knowledge and power. This framework for understanding the way specific representations can garner power is instructive in the context of UAD. It can be used to analyze the way these products are represented and their claims to reproduce the "sound" of analogue technologies. UAD's discursive practices – and the recurring descriptors used to identify the product – can be understood as a process that is mobilized in order to reinscribe and perpetuate a value system within the field. Over the past ten years, the Internet has become the dominant discursive space of pop music production. For Cecilia Suhr (2012), new media, such as social media and the Internet, have become an important site of texts and their political negotiations. The UAD website provides a site to participate in this space. Its descriptions and photographs of its products provide a site to negotiate the political field.

The relationships between these texts and the practice of music production can be understood through the work of Pierre Bourdieu (1984, 1990). In his work, Bourdieu introduces a range of perspectives and terms for understanding fields. Bourdieu uses the term "cultural field" to describe a particular "site of cultural practice" (Webb 2002: 22; Bourdieu 1990). Cultural practices relate to the behaviors of people or, as he writes, "agents". These behaviors produce hierarchies of power that are determined by economies of capital. This capital can be economic, cultural, or social. For instance, economic capital is straightforward: it refers to one's spending capacity. Cultural capital lies in patterns of particular consumption: some products are valued more highly than others (Webb 2002: 22). Within a field, specific knowledge of what is considered "tasteful" exists within economies of value. The ability to identify and practice these attributes results in cultural capital for agents. "Social capital" is another important element of Bourdieu's work. It refers to the degree of mobility agents have within a social field; agents establish – and have access to – social networks through their status, practice, and the cultural capital they acquire.

Bourdieu's work has been adapted to music production. It has been used to understand specific economies of value within home studios (Cole 2011). Further, his work has been used to understand the political negotiations for status and reputations within the field of advertising production (Taylor 2012). Bourdieu's foundational work is also particularly relevant for understanding economies of value around analogue technologies and, more important, how they can be mobilized for political gain. The respective work of these theorists brings into focus how cultural capital associated with analogue is mobilized in order to negotiate value for digital technologies. These negotiations demonstrate the degree to which consumers of audio technologies are not only dealing with production tools but also with political cultural goods. Using this framework, it is possible to understand the ways in which sound and hardware are discussed and represented within the context of UAD.

## RECORDING TECHNOLOGIES IN POP MUSIC PRODUCTION

The emergence of UAD plug-ins can be situated within a number of key developments in music production during the second half of the 20th century. During this time, recording technologies have increasingly mediated the sound of pop music. Originally, obsolete equipment from radio production was repurposed for use in music recording studios. Since the 1960s, however, large studios have developed, and as a result, particular technologies have dominated – and become synonymous with – recording (Cunningham 1999). In these spaces, for example, microphones, speakers, and consoles have become crucial cultural objects within the field. The use of these technologies has transformed pop music. For Theodore Gracyk, it resulted in new styles: "Rock is a tradition of popular music whose creation and dissemination centres on recording technology" (1996: 1). The use of recording technologies has also further distinguished between performance and recording. As Paul Théberge states, "Sound recording allows for the musician to distance themselves from the act of performance and to create 'impossible music', that is, music that could not otherwise be conceived or performed" (Théberge 1997: 216). Further, Phillip Auslander (2008) argues that what constitutes a "live performance" has shifted because of the use of digital technology and audience expectations.

In recording studios, signal processors form one of the most important technologies. They are used to filter audio via processes such as equalizing frequencies, compressing and gating dynamic range, and providing the sound of space through reverb. Their influence on recording has been demonstrated within a number of recent discussions. For instance, the debate known as the "loudness wars" raised concerns about the degree to which the dynamic range of a recording is reduced by compression (Devine 2013). There has also been debate on the influence of pitch correction vocals on performance. Since its uptake in the 1990s, the digital signal processor "auto-tune" has led to concerns about the degree to which a singer's intonation has become redundant (Hughes 2015). Here, signal processors are ostensibly compromising highly valued aspects in pop music.

As these debates within pop music production demonstrate, the uptake of digital technologies has considerably shaped the means of filtering audio in music production. For Tim Taylor, the advent of digital technology "marks the beginning of what may be the most fundamental change in the history of Western music since the invention of music notation in the ninth century" (2001: 3). It has dramatically changed workflow options for producers. Taylor states, "It is now possible to create entire worlds of sound all by yourself with your computer; it is no longer necessary to be with other people. Music as a social activity is becoming a thing of the past for many of these musicians" (2001: 139). Tim Warner (2003) notes that digital technologies facilitate "infinite copying with no deterioration of quality; non-destructive, software based editing" (2003: 20–21). These technologies offered alternative cheaper and miniaturized tools for music production (Moorefield 2005). This trend continued with the increase in computer-processing power of the 1990s. It has also led to an increase in home studios (Cole 2011; Tandt 2004).

The development and uptake of software plug-ins to filter audio is also a key element of the "digital revolution". Plug-ins offer additional ways to filter audio in digital audio workstations. They differ significantly from analogue hardware. For example, plug-ins are a relatively cheap option for consumers. As a result, they have shaped and enhanced the field of home recording (Théberge 2012). Plug-ins also offer the user "total recall", a characteristic of some digital technologies that allow users to move to another project on their device and then return to the original project later. Conversely, analogue technologies are "destructive" in that they require the user to commit to settings prior to moving on to another project. Their capability for multiple concurrent usage is another important characteristic of plugins. Digital plug-ins can be copied across multiple channels and used individually. Conversely, hardware compressors can only be used on one instrument at a time. Here, the audio effect must be "destructively" recorded before the compressor is available for another instrument. The functionality of digital plug-ins, however, is determined by the host computers' processing power. As the processing power of new computers has exponentially increased in recent years, the functionality of plug-ins has also increased. In addition, auxiliary hardware processing solutions have come to market to reduce the reliance of plug-ins on computer processors. Audio interfaces such as those by Metric Halo, for example, offered a processor that ran a suite of plug-ins. Also, digital hardware signal processors have provided their own processing power to reduce the stress on the processor of the computer while maintaining functionality. For example, some Lexicon hardware reverb units have their own processing power, but allow control and functionality within a digital audio workstation.

## "VINTAGENESS" IN PLUG-INS

In recent years, digital plug-ins have been implicated within a broader cultural shift back to analogue. This trend has occurred across a number of media technologies. Pop culture texts have argued that analogue dominated the 2000s. Music journalist and author Simon Reynolds, for example, claims that "instead of being the threshold to the future, the first ten years of the 21st century turned out to be the 'Re' Decade. The 2000s were dominated by the 're' prefix: *revivals, reissues, remakes, re-*enactments. Endless *re*trospection" (2011: xi). Renewed interest in analogue technologies has also been observed by LJ Rich (2010) in *Analogue Gadgets Back in Fashion in a Digital Age*. The academy, however, has taken a more cautious and case-by-case approach to examining the recent interest in vintage cultural objects. In her work, Elena Caoduro (2014) examines the reproduction of analogue elements in digital photography on social media. Music recording formats is another key area of this change. Roy Shuker (2013), for example, analyzes the recent renewed interest in vinyl record. Music production has also experienced a trend back to analogue. In her work, Samantha Bennett (2012) explores the recent re-uptake of previously disused analogue technologies in recording studios.

Software plug-ins have also continued this trend within music production. In recent years, recording technology companies have developed renderings of vintage technologies in digital plug-ins. Examples of this trend include UAD and Waves. However, some hardware users question the quality of the emulations and the extent to which they can reproduce the highly valued aspects of hardware. For Carsten Kaiser (2017), distaste toward these plug-ins compromises the often celebrated "social equality" brought about by their relative affordability. Despite the degree of distaste toward emulations, however, UAD is used across a range of professional studio contexts. Despite the relatively few large recording studios in Sydney, for example, both Studio 301 and Track Down advertise UAD plug-ins in their list of equipment. UAD is advertised alongside their hardware (Studios 301 n.d.). This trend is also present in some of the smaller yet professional spaces. The small recording studio, The Vienna People, for example, also include UAD in their advertised list of equipment. These examples demonstrate that, despite their extensive hardware, large studios offer and advertise UAD. This trend suggests that the politics of vintage emulations is not only complex; it is also not a binary of hardware or software contexts.

## THE POLITICS OF DISCUSSING AND REPRESENTING SOUND

The success of recording technologies that reproduce analogue sounds is complex. Although the ability of digital technologies such as UAD to accurately reproduce the sound characteristics of analogue technology might be a rationale for its success, we must also consider the spatial composition of the field and the way that consumers perceive sound. In the first instance, the spatial composition of the field is relevant to the reproduction of digital technologies for use in music production. Since the original analogue technologies were introduced, the industry has changed. As large commercial studios dominated the 1960s, successful musicians began to use their economic and cultural capital to build their own studios with select parts of commercial equipment (Théberge 2012). However, as digital technologies developed in the 1980s, the politics of access to recording technologies changed once again. A wider range of musicians began to access cheaper digital technologies. As a result, pop music is presently produced across a wide range of recording studio contexts. Although large studios are still common, shifts in the politics of access suggest that a lot of music is also being produced in small-scale production circumstances using digital technologies. Further, as it is now four decades since music producers first adopted digital technologies, many contemporary musicians have not worked with analogue technologies. These developments suggest that many in the field of audio consumers are not equipped to make productive comparisons between hardware and UAD plug-ins.

In the second instance, the ways in which people perceive sound is also relevant to the production of digital technologies. As it is very difficult to understand how people perceive sound, it is difficult to argue that the reproduction of analogue technology is exclusively based on its sound characteristics. Within the discursive space of the field, people can claim whatever they wish about sound. Few are placed in testing conditions under which their perception of sound is "proven". Further, those who are put under testing conditions often reject the results (Crowdy 2013). This reaction aligns with that of consumers who are presented with scientific evidence about the ability of high-fidelity technologies to enhance sound. According to Perlman (2004), they also reject the results. The perception of sound is also culturally specific. Throughout the Middle Ages, for example, analysis of Western music culture emphasized melody in music to the extent that timbal variation was often ignored. For Charles Burnett, pitch is the primary "differentiating factor in sound – rather than say, brightness, volume or sweetness" (1991: 48). Cornelia Fales's (2002) work is also important in understanding how people perceive sound differently. In one instance, an ethnomusicologist's emphasis on the melody led to a recording technique that overlooked crucial "whispering" timbral qualities of the music. These ideas offer a number of relevant points for how we can think about signal processing. First, it can be situated within Gracyk's (1996) work on recording styles and music being the product of studio manipulation. Against the backdrop of a plurality of production spaces, however, these factors highlight the extent to which we should avoid an assumption that there is a universal cognitive perception of sound. Rather, it is culturally and historically specific.

Coupled with the work of Bourdieu and Hall, these factors challenge the role that the accurate sound reproduction of vintage hardware plays in negotiating the political field. I want to stress, however, that the sound produced still plays a role in reproducing vintage analogue, as agents within the field who produce audio indeed have listening skills. Bourdieu acknowledges a certain degree of non-transferable knowledge is required to work within a specific field. But the foregrounded aspects suggest that there are possibly a range of other important factors involved in how UAD negotiates this complex field.

## STRATEGIES FOR REPRODUCING VINTAGE

### Hardware

UAD deploys a number of strategies in order to reproduce the sound of vintage analogue technologies within digital contexts. An important element of these strategies is the company's long-standing industrial links to signal processing hardware production. This link is complex for five reasons. First, within UAD's plug-in library, some signal processors are no longer available in their original form. For instance, the Fairchild is no longer available as a new hardware device. Second, the majority of the plug-ins offered by UAD are also available as hardware "clones" by various companies. A hardware "clone" reproduces the components of an original device. Markets for clones exist due in part to the unavailability of – or expense associated with – the original product. Pultec and Fairchild signal processors are cloned by a range of companies who did not produce the original. Third, the extensive second-hand market is another important element in the complexity of the field. In her work, Bennett (2012) outlines an emergent market of vintage analogue recording technologies. Until recently, these technologies were rendered obsolete by the so-called "digital revolution". This trend is particularly prominent with large recording consoles. The main focus of the 2013 film *Sound City*, for example, is a Neve console. The film details the history and then closure of the studio, after which the console was moved into Dave Grohl's private studio. Similarly, the sale of the recording console used by Pink Floyd on their album *Dark Side of the Moon* recently attracted media attention (Planet Rock 2017). Fourth, some signal processors by UAD are still remodeled and offered by their original companies. These include SSL, Neve, API, Ampex, Manley, Lexicon, and Pultec. Further, some of these companies have changed since the 1960s. "Neve Electronics", founded in 1961, for example, merged with AMS in 1992 to become "AMS Neve". Other companies have remained more or less in operation since the 1960s. For example, API have remained in production since 1969. Fifth, in contrast, some companies responded to the renewed interest in vintage technologies by reissuing their products more. In 2000, for example, Pulse Techniques developed a reissue of the PULTEC EQP-1 (Pulse Techniques 2011).

Universal Audio's associations with hardware is another crucial strategy used to promote its role in reproducing analogue technologies for digital contexts. For instance, it has reissued the "Teleronix LA2A" and "1176". Following its role in shaping the technological landscape of recording studios from the mid-1960s to the 1970s, Universal Audio was reformed by the founder Bill Putnam's sons in 1999. This has resulted in a transfer of cultural capital to the Teleronix LA2A and 1176. Although the political economies of the field are complicated by notions of value around second-hand yet original vintage technologies, cultural capital is nonetheless linked with this remodeled hardware. Here, Universal Audio mobilizes the cultural capital associated with its hardware into digital contexts. These aspects put UAD at a distinct advantage compared to other companies, such as WAVES, which does not produce hardware.

## Research

In line with broader claims made within the field of music production, UAD also attributes its ability to reproduce vintage technologies within digital contexts to its use of university-based research and ground-breaking science. The use of scientific research in the development of music products is common. For example, the Lossy Compression format "MP3" was developed at the Fraunhofer Society in Germany during the 1990s (Collins & Young 2014; Sterne 2012). Universal Audio draws similar links to university research. According to the Universal Audio website, UAD is the result of work by scientists from Stanford University. The company claim that they have developed a distinct approach to emulating analogue technology. This is called "circuit modelling":

> Unlike other plug-in makers, Universal Audio constructs their software using circuit modelling, rather than circuit simulation. With circuit simulation, a number of test signals are simply run through the unit being modelled, and a mathematical model is then developed to mimic these changes. UA prefers to follow the more rigorous scientific approach of circuit modelling by fully rebuilding all elements of

each circuit – whether transistor, transformer, or tube – in the digital world. The results are the best-sounding and most faithful emulations of the subtle nuances of classic analog hardware available.

(UAD 2017)

In this passage, UAD argues that the process of separating the individual components of a hardware system allows a more accurate reproduction of the sound within digital contexts. Their reference to mathematics and scientific practice provides some evidence that they have accurately – and faithfully – reproduced analogue sounds. However, as foregrounded earlier with Crowdy (2013) and Perlman (2004) references to science are not always a productive strategy for shaping the views of participants within pop music discourse. In line with understandings of cultural capital as a political tool, agents tend to reject scientific findings in favor of their own ability to hear what they are looking for. Despite this contradiction, the way that universal draws links to university research is useful in order to demonstrate the extent to which the development of the product is linked with the original hardware.

## Skeuomorphism

In order to further strengthen its claims that UAD plug-ins faithfully reproduce analogue sounds, Universal Audio also utilizes the physical appearance of the product. Plug-ins are made to look like vintage analogue technologies. This approach is known as "skeuomorphism", and it is most overtly used within digital photography (Caoduro 2014). For instance, Caoduro argues that the use of Instagram filters that reproduce vintage photographic styles is a form of skeuomorphism that reproduces a "self-conscious fetish for cultural objects from the immediate past, and the imitation of the style, design, and sound of analogue technologies" (2014: 5). It is also common in digital music production contexts. In their work, Adam Bell et al. (2015) examine the use of skeuomorphism in digital audio workstations. In contrast to Caoduro, they argue that that these symbols are deployed to ease users away from the traditional form of analogue hardware. For Bell et al., however, digital environments require new "interface metaphors" to accommodate for the way in which users work in contemporary contexts. Despite significant differences in the way these technologies are used, skeuomorphism is at work in this context as UAD plug-ins reproduce much of the form of the original hardware. This reproduction is emphasized in the front panels. In the analogue technology, the controls were located in order to optimize human hand gestures. In the UAD skeuomorphic design, however, the controls are typically operated by a hand-driven computer mouse. Yet, Universal has ensured that they reproduce the visual form of the hardware. The LA2A compressor, for example, features a number of surface scratches. Although scratches typically occur on technology after extensive use, they are used here to highlight the vintage age of the original hardware. This practice also aligns it with the cultural capital associated with the original versions of the hardware.

The representations of the front panels reproduce the visual appearance of distinct cultural objects and spaces associated with music production. Although large mixing consoles tend to dominate pop culture representations of typical mixing rooms in large studios, "racks" of the signal processors also play a role. In *Sound City*, for example, signal processors are shown as part of the broader focus on the studio and its mixing console. The association of these panels with large studios mobilizes cultural capital within digital contexts. Taking into account the problems associated with understanding how people perceive sound, visual reproduction functions as an effective – and relatively simple – strategy to reproduce vintage analogue. Here, aspects of user interface functionality are traded for the visual aspects of the hardware.

## Analogue Technology Discourse

The most crucial part of UAD's reproduction of analogue technologies is its placement within discourses of sound. The language used to describe sound plays an important role within the political economy of pop music production. As Thomas Porcello (2004) argues, distinct languages and vocabularies about sound are deployed and taught within large studio contexts. The words that one uses to describe specific sounds are judged and then used to make distinctions between "professionals" and "novices". Yet, Crowdy (2013) argues that these distinctions have become so nuanced that they are often inaudible. The work of Louise Meintjes (2012) further demonstrates the political discourses of music production. She contests that studio technology is "fetishized". "Fetish" is a particularly productive term for understanding the politics of sound and its associated technologies. The term highlights the ways in which discussions about technologies are not only a productive means for communication in collaborative contexts; they also function as powerful political tools. However, the political dimensions of music production have often worked against new technologies. During the 1980s, for example, the introduction of digital technologies, such as MIDI, caused upheaval within the field and sparked debates about authenticity, musicianship, and the computerization of music (Théberge 1997; Taylor 2001).

UAD deploys the discourse of analogue to reproduce vintage analogue technology within digital contexts. The discourse of vintage analogue technology can be frequently observed in discussions of music production. In these texts, recurring terms tend to dominate discussions and convey knowledge. For example, terms such as "smooth", "sheen", "glue", and "musical" are frequently used. One of the most common terms is "warmth". In recent years, it has been used within the academy as academics demonstrate a surge in interest in record production. Samantha Bennett (2012) and Zagorski-Thomas (2014: 70), for example, use "warmth" as an analytical means of describing specific sound characteristics associated with analogue technologies. Their usage of the term stems from its extensive use within industrial discussions. In industry, the term is deployed as a characteristic that can be added to the mixing process. For example, in their blog, "The quick guide to warming up a digital", *Music Radar* uses the term as a verb to describe a range of techniques that can be applied by using specific analogue recording technology. For the author, this process will result in specific sound qualities. "Warmth" is also discussed as a quality in music production that is mysterious and elusive. In his article, "Analogue Warmth" (2010) for *Sound on Sound*, Hugh Robjohns uses the term "mysterious" to describe the sound characteristics of constructing and explaining warmth.

UAD reproduces the discourse of analogue technology in their descriptors of the sound characteristics of their plug-ins. As demonstrated in Table 8.1, a number of terms recur in descriptions of plug-ins. The terms "warmth" or "warm", for example, occur in eight of their plug-in descriptions, including "Fatso Jr/Sr Tape Sim & Compressor", "Studer A800 Multi-channel Tape Recorder", and "Lexicon 224 Digital and Reverb EP-34 Tape Echo". UAD also uses the term "Glue". It is used to describe three UAD plug-ins including "SSL G Series Bus Compressor" and "Studer A800 Multichannel Tape Recorder". Similarly, "smooth" is used to describe three plug-ins, including the "Neve 33609/33609SE Compressor". "Musical" is one of the more peculiar and loaded terms. Despite its typical associations with musicological aspects of instruments, such

**Table 8.1** Descriptors of UAD Plugins from the UAD Website

| Plug-in | Term used in description |
| --- | --- |
| Magnetic Tape Bundle | Warmth |
| Fatso Jr/Sr Tape Sim & Compressor | Warmth |
| Studer A800 Multichannel Tape Recorder | Warmth |
| Lexicon 224 Digital Reverb EP-34 Tape Echo, | Warmth |
| Sonnox Oxford Inflator | Warmth |
| Fairchild Tube Limiter Collection | Warmth |
| Oxide Tape Recorder | Warmth |
| Precision Buss Compressor | Glue |
| SSL G Series Bus Compressor | Glue |
| Studer A800 Multichannel Tape Recorder | Glue |
| Summit Audio TLA100A Compressor | Smooth |
| Neve 33609/33609SE Compressor | Smooth |
| Fuch Overdrive Supreme 50 Amplifier | Smooth |
| Engl E646 Limited Edition | Musical |
| Fender '55 Tweed Deluxe | Musical |
| Sonnox Oxford Inflator | Musical |
| Galaxy Tape Echo | Musical |
| Neve 1081/1081SE Classic Console EQ | Musical |
| Studio D Chorus | Musical |
| API 500 Series EQ Collection | Musical |

as pitch and rhythm, "musical" is also used to describe plug-ins such as "Neve 1081/1081SE Classic Console EQ" and the "API 500 Series EQ" collection. As demonstrated by the recurring use of these terms, the discourse of analogue technology is powerful. It is an effective way of associating with analogue technology within digital contexts. Consequently, it is put to work to align the sound characteristics of UAD plug-ins with the cultural capital associated with the emulated hardware.

## Endorsements

Artist endorsements of UAD plug-ins also deploy the discourse of analogue technology. Endorsements are common within the music industry. Artists will often endorse guitars, amps, and keyboards in order for payment or the use of endorsed instruments. UAD uses this approach as a political tool. In their promotional material, UAD quotes from a number of musicians and producers who endorse specific plug-ins. This approach contains a quote, a picture of the person, his or her name, and some of the high-profile people with whom he or she has worked. As demonstrated in Table 8.2, UAD redeploys the terms "warm", "glue", and "musical". Artists also use these terms. For example, Richard Devine provides the following endorsements of the FATSO: "When you need something to really punch but still have warmth, the UAD FATSO is my weapon of choice". Here, Devine takes a highly valued aspect of analogue recording

**Table 8.2** Artist Endorsements of UAD Plugins from the UAD Website

| Artist | Plug-in | Descriptor | Quote |
| --- | --- | --- | --- |
| Vance Powell (Jack White, The Dead Weather, The Raconteaurs) | LA2A | Warm | This sounds exactly like I hoped it would – real, warm, and 100% fantastic. |
| Dean Coleman | FATSO | Warm | The UAD FATSO really brings back the bottom-end warmth to a vocal that's lost its way. |
| Richard Devine | FATSO | Warm | When you need something to really punch but still have warmth, the UAD FATSO is my weapon of choice. |
| Jacknife Lee (R.E.M., The Cars and Snow Patrol) | Oxide Tape Recorder | Warm | It's simple to use with endless possibilities, and its warm and crunchy while keeping the transients intact. |
| Glen Nicholls (The Crystal Method, Nine Inch Nails, No Doubt) | Pultec Passive EQ Collection | Warm | It has heaps of warmth and a nice open clarity on the top end. It's beautiful on vocals and guitars. |

*(Continued)*

**Table 8.2** (Continued)

| Artist | Plug-in | Descriptor | Quote |
|--------|---------|-----------|-------|
| Jimmy Douglass (Pharrell, Timbaland, John Legend) | Neve 88RS | Warm | It has all of the warmth and color of the analogue console, and its mic preamp is simply as sweet as it gets – I love it for vocals. |
| Joey Waronker (Atoms for Peace, Bec, Other Lives) | Neve 88RS | Warm | Just running signals through the UAD Neve 88RS plug-in without any dynamics or EQ engaged, adds a beautiful, warm character. |
| Joe Chiccarelli (Jason Mraz, Bec, The Strokes, Morrissey) | EMT 140 states | Warm | The EMT 140 plug-in is perfect to add some warmth and size to a lead vocal. |
| Ross Hogarth (Van Halen, Ziggy Marley, Motely Crue) | Studer A800 | Musical | I've been really loving the Studer A800 on bass, kick, overheads, horns, and strings. It adds a tonal color that is very musical. |
| Darrell Thorp (Beck, Radiohead, Charlotte Gainsbourg) | Studer A800 | Musical | The Studer A800 plug-in is tape in a box. Tracks that are thin and dull become full of life and tone . . . so creative, and musical. |
| Ariel Borujow (Chromeo Mac Miller, Madonna) | Neve 33609 | Smooth | I always reach for the Neve 33609 plug-in to smooth out background vocals. It's just like the real thing! |
| Jason Schweitzer (Eminem, Busta Rhymes, Ciara) | Pultec Passive EQ | Smooth | The UAD Pultec Passive EQ Collection has a very smooth and analogue sound. |
| Jacknife Lee (R.E.M., The Cars, Snow Patrol) | Pultec Passive EQ | Smooth | The UAD Pultec Passive EQ Collection sounds smooth and lovely. Use them gently or slam them – they still sound good. Another gem. |

and claims that he uses UAD – not the original hardware – to reproduce this aspect. This functions as a powerful endorsement. Once again, these endorsements demonstrate the use of cultural capital – associated with analogue technology – to legitimize digital contexts.

However, another aspect of cultural production is also present in artists' endorsements. The artists with whom these producers have worked provide social capital. Their social capital within the field is now tied to particular products. Their endorsement of these plug-ins suggests that UAD plugins are used to produce these artists. This functions as a powerful endorsement for consumers who wish to reproduce these aspects in their sound. Here, the artist, the producer, and the original hardware are implicated in a political negotiation. UAD mobilizes the social capital associated with each producer and then converts it into cultural capital for its plug-ins.

## ANALOGUE AND THE FIELD OF POP MUSIC PRODUCTION

The tools that UAD mobilizes in order to reproduce vintage analogue technology exposes a number of important aspects of the field of music production. First, it reveals recent cultural practices associated with music production. The large range of UAD signal processors demonstrates the extent to which specific production tools continue to shape the sound of pop music. It highlights that pop music continues to be, as Gracyk (1996) argues, a "recording style". In their range of plug-ins, for example, UAD offer a number of compressors; each with an ostensibly nuanced approach to filtering sound. In this context, compressors are not only used to manipulate the dynamic range of audio. Each specific compressor ostensibly offers a specific approach to shaping sound. UAD also highlights the pervasive use of analogue technology across a range of production contexts. Within these contexts, UAD's approach to reproducing the sound challenges the analogue/digital divide that has shaped notions of legitimacy of production spaces over the past ten years.

Second, and more crucially to the key argument of this chapter, UAD brings into focus the complex politics that make up the field of music production. It highlights the role of recording technologies in negotiations of power. The specific ways in which UAD reproduces analogue technology, for example, demonstrates that the complex politics of legitimacy that have historically surrounded digital technologies be must negotiated. However, in doing this, UAD also highlights the continuation of fetishizing analogue technologies within the aesthetics of pop music and the practice of its production. This fetishisation demonstrates that seemingly contradictory attitudes to technology are put to work to build status around a particular cultural good. Here, particular analogue technologies hold considerable cultural capital within the field. Cultural capital is acquired through essentalizing these technologies. Through its use of hardware, research, skeuomorphism, discourse, and endorsements, UAD mobilizes cultural capital and puts it to work to shape value around digital contexts.

## REFERENCES

Auslander, P. (2008) *Liveness: Performance in a Mediatized Culture*, London: Routledge.

Bell, A., Hein, E., & Ratcliffe, J. (2015) "Beyond Skeuomorphism: The Evolution of Music Production Software User Interface Metaphor," *Journal of the Art of Record Production* 5. Available from: http://arpjournal.com/beyond-skeuo morphism-the-evolution-of-music-production-software-user-interface-meta phors-2/ [Accessed 8 Feb 2017].

Bennett, S. (2012) "Endless Analogue: Situating Vintage Technologies in the Contemporary Recording & Production Workplace," *Journal of the Art of Record Production* 7. Available from: http://arpjournal.com/endless-analogue-situat ing-vintage-technologies-in-the-contemporary-recording-production-work place/ [Accessed 15 Jul 2017].

Bourdieu, P. (1984) *Distinction: A Social Critique of the Judgement of Taste*, Cambridge, MA: Harvard University Press.

Bourdieu, P. (1990) "The Intellectual Field: A World Apart," in *In Other Words: Essays Towards a Reflexive Sociology*, Cambridge: Polity, 140–149.

Burnett, C. (1991) "Sound in the Middle Ages," in C. Burnett, M. Fend, & P. Gou (eds.) *The Second Sense: Studies in Hearing and Musical Judgement From Antiquity to the Seventeenth Century*, London: University of London, 43–69.

Caoduro, E. (2014) "Photo Filter Apps: Understanding Analogue Nostalgia in the New Media Ecology," *Networking Knowledge: Journal of the MeCCSA Postgraduate Network* 7(2). Available from: http://ojs.meccsa.org.uk/index.php/ netknow/article/view/338 [Accessed 8 Sep 2017].

Cole, S. J. (2011) "The Prosumer and the Project Studio: The Battle for Distinction in the Field of Music Recording," *Sociology* 45(3), 447–463.

Collins, S., & Young, S. (2014) *Beyond 2.0: The Future of Music*, Bristol: Equinox Publishing.

Crowdy, D. (2013) "Chasing an Aesthetic Tail: Latent Technological Imperialism in Mainstream Production," in S. Baker, A. Bennett, & J. Taylor (eds.) *Redefining Mainstream Popular Music*, London: Routledge.

Cunningham, M. (1999) *Good Vibrations, a History of Record Production*, London: Sanctuary.

Daley, D. (2011) "Analog Recording Makes a Comeback," *Variety*. Available from: http://variety.com/2011/more/news/analog-recording-makes-a-comeback-1118029668/ [Accessed 15 Aug 2017].

Devine, K. (2013) "Imperfect Sound Forever: Loudness Wars, Listening Formations and the History of Sound Reproduction," *Popular Music* 32(2), 159–176.

Fales, C. (2002) "The Paradox of Timbre," *Ethnomusicology* 46(1), 56–95.

Gracyk, T. (1996) *Rhythm and Noise: An Aesthetics of Rock*, Durham, NC: Duke University Press.

Hall, S. (2013) "The Work of Representation," in S. Hall, J. Evans, & S. Nixon (eds.) *Representation*, London: Sage.

Hughes, D. (2015) "Technologized and Autonomized Vocals in Contemporary Popular Music," *Journal of Music, Technology and Education* 8(2), 163–182.

Kaiser, C. (2017) "Analog Distinction – Music Production Processes and Social Inequality," *Journal on the Art of Record Production* 11. Available from: http:// arpjournal.com/analog-distinction-music-production-processes-and-social-in equality/ [Accessed 5 Jul 2017].

Meintjes, L. (2012) "The Recording Studio as Fetish," in J. Sterne (ed.) *The Sound Studies Reader*, New York: Routledge.

Moorefield, V. (2005) *The Producer as Composer: Shaping the Sounds of Popular Music*, Cambridge, MA: The MIT Press.

Perlman, M. (2004) "Golden Ears and Meter Readers the Contest for Epistemic Authority in Audiophilia," *Social Studies of Science* 34(5), 783–807.

Planet Rock. (2017) "Dark Side of the Moon Recording Smashes World Record at Auction," *Planet Rock*. Available from: www.planetrock.com/news/rock-news/pink-floyd-the-dark-side-of-the-moon-recording-console-smashes-world-record-at-auction/ [Accessed 8 Nov 2016].

Porcello, T. (2004) "Speaking of Sound: Language and the Professionalization of Sound-Recording Engineers," *Social Studies of Science* 34(5), 733–758.

Pulse Techniques. (2011) "The Back Story," *Pulse Techniques*. Available from: www.pulsetechniques.com/about-us [Accessed 4 Jan 2017].

Reynolds, S. (2011) *Retromania/ Pop Culture's Addiction to Its Own Past*, London: Faber and Faber.

Rich, R. J. (2010) "Analogue Gadgets Back in Fashion in a Digital Age," *BBC Click*. Available from: http://news.bbc.co.uk/2/hi/programmes/click_online/8710938.stm [Accessed 7 March 2016].

Robjohns, H. (2010) "Analogue Warmth | Sound on Sound," *Sound on Sound*. Available from: www.soundonsound.com/techniques/analogue-warmth [Accessed 20 Jul 2017].

Shuker, R. (2013) *Wax Trash and Vinyl Treasures: Record Collecting as a Social Practice*, Surrey: Ashgate.

Sterne, J. (2012) *MP3: The Meaning of a Format*, Durham, NC: Duke University Press.

Studios 301. (n.d.) "Castlereagh Studios," *Studios 301*. Available from: https://studios301.com/about-studios-301/castlereagh-studios/ [Accessed 8 May 2017].

Suhr, C. (2012) *Social Media and Music: The Digital Field of Cultural Production*, New York: Peter Lang.

Tandt, C. D. (2004) "From Craft to Corporate Interfacing: Rock Musicianship in the Age of Music Television and Computer-Programmed Music," *Popular Music and Society* 27(2), 139–160.

Taylor, T. D. (2001) *Strange Sounds: Music Technology and Culture*, London: Routledge, 2001.

Taylor, T. D. (2012) *The Sounds of Capitalism: Advertising, Music, and the Conquest of Culture*, Chicago: University of Chicago Press.

Théberge, P. (1997) *Any Sound You Can Imagine: Making Music/Consuming Technology*, Middletown, CT: Wesleyan.

Théberge, P. (2012) "The End of the World as We Know It: The Changing Role of Studio in the Age of the Internet," in S. Frith & S. Zagorski-Thomas (eds.) *The Art of Record Production; an Introductory Reader for New Academic Field*, Surrey: Ashgate.

UAD. (2017) "UA's Art and Science of Modeling UAD Plugins, Part 2," *Universal Audio*. Available from: www.uaudio.com/blog/ask-doctors-ua-modeling-plug-ins/ [Accessed 2 Sep 2017].

Warner, T. (2003) *Pop Music – Technology and Creativity: Trevor Horn and the Digital Revolution*, Surrey: Ashgate.

Webb, J., et al (2002) *Understanding Bourdieu*, London: SAGE.

Zagorski-Thomas, S. (2014) *The Musicology of Record Production*, Cambridge: Cambridge University Press.

# 9

## Producing Music for Immersive Audio Experiences

Rob Lawrence

### INTRODUCTION

The desire to immerse audiences in sound and music is not a new one. Yet one could conceivably, and easily, contend that the ability to do so is becoming a more important endeavor. For the beginner, and those new to the field of music production, the subject of convincingly immersing audiences in sound and music can be a complex pursuit. However, as will be revealed in this chapter, for the producer with a more traditional background in audio production, it need not be an arduous one. Many academics, researchers, audiophiles, and producers consider the field of spatial audio as one of the final frontiers for high-quality sound reproduction and recorded sonic immersion (Rumsey 2013; Van Baelen 2013a).

One of the biggest challenges for those new to the field of spatial audio is choice – choice of how to create an experience given the vast variety and growing number of formats, standards, and configurations, which, understandably, deter as many as the subject inspires. For others, it is a world of audio and music production that is still relatively unknown. This is supported by the many undergraduate curricula yet to adopt the subject beyond 5.1 and 7.1 as advanced surround-sound configurations. Meanwhile the number of spatial audio capable sound studios, research laboratories, and auditoriums that offer such experiences is increasing.

### BEYOND STEREO

A sweeping glance through academic textbooks, papers, and various historical audio societies' archives over the last few decades and one would be forgiven for forming the firm impression that when "3-D sound", "immersive audio", or "spatial audio" is referenced, what is being referred to is the reproduction of surround sound with height channels. This is only partly true (Nipkow 2017). What is known, however, is that there is a growing interest among composers, music producers, and music recording companies in increasing the details in a sound experience and creating a sense of envelopment for the listener, by adding elevated loudspeakers to some performances (Barbour 2003; Van Baelen 2013a). Yet, to broadly brush the world of spatial audio as only being limited to loudspeaker arrays, systems that employ height channels, would be incorrect (Nipkow 2017). "Binaural", "360 sound", "VR", and "AR" are increasingly popular terms in the field too and often refer to immersive audio experiences using headphones. The increasing use of these terms today reflects the popularity of modern listening choices and the new experiences they may offer. What is clear is that the field and subject of spatial audio is as broad as it is deep, and it is expanding. With the accelerating convergence of technology, research, and practical know-how and microphone configurations and head-tracking software and hardware, not forgetting the latest loudspeaker systems and file formats, which we will briefly explore, the width and depth of the subject are rapidly flourishing.

Before we begin to consider creating a spatial audio mix, a brief reflection on what we mean by "spatial audio" may be helpful. Although a deep historical introduction to the subject is not required here and is available elsewhere, let us review what we mean by these terms and their associated expressions in the context of creating an immersive audio music mix.

Generally speaking, immersive audio experiences, outside of theaters and theme parks, fall into two broad categories: headphones (often referred to as "360" or "binaural" experiences) and loudspeakers ("immersive audio" or "spatial audio" experiences.) From this point, there are a number of branches one can further explore. For headphones, we can be concerned with virtual reality (VR) and augmented reality (AR) experiences which can rely upon a head-tracking system for realism and "360" experiences, which do not (Ammermann 2017). For loudspeaker systems, particularly in the cinema and home, we are often dealing with hybrid systems, a combination of object- and channel-based technology. The fundamental difference between these two technologies is how and where the sound is rendered. For example, in cinema, Dolby Atmos mainly employs object-based technology to add the height dimension combined with individually amplified loudspeakers (up to 64 channels in a single auditorium, with subwoofers, are known), whereas Auro Technologies' Auro-3D is fundamentally a channel-based format adding object-based technology on top of their channel-based format, based upon three layers (see Figure 9.1). This system, called AuroMax, creates up to 27 different "zones", which are forming part of the upcoming SMPTE standard for Digital Cinema. Auro 9.1, Auro 11.1, or Auro 13.1 are compatible speaker layouts also used for home. Both Atmos and Auro-3D offer audiences a sense of immersion in cinematic environments, using height loudspeaker arrays in a variety of different ways.

For the sake of brevity, we will limit our exploration here to the more common experiences of music: the headphone, the loudspeaker, and the club. However, we will remind ourselves: how we experience sound in the natural world offers an opportunity from which to base good mixing decisions because of the way we naturally experience sound.

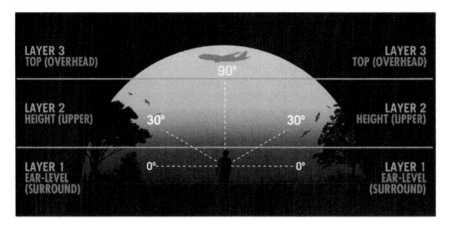

**Figure 9.1** Auro-3D with height and top (overhead) layer
Source: Auro Technologies 2010

Everyday life is full of numerous sound events and sources that surround us: the blending of various cues of individual elements create a general perception of space or "spaciousness" (Rumsey 2013). These cues often form the objects component in a spatial audio production, such as a bird singing or a helicopter passing overhead. Yet such cues are combined with a variety of other sounds originating from multiple sources, creating the diffuse ambience of what we might call the "sound of our environment".

Indoor environments and their corresponding surfaces, conversely, create a different sense of space. These enclosed environments offer their own unique character from which the listener derives his or her own sense of position and experience in a place. It is often the mostly early reflections that dominate such a space, forming a large part of the characteristics of a room offering a quality which is distinctly different from that which is outdoors (Rumsey 2013).

For some time now, there has been ongoing dialogue and debate about the specific aims of recording and reproducing sound. What are the aims of recording sound? On the one hand, there is the opinion that the aim of creating an immersive experience is to recreate a recorded performance within a particular space: that is to create a believable reproduction from a recording and a sense of "being there" (Rumsey 2013). On the other hand, there are those who claim the aims of reproducing sound are only worthwhile when creating a new experience altogether, one that could be synthetic, augmenting the artist's original intentions (Milner 2009). Either way, a common belief is that whatever outcome is realized, and no matter what process or workflow is followed, or adopted, the final result will always, ultimately, be an illusion (Nipkow 2017; Zielinsky and Lawrence 2018).

## MIXING WITH THE END IN MIND

Aspiring music producers, armed with traditional music production skills, experience, and knowledge, honed with monophonic and stereophonic formats, are rightfully asking, "What is involved in producing for 5.1 and beyond and for VR and AR and other applications that involved head-phones beyond stereo?"

Before an immersive audio mix begins, there are three core considerations. First, what are the objectives of the mix concerned? This will include the artistic aims and intentions as well as the final reproduction format, if known at this stage. To achieve a correspondingly convincing result, a sophisticated sound concept is necessary (Nipkow 2017). Second, what tools and resources are needed to achieve these objectives? In order to create an extraordinary experience, it has to be produced with a high degree of quality (Nipkow 2017). Hardware and processing power are one aspect; soft skills are another – what talent, knowledge, and experience are required? (Llewellyn 2017). Third, which of these identified resources will be available, during the production time line, to achieve the final production on time and in the variety of formats requested? Music producer, Gregor Zielinsky, suggests that most current productions will require, at the very least, a stereo and likely a 5.1 mix too (Zielinsky and Lawrence 2018).

The importance of having a conceptual idea of how the music is to be presented in the final format is not to be understated: there should be some kind of artistic intention (Lee 2017a). As with music presented in stereo and other legacy formats, the experience needs to be plausible (Ammer-mann 2017). Whatever the intentions of the music, the listener needs to be convinced. Therefore, when mixing music, the producer is often best informed in making critical mixing decisions by beginning with the listener in mind.

Giving consideration to where listeners will be, what their environment is likely to be like, when they are listening, and how and where they are experiencing the music can go a long way to assist both the recording and mixing process in spatial applications. The sample frequency applied during the workflow up to the point of mastering and distribution is another important consideration for the quality of the overall sound experience (Van Baelen 2018). As with almost all music production, the final format, or environment, is often not known, yet any clarification will benefit the music producer. A film being produced in Auro-3D would be a good example. If it is known that most Auro-3D-capable auditoriums are Auro 13.1, with easy Auro 11.1, 9.1, and 7.1 surround down-mix possibilities in the playback devices, the producer can begin there. "If you're doing stuff in game engines or you're doing stuff in objects there's a lot more reliance on monophonic sounds and it's kind of a different kettle of fish really" (Llewellyn 2017). Music for broadcast is another example, particularly with the increasing popularity of soundbar devices (Walton et al. 2016). If aiming for the mass domestic market, soundbars are "a very important part of bringing sound to the audience" (Ammermann 2017). In the United States, 5.1 is broadcast natively on terrestrial television (National Association of Broadcasters 2008). Other examples would include listeners who enjoy podcasts on their headphones whilst experiencing their daily commute or jog, offering the producer binaural, 360, stereo, or mono opportunities. Another common example is creating electronic music for clubbers on the dancefloor of a nightclub. In the case of club music, the production would have to be created specifically, and exclusively, for the 3D audio live situation; for the consumer at home, it would then require a separate mix (Nipkow 2017). "Every 3D production, no matter what the format is, must always be compatible with stereo and 5.1" (Zielinsky and Lawrence 2018).

Elements recorded with traditional monophonic and stereophonic techniques are as valid as those from single and multi-channel synthetic sources (Ammermann 2017; Lee 2017a; Nipkow 2017). "3D is only a further extension of traditional music production" (Zielinsky and Lawrence 2018). For live performances, a producer may receive multiple tracks recorded using the many spatial audio microphone techniques and approaches employed, which aim to reproduce an impression of the performance space. Many 5.1 albums, for example, are live albums (Lee 2017a). The study of ambisonics offers those wishing to capture sounds in space a higher degree of scalability and flexibility, beyond the recording stage, and throughout the production process. Ambisonics inherently includes height information, and depending upon how many channels are utilized, it is possible to reproduce in an increasing number of dimensions and at higher resolutions (Frank et al. 2015). However, this flexibility has challenges, and one is the potential for a less natural timbre introduced by phasing issues inherent to the concept (Van Baelen 2018).

Whilst 360 localization is interesting, the modern producer has to seriously ask, "What is it the listener is really wanting to experience?" (Lee 2017a). The expectations of listeners are important. For example, with modern dance music, a feeling of being immersed in the music is often the purpose. Sometimes the aim is to make a sound experience "more interesting" (Lee 2017a). For other producers, the attraction of producing music in spatial audio is a new opportunity to "create a large version of your idea" (Ammerman 2017). Whatever the intentions, being clear on the production objectives of the outcome greatly assists with decisions throughout the mixing process.

## MIXING ENVIRONMENT

Mixing with the end in mind, it is essential to replicate an environment that, to some degree, emulates where the audience will be when they experience the final production. At the very least, setting up a studio environment that enables a producer to make critical decisions allowing the final mix to translate well between a variety of systems, as with stereo, is important. As we will see, there are strong similarities between the many decisions that need to be made around mixing aspects of creating music in spatial audio, with a few additional aspects to consider. For those new to mixing in spatial audio, there are two trains of thought on where to begin – first, with what you've got (e.g., headphones for binaural and 360 applications), and second, by augmenting an existing loudspeaker setup with additional loudspeakers.

A majority of independent music producers today prefer the relative simplicity of an Auro 9.1 loudspeaker configuration in the mixing studio. This format was introduced and popularized by Auro Technologies' Wilfried Van Baelen, who introduced the Auro-3D format in 2006, augmenting the ITU-R BS.775–1 5.1 loudspeaker specification by adding four additional height channels to reproduce sound from an elevated layer (Theile and Wittek 2011; Van Baelen 2006, 2018). The benefits of height channels in such a configuration include the opportunity to reproduce early reflections from above, emulating a natural occurrence in a room and the ability to immerse, or engulf, a listener into a spatial listening experience offering more depth and transparency and an improved harmonic reproduction resulting in a more natural timbre. Timbre is still the most dominant parameter in the judgment of a natural sound reproduction (Van Baelen 2018). Atmos also adds height; however, there are numerous differences between the formats for the music producer. Until recently, Atmos was limited to the big screen and only a few studios were equipped with the required configuration and number of loudspeakers. Although Dolby greatly assists engineers with plug-in software that easily allows for the placement of objects, the opportunity to work in an environment that goes toward accurately assessing the final production is limited for most independent music producers without large budgets. That said, both Atmos and Auro-3D offer the music producer the opportunity to create music with great spatial diffusion particularly of the reverb components of a mix (Theile and Wittek 2011).

There are other limitations to all current formats to be aware of too. Such limitations chiefly include the ineffectiveness of panning between the horizontal and elevated layers, as well as the overuse of phantom sources, which can create unusual coloration of sound elements (Lee 2017a and b; Nipkow 2017; Theile and Wittek 2011).

A popular place to start for live loudspeaker music reproduction and mixing is using an Auro 9.1 setup, which involves an easy placement of each loudspeaker between ear-level and Auro-3D's height layer (x/2 in Figure 9.2a). An advantage being that there is no need, in small rooms, to add overhead speakers (Wilfried Van Baelen 2013b). Another place to start is with Sennheiser's AMBEO Cube: "I always start with a normal stereo production, [setting] up my remaining AMBEO Cube around it" (see Figure 9.3 and Zielinsky & Lawrence 2018). Loudspeaker calibration is critical; checking for consistent intensity and frequency responses from each loudspeaker will often help the producer to avoid making poor-quality mixing decisions (Ammermann 2017). LFEs for music mixing are frequently employed by immersive audio music producers (Ammerman 2017; Lee 2017a). However, there is ongoing debate as to whether to angle

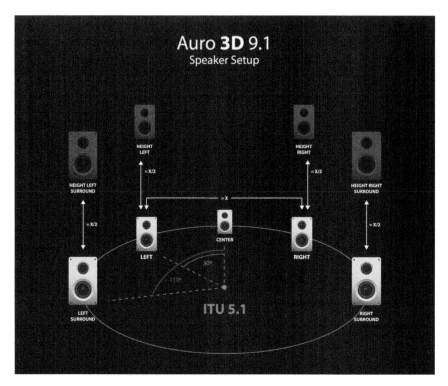

**Figure 9.2a** Auro-3D 9.1, the most efficient speaker layout able to reproduce a 3-D space around the listener while based upon, and backward compatible with, ITU-R BS.775–1

Source: Van Baelen 2006

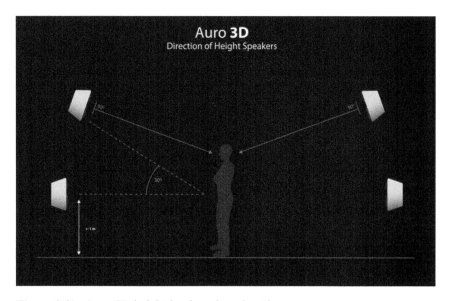

**Figure 9.2b** Auro-3D height loudspeaker elevation

Source: Van Baelen 2006

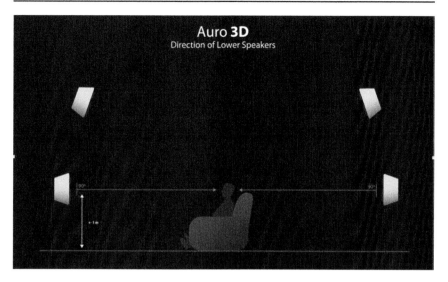

**Figure 9.2c** Auro-3D, horizontal layer relationship to listener

Source: Van Baelen 2006

the height loudspeakers directly to the sweet spot or not. Tilting the height speakers in such a way that the crossing point is just above the listener's head not only creates a larger sweet spot but also offers a more natural immersive sound experience (Wilfried Van Baelen 2006, 2018). Another deliberation is the use of the center loudspeaker and for what? Traditionally, the center speaker is introduced for film dialogue and narration. In Auro-3D 13.1 cinema configurations, for example, film dialogue is reproduced using the 13th directly overhead, or "Voice of God", loudspeaker. Music producers are also occasionally known for using the center speaker for "centralized" sound sources, such as vocals, snare drums, and certain frequencies of kick drums (Ammermann 2017). For most producers using height channels, the important part of an Auro 9.1 array, for example, is the eight left and right loudspeaker channels creating Auro-3D's coherent "vertical stereo field" all around the listener (Wilfried Van Baelen 2006, 2018). The opportunity here is to experiment whilst ultimately trusting your ears (Senior 2011). Although many producers may be tempted to create a cinema environment, Ammermann (2017) cautions not to "create a cinema environment. . . [but to] create a near-field situation" for good music mixing decisions.

For the producer interested in targeting headphone listeners, good-quality mixing headphones are a good place to start. Modern popular choices include Beyer Dynamic DT-1770 and DT-1990 models, as well as AKG K-812 and Shure SRH-1440 headphones, all of which are available for under $1,000. Bear in mind that one of the key differences between creating an immersive audio experience for music and for virtual reality applications and gaming, in headphones, also reflect some of the key differences in creating a binaural experience and a 360-degree experience

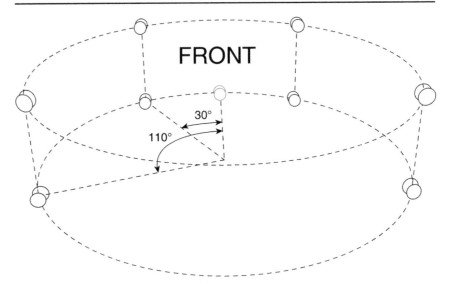

**Figure 9.3** Sennheiser AMBEO for Loudspeakers

Source: Sennheiser 2016

(Ammermann 2017). The former is a static (linear) experience, and the latter is dynamic and involves head-tracking. The production and rendering
processes for each differ too, which is important to note when embarking
on any project. A binaural experience relies upon the linear (re)produc-
tion of sound. VR and AR applications and 360 experiences, are rendered
using object-based game engines, whilst interactive, often involve music
beds (Ammermann 2017). "[The] linear stuff lends itself very well to a lot
of traditional [production] techniques" (Llewellyn 2017). Those interested
in dynamic and interactive applications may wish to invest further using a
headphone system that involves head-tracking such as the Smyth Realiser
A8 or the more recent A16 (see Figure 9.3).

## MIXING WORKFLOW

Having outlined the importance of having clear objectives for a mix and
having given some thought toward what is required to achieve a mix, this
section offers the producer ideas toward creating a mix workflow. Further
exploration and experimentation is encouraged.

Once the artistic intentions of the production are known, it is then the
responsibility of the producer to interpret these aims, in a technical sense,
toward achieving a finished music production. This stage often includes
the mastering and delivery of the music material in the final format(s)
required. There is no one single approach toward achieving this; as with
all music production, producing a mix in a spatial audio format is rarely
different, in so much as it is an art as much as it is a science. Mixing decisions are often subjective, if not solving problems. Creative exploration

will serve the aspiring producer well, as trial and error often lead to exciting, new, entertaining, and engaging final mixes (Ammerman 2017).

Producers and studios have their own unique variety of workflows. These workflows often comprise a preferred, and individual, method which involves many thousands of subconscious, time-dependent, material-related decisions. Presented here are some of the more common themes that overarch the subjective mixing decision process where a number of spatial audio producer workflows have been explored. The approach that follows summarizes some of the broader considerations taken into account at the various different stages during the final mix of a spatial audio music mix.

By breaking down any music mix into the textural components, or sonic qualities of a sound, it is possible to observe subjective mixing decisions in an objective way. For more detail on how the spatial attributes of multidimensional sound can be assessed and evaluated, the AES 16th international conference paper, *Methods for Subjective Evaluation of Spatial Characteristics of Sound*, is recommended (Bech 1999). One approach is to organize a set of attributes that allows the producer to analyze, or at least consider, sonic components and qualities in a way that assists with the mixing decision process. Table 9.1 offers broad definitions that consider timbral and spatial qualities of a sound (Corey 2010). Let us briefly explore each of these elements and how to consider them from a practical, spatial-audio mix perspective.

Approaching a mix from a timbral and spatial definition perspective offers the modern producer the opportunity to consider both creative and technical aspects in a unified way. Professional producers often make decisions around EQ, modulation, and dynamics, for example, as a whole, rather than in isolation. The approach outlined here considers this.

If *Spectral content* concerns the frequencies present in a sound, an aim for an immersive audio mix could be, then, to increase the clarity of the overall sound image and, therefore, an improvement upon the impression when compared to reproducing the same sounds in legacy formats. This is

**Table 9.1** Timbral and Spatial Definitions

| Attribute | Definition |
| --- | --- |
| Spectral Content | "All frequencies present in a sound". |
| Spectral Balance | "The relative balance of individual frequencies or frequency ranges". |
| Amplitude Envelope | "Primarily the attack (or onset) and decay time of the overall sound, but also that of individual overtones". |
| Spatial Extent | "Width and depth of the sound stage". |
| Spaciousness | "The perception of physical and acoustical characteristics of a recording space". |

Source: Adapted from Corey 2010.

often achieved through a balance of equalization (EQ), the placement, panning, and perceived spatial positioning between elements (often involving delays and reverberation), creating the sonic characteristics of a perceived space or "sound stage".

Closely related, the *spectral balance* of a mix refers to the relative differences in power of each frequency, or frequency bands within an audible range (Corey 2010). Supposing *spectral content* is the rainbow, *spectral balance* could be considered as the contrast between each color (or frequency heard in audio terms). Timing between elements is critical for clarity, as the onset of each sound element is often crucial for listeners being able to identify individual elements in a mix. Equally, the timing and delivery of specific elements within a mix can contribute to an overall soundscape or texture, allowing sounds to blend; in such instances, the clarity of the attack of each element is less critical whilst their combination (the sum of their parts) can contribute to an overall onset of a sound texture (Corey 2010).

The *amplitude envelope* is concerned primarily with the definition of a sound element within a mix. An element could be either a single sound source or a number of sounds blended or manipulated together toward creating a single element. In traditional mixing terms, the latter could be thought of as tracks of sounds grouped together. The signature sound, or sonic characteristic of each element, often relies upon a clear *attack* component of the attack, decay, sustain, and release (ADSR) envelope profile. The skillful control of the dynamic range for each element and for the overall image will result with the definition required for a cohesive mix. That is not to say that equalization and reverberation are not as important, or indeed starting points, but more emphasis is placed on careful dynamic control with immersive audio mixes. Gregor Zielinsky uses a trick when mixing for immersive audiences: room signals are compressed using a side-chain input, which reacts to specified loud signals by reducing room reverb (Zielinsky and Lawrence 2018). During a piece when, for example, a loud bass instrument or snare signal exceeds the set threshold (whilst creating extreme reverb from behind the listener's position), the side-chain input triggers the compressor to duck the room signals. Consequently, the louder the music becomes, the more the room components are reduced. Ideally, the listener experiences this effect subconsciously.

Finally, the *spatial extent* of a sound, or a mix, refers to the perceived width and depth of the sound stage, which is separate from the individual sound elements, or sound sources, presented in a mix (Corey 2010). This attribute closely relates to that of *spaciousness*, which corresponds to the perceived physical and acoustical characteristics of a completed production. Corey (2010) postulates that with only two loudspeakers, it is difficult to realize a sense of true envelopment, engulfment, or immersion. For this reason, one may argue that the spatial characteristics in an immersive music mix require even more detailed attention, at the mix stage, than a mix created for stereo; the implication is that sensitivity to the element of *spaciousness* has greater importance.

Seeking the most appropriate timbral and spatial definition for each element, including those that contribute principally toward a sense of space, such as natural room tones, reverbs, and delays, offers a good place to express the producer's creativity for a spatial audio music mix. Often, it is the grouping of sources toward creating textures and new sound elements that can separate an immersive mix from a stereo one. The overall resultant effect becomes more noticeable when switching from a spatial audio mix back to stereo, rather than vice versa (Lawrence 2013). The style of the material and the nature of the audience are important factors when decisions are made toward achieving a sense of space and a mix that achieves the artistic aims of a production (Corey 2010).

Where height channels are concerned, Lee (2017a) suggests addressing elements and sound sources on the horizontal plane first, as mixing in 5.1 is easier. Depending upon the material and the objectives concerned, the producer may or may not choose to begin with a stereo mix. An increasing number of music producers prefer their stereo mixes when they originate in a multichannel configuration such as refers to Auro-3D 9.1 (Ammermann 2017; Llewellyn 2017). Where a recording has been made specifically for a spatial-audio mix, there is the opportunity to down-mix later (Llewellyn 2017). For example, with a classical or orchestral recording, where the objective is to create a sense of being in a room, it may be that the microphones were arranged in such a way that it is anticipated that they will discretely map to specific channels, with spot microphone sources to add clarity and detail (Nipkow 2017; Zielinsky and Lawrence 2018). Sennheiser's AMBEO Cube and AMBEO Square, Schoeps's ORTF-3D, the PCMA-3D, and the Williams STAR surround microphone arrays are examples of microphone array configurations often used to achieve these types of production (Lee and Gribben 2014; Posthorn 2011; Schoeps 2018; Sennheiser 2016; Williams 2004).

Localization of specific instruments, particularly soloists, may be important, depending upon the aims of the project, although there is much deliberation about the importance of accurate localization of specific sources in both stereo and immersive music mixes. More broadly, a sense of envelopment can depend upon specific sound cues, which may also rely upon localization to offer the listener a convincing experience (Lee 2017a). For this reason, the importance of localization rests upon the aims of the material. If using height channels, the elements reproduced in the height dimensions will need to be clearly defined to be effective (Lee 2017a). This emphasizes the importance of well-recorded material (Zielinsky and Lawrence 2018); artifacts are amplified with multichannel loudspeaker arrays. Ambient sounds, reverbs, and early and late reflections are frequently a first candidate for heights channels. Lee (2017a) cautions against using time delays when employing height channels, with material intended for the horizontal plane, as this approach rarely renders a successful result.

For popular and electronic music material, and where the original multitrack recording of a stereo mix is available, an immersive mix may begin with the two stereo channels with specific elements being split out into other channels toward a new artistic intention. This is a less-often-adopted approach with jazz, classical, and other styles of music; however, there are tools and possibilities for upmixing two-channel audio to multi-channel arrays (Llewellyn 2017). For every mix scenario, the key question for the producer to ask is "What's the best way to render what it is you're trying to create?" within the constraints of the final format and beginning from there (Lee 2017a).

## ADDITIONAL MIX CONSIDERATIONS

Spatial audio software plug-in tools are available for most DAWs. These are designed to manipulate the spatial position of elements for most 3-D audio applications. For the spatial audio producer, such tools are essential (Ammermann 2017). Once installed, these tools enable the producer to experiment and explore a combination of sound elements together quickly and easily. As with other software, much of the complexity, particularly in relation to phase and amplitude, is removed, allowing the producer to simply get on with the craft of creating a mix. Such tools offer a variety of output modes. The Spatial Audio Designer (Figure 9.4) is one example, and the Auro-3D Creative Tool Suite (Auro Technologies 2018) is another. These allow the producer to check compatibility across multiple output formats, both quickly and frequently, and at the touch of a (often virtual) button. Comparing a 5.1 mix to an 11.1 mix and back again becomes routine practice as does checking for mono compatibility with a stereo mix. The predominant concerns are always issues related to phase and balance.

From the perspective of hardware, employing height channels benefits immersive listening experiences and need not be a complicated addition to a stereo or 5.1 configuration (Theile and Wittek 2011). Once you know what to do with both the height and surround-sound channels, it is possible to offer a deeper spatial impression (Ammermann 2017; Lee 2017a). Bear in mind, however, that more channels will introduce "problems with coloration" (Lee 2017a). According to Lee (2017a) and Ammermann (2017) decisions around what elements to place and where are fundamentally governed by the frequency spectrum of the sound sources presented in the original music material. For example, it is well known that low-frequency sound sources are challenging to localize (Hill 2014). A common practice among immersive audio professionals is to reduce lower frequency information for height channels whilst boosting higher frequencies. This does not need to be done in an extreme manner; in all channels, Lee (2017a) suggests that frequencies above 5 kHz can often improve the perceived vertical spread, although this happens to be where the upper frequencies for classical instruments roll off (around 5 to 6kHz). We also tend to perceive higher frequency sounds as arriving from elevated positions (Lee 2016). The importance of researching and better understanding the psychoacoustic effects is evident. Auro-3D creator Wilfried Van Baelen (2018) also suggests that our human brain is very sensitive to spatial

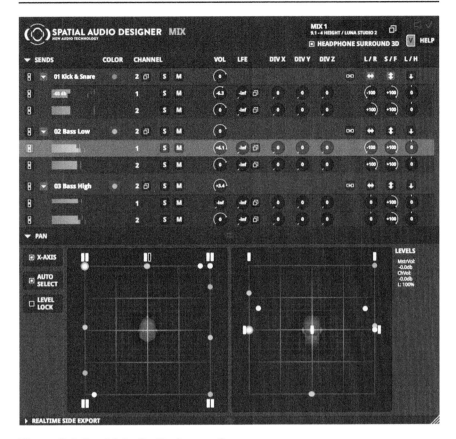

**Figure 9.4** Spatial Audio Designer software

Source: New Audio Technology 2018b

reflection modes between approximately 200 and 800Hz. This implies a less successful reproduction of natural immersive sound experiences with loudspeaker systems used in height arrays which struggle to reproduce frequencies down toward 200 Hz.

Another consideration is the phantom image elevation effect (Lee 2017b). The effect suggests that as the width between loudspeaker sources is increased, the degree of perceived elevation of correlated sources rise. The sound of a single instrument, such as a trumpet, would be a good example. Where the phantom signal of the recorded trumpet is placed in both left and right channels, as the width between the loudspeakers, on a horizontal plane, reproducing the instrument is increased, the perceived position of the trumpet player, in relation to the listener, elevates. Similarly, panning to the rear speakers changes the listener's perception of where a source sound is arriving from; the source of a single sound could appear somewhat above or slightly to the rear of the listening position (Lee 2017b). Continuous panning between front and rear loudspeakers is not recommended because of sounds dithering, at different points for different listeners, between front and rear sources. Many engineers expect to be able to create phantom sources in the vertical axis as with the horizontal axis. This is not true, as phantom sources are created on the horizontal axis because of time and level differences between each ear. We do not have an ear on the top of our heads to accommodate vertical spread (Van Baelen 2018).

Finally, beyond a height loudspeaker elevation angle of more than 40 degrees, vertical coherence is lost. This coherence, in the vertical axis and around the listener, is vital to creating a natural immersive sound experience (Van Baelen 2013b, 2018). False listener expectations with height channels are not uncommon (Lawrence 2013). Ultimately, the producer must trust his or her ears for final mixing decisions (Senior 2011). Awareness of such challenges and theories is useful, as they offer the grounds upon which critical mixing decisions can be made.

The role of dynamic control (compression, expansion, and limiting) in the discipline of music mixing is fundamental to the overall impression. The successful management of dynamics, throughout a musical piece, for the variety of sources and elements, will often determine the success of the final result; too much or too little dynamic control may reveal an amateur sounding mix. The more loudspeakers used to reproduce sound, the more the skills of dynamic control are noticed. The effect is less evident with headphone applications. For individual sources, solo elements and instrumentation, dynamic control, within a mix, can be approached as one would with a 5.1, stereo, or mono production. The subtleties, so far as spatial audio material is concerned, are more to do with how elements are creatively blended together. For example, master bus compression may or may not be used in a traditional stereo mix and at the mixing or mastering stage; often such decisions are down to the style of music presented and the taste of the artist, producer, or mixing engineer (Senior 2011). However the recommendation for producers fresh to mixing spatial audio is to leave master bus compression for now. At the very least, master bus compression is complicated and is not recommended if a number of different mixes are required for different formats at different points during the mixing stage of a project (Senior 2011).

As far as creating a sense of space is concerned, there is a conflict in the field of spatial audio (Zielinsky and Lawrence 2018). On the one hand, with indoor recordings, we want to represent a room – a three-dimensional space. On the other hand, we want to represent the warmth, close proximity, and detail of the sound sources or instruments present, in a manner that listeners used to stereo reproduction are familiar with. As with all music production opportunities, *a sense of space or spaciousness is often an illusion* (Zielinsky and Lawrence 2018). Therefore, when adding spatial content to immersive audio music mixes, there are three specifc considerations. First, what is the spatial illusion? Second, what will be the final reproduction format toward achieving the illusion (if known?). Third, what tools are required to achieve the aims in mind? Those seeking to accurately replicate a true sense of space, e.g., a concert hall, in a spatial audio music production are at the risk of chasing an elusive aim.

Until the introduction of 3-D reverb plug-ins, early spatial audio productions were mixed using conventional multiple stereo reverbs on multiple channels and often still are (Llewellyn 2017). For example, in an Auro 9.1 array, four reverb units could be employed to manage the depth and reflections in four separate zones: front horizontal, front height, rear horizontal, and rear height. To break this down further, a stereo reverb would be assigned to the front, horizontal plane, the left and right pair of loudspeakers (L and R); another unit to the rear left and right pair of loudspeakers (Ls and Rs); a third unit to supply the upper front pair (HL and HR); and a fourth to supply the upper rear pair (HLs and HRs.) This approach of using stereo reverb units as separate elements offers greater control (Ammermann 2017). Disney re-recording mixer David Fluhr has successfully employed such approaches commercially. Fluhr is known for his success as a re-recording mixer with films such as *Frozen* and is recognized as having received multiple awards and nominations (Disney 2018). 3-D reverb plug-ins have made life easier for music producers, mixing for both headphones and loudspeaker systems (Ammermann 2017). Not only do these new tools offer benefits such as recall and automation; such software is becoming essential when working between studio facilities and locations.

From an audience perspective, sounds perceived in space are not only subjective to each individual listener; the listener's perceived position and the process in which the brain processes, and focuses in on, specific sounds is a complex one. The way we experience sound is beyond the mechanics of simply hearing a sound in the space we perceive. For example, the visual cortex plays an important part, as demonstrated by the McGurk effect phenomenon (McGurk and MacDonald 1976). Thus, the reception by a listener of a musical performance cannot be objectively determined at the mixing stage (Zielinsky and Lawrence 2018). Knowing this, the producer is offered an opportunity: to mix toward an illusion that accommodates the artistic intentions of the musical piece created toward an ideal. This might be the ideal listening position at a concert or in the center of an unworldly experience in other applications, for example. To illustrate another case, the room in rock, pop, and electronic music material may or may not have little to do with the overall impression created (Lee 2017a; Zielinsky and Lawrence 2018). Where a room and live instrumentation has been recorded, a good quality recording of a decent sounding room is paramount for the reasons stated earlier (Zielinksy and Lawrence 2018). Where height channels are available, one opportunity is to "engulf [the listener] from the top" (Lee 2017a). Although pure reverb sounds are "not very bright", such elements can contribute to the perceived depth of an experience whilst offering space to upper harmonics and other high-frequency detail (Lee 2017a). The perceived distance and position for a listener, from a specific source, whilst it may vary for each participant, can be created, to a degree, by careful use of early reflections affected by delayed signals: "A ceiling or floor reflection can really affect the sound" (Lee 2017a). The key here is experimentation whilst keeping these core principles in mind (Ammermann 2017; Lee 2017a).

## DELIVERY

Familiarity with emerging and modern audio standards, particularly MPEG-H,[1] and workflow concepts that include Dolby's DCP and tools such as Auro-3D's encoding system, which allows the dynamic down-mix of all 3D channels into a standard 5.1 PCM master, are helpful for those interested in learning more about what is required for modern final productions in broadcast, music, cinema, and beyond. Not knowing the final format ought not to deter the producer from making good decisions based upon ideas validated through testing and experimentation. Understanding more about what is required to satisfy the mastering stage, as in any mix situation, is vital.

## FINAL THOUGHTS

For producers interested in new avenues of creative exploration, spatial audio may have the answer. "Nobody really understands that there is an immersive market [although] VR audio is helping to bring immersive audio to the masses" (Ammermann 2017). There has never been a better time for the modern music producer to start learning how to expand his or her repertoire, building upon existing talents, knowledge, and experience and creating new and innovative productions with a unique perspective. "The good news is you don't need that much stuff" and herein lies the opportunity to "conquer a new market [without a large studio]" (Ammermann 2017). The opportunities for the aspiring music producer are both exciting and enticing, although not without challenges. These challenges ought to be viewed as creative possibilities. The promise is, however, that the underlying approaches for spatial audio allow the creation of something new in a market that is growing in momentum. The perspective that "everybody [in professional audio] is an expert in stereo" suggests that, for the more ambitious and adventurous producer, particularly those looking to differentiate their talents in the world of mixing music, spatial audio offers "more opportunities to become a specialist" (Ammermann 2017). The growing number of advanced formats for spatial-audio reproduction is worth exploring. These advanced formats primarily include the well-researched and documented subjects of wave field synthesis (WFS) and ambisonics. *Spatial Audio* (Rumsey 2013) is a good general starting point and for references and reading. The vast array of possibilities may seem daunting at first for some; it need not be.

Learned production techniques and approaches are still relevant, since spatial audio is an extension of stereo (Nipkow 2017). Depending on the material being produced, creating a good sounding quad mix can sound much better than an approach diving straight into ambisonics, as well as be a less arduous journey for those starting out (Ammermann 2017). Unless you are aiming for a piece that is entirely experimental, you can always derive what you need for stereo from what is required to achieve 3-D (Zielinsky and Lawrence 2018). Modern DAW plug-ins, such as the Spatial

Audio Designer (see Figure 9.4. and New Audio Technology 2018a), and the Auro-3D Creative Tool Suite (Auro Technologies 2018) allow music producers to create content, as well as monitor, in surround and 3-D, quickly and easily. Such products also offer content creators the option to down-mix material (Llewellyn 2017). The Auro-Matic Pro Plugin (see Figure 9.5.) offers producers the opportunity to create natural-sounding 3-D reflections from each source sound regardless of mono, stereo, or surround origins.

Curiously, dummy heads, such as the Neumann KU-100, are being experimented with, to record music rendered through multichannel loudspeaker arrays and create a version for listeners interested in binaural versions and with positive results. Such examples demonstrate some of the creative possibilities available for music producers. Lee (2017a) postulates that "as a producer you should be open to new things," emphasizing the importance of pushing the boundaries and not being shy about introducing and trying out new elements. "Binaural [listening] is popular these days", and experimentation need not have purely immersive aims. "Binauralization" purely for stereo effects or headphones may offer rich, new, and exciting textures for all reproduction formats. Spatializing elements, simply for effect, can be fun and produce unique results. The bonus of such creativity is that when reproduced in the initially anticipated format (i.e., headphones with binaural), the listener is gifted with an exciting immersive moment unavailable elsewhere. Such ideas are not new; for decades, this has been a feature of everyday pop music production (Lee 2017a). An analysis of the productions created by Tchad Blake, for example, are good

**Figure 9.5** Auro-3D Auro-Matic Pro plug-in

Source: Auro Technologies 2018

audio references for those interested in getting started. Blake, well known for placing small microphones in his ears and using these recordings in his music productions, offers artistic inspiration for the aspiring producer (Tingen 1997). As headphone and head-tracking technology evolve, one only has to look at the glowing testimonials of products such as the Smyth Realiser A8 (Smyth Research 2018) to appreciate what the new opportunities are, and will become, when mixing music for spatial applications and almost anywhere.

Beyond adopting new approaches whilst learning how new and emerging tools work, the challenges for producing music in spatial audio are mostly technological and commercial. Although the latest software will perform most of the heavy lifting, it is processing power, not hardware, that producers will require more from in the future, particularly where AR and VR experiences are concerned (Ammermann 2017; Lee 2017a and Llewellyn 2017). From a commercial perspective, the challenges are more specific to the economics and marketing of how to entice listeners to invest in new sonic opportunities. At the end of it all, the responsibility for the music producer is to convince the listener of the true added value offered by spatial audio formats. Ammermann (2017) suggests that one of the biggest hurdles for spatial audio is that "you have to prove to your client that they can earn money with [what is produced]". For that reason, the success of the immersive audio industry relies upon not one but at least three key concerns: first, improving the technical delivery from concept to listener; second, creating new ideas for aesthetic uses of height channels and binaural applications (Theile and Wittek 2011); third, developing sufficient experience to establish a workflow that allows delivery in all of the main formats simultaneously (Van Baelen 2018). According to Lee (2017a), you can learn a lot by playing with sounds and experimenting – "things become a lot simpler when you try these things" – implying that future success in this segment of the audio and music industry relies upon, more than ever, the courage and creativity of ambitious music producers.

## CONCLUSION

Mixing music with the aim of creating an immersive listening experience need not be a laborious challenge but can be a rewarding investment. The modern producer has a great deal of research, and recorded work, from which inspiration and direction can be sought toward grasping a feel for what is involved, taking only what is needed, indeed, what one is interested in or curious about, to get started.

All of the producers interviewed for this chapter claim that their stereo mixes "sound better" when they originate in an immersive audio format. This opinion is shared among other spatial-audio music producers working with a unified sense that sound elements are easier to place and hear when mixing with surround or spatial loudspeaker arrays. Yet, relatively few professional modern music producers consider, or practice, beyond stereo, and if they do, it is limited to 5.1 or 7.1 surround formats. Despite the practical, economic, and partly academic reasons for this, the number of those attending spatial-audio conferences, such as the International Conference on Spatial Audio (ICSA) and conventions held by the Audio Engineering Society (AES) is increasing.

As is true for so many fields of artistic and scientific endeavor, it is always important to follow instinct and curiosity. Those interested may soon find that their exploration of spatial audio is a rich endeavor filled with new research, information, and experiences, all of which are being added to every day. The possibilities are seemingly infinite, and the future is compelling. Music experienced in spatial audio may be the final frontier for producers and listeners alike. What is evident, however, is that the importance of multichannel surround sound and 3-D audio in music for the 21st century is not to be understated.

## NOTE

1. An audio coding standard designed to support coding audio as audio channels, objects, and higher-order ambisonics (HOA).

## REFERENCES

Ammermann, T. (2017) Conversation With Rob Lawrence. 16 November.

Auro Technologies (2010) Auro-3D/Auro Technologies: Three-Dimensional Sound. Available at: www.auro-3d.com/system/concept [Accessed: 19 May 2018].

Auro Technologies (2018) Auro-3D/Auro Technologies: Auro-3D® Auro-Matic Pro Plug-In. Available at: www.auro-3d.com/system/creative-tool-suite/ [Accessed: 19 May 2018].

Barbour, J. L. (2003) Elevation Perception: Phantom Images in the Vertical Hemi-Sphere. In: *AES 24th International Conference: Multichannel Audio, The New Reality*, 26–28 June. Available at: www.aes.org.sae.idm.oclc.org/tmpFiles/elib/20130928/12301.pdf [Accessed: 28 September 2013].

Bech, S. (1999) Methods for Subjective Evaluation of Spatial Characteristics of Sound. In: *16th International AES Conference: Spatial Sound Reproduction*, March 1999. Available at: www.aes.org.sae.idm.oclc.org/e-lib/inst/download.cfm/8010.pdf?ID=8010 [Accessed: 24 September 2013].

Corey, J. (2010) *Audio Production and Critical Listening*. Burlington, MA: Focal Press.

Disney. (2018) David Fluhr Presented With JHW Award. [Online] Available at: www.disneydigitalstudio.com/david-fluhr-presented-with-jhw-award/. [Accessed: 4 May 2018].

Frank, M., Zotter, F. and Sontacchi, A. (2015). Producing 3D Audio in Ambisonics. In: *AES 57th International Conference*. [Online] Hollywood, CA: Audio Engineering Society. 6–9. Available at: www.aes.org/e-lib/browse.cfm?elib=17605 [Accessed: 4 May 2018].

Hill, A. (2014) *Low Frequency Sound Localisation Research* [Presentation], 30 June.

Lawrence, R. (2013) *Evaluating Immersive Experiences in Audio for Composition*. BSc (Hons). Oxford: SAE.

Lee, H. (2016) Perceptual Band Allocation (PBA) for the Rendering of Vertical Image Spread With a Vertical 2D Loudspeaker Array. *Journal of the Audio Engineering Society*, 64(12), 1003–1013.

Lee, H. (2017a) Conversation With Rob Lawrence. 15 November.

Lee, H. (2017b) Sound Source and Loudspeaker Base Angle Dependency of Phantom Image Elevation Effect. *Journal of the Audio Engineering Society*, 65(9), 733–748.

Lee, H. and Gribben, C. (2014) Effect of Vertical Microphone Layer Spacing for a 3D Microphone Array. *Journal of the Audio Engineering Society*, 62(12), 870–884.

Llewellyn, G. (2017) Conversation With Rob Lawrence. 30 October.

McGurk, H. and MacDonald J. (1976) Hearing Lips and Seeing Voices. *Nature*, 264(23/30), 746–748.

Milner, G. (2009) *Perfecting Sound Forever: The Story of Recorded Music*. 1st ed. London, UK: Granta Books.

National Association of Broadcasters. (2008) Preparing for the Broadcast Analog Television Turn-Off: How to Keep Cable Subscribers' TVs From Going Dark. [Online] Available at: www.nab.org/documents/resources/paperGoldman.pdf [Accessed: 4 May 2018].

New Audio Technology. (2018b) SAD-2015 [Online]. Available at: www.newaudiotechnology.com/en/products/spatial-audio-designer/ [Accessed: 9 February 2018].

Nipkow, L. (2017) Book Contribution. 11 November. [email].

Posthorn. (2011) Posthorn the "Williams Star" Surround Microphone Array. [Online] Available at: www.posthorn.com/Micarray_williamsstar.html [Accessed: 5 February 2018].

Rumsey, F. (2013) *Spatial Audio*. 1st ed. Burlington, MA: Focal Press.

Schoeps GmbH. (2018) Plug-and-Play Setup for 3D Audio Ambience Recordings ORTF-3D Outdoor Set – Overview – SCHOEPS.de. [Online] Available at: www.schoeps.de/en/products/ortf-3D-outdoor-set [Accessed: 22 January 2018].

Senior, M. (2011) *Mixing Secrets for the Small Studio*. 1st ed. Burlington, MA: Focal Press.

Sennheiser Electronic GmbH & Co. KG | United Kingdom, (2016) x1_desktop_ambeo-blueprint-loudspeaker-7 [Online]. Available at: https://en-uk.sennheiser.com/img/10419/x1_desktop_ambeo-blueprint-loudspeaker-7.jpg [Accessed: 9 February 2018].

Smyth Research. (2018) Smyth Research. [Online] Available at: www.smyth-research.com/indexA8.html [Accessed: 19 January 2018].

Theile, G. and Wittek, H. (2011) Principles in Surround Recordings With Height. In: *AES 130th Convention*, London, UK, 13–16 May 2011. Available at: www.aes.org.sae.idm.oclc.org/e-lib/browse.cfm?elib=15870 [Accessed: 9 August 2013].

Tingen, P. (1997) TCHAD BLAKE: Binaural Excitement. *Sound on Sound.* December 1997. Available from: www.soundonsound.com/sos/1997_articles/dec97/tchadblake.html [Accessed: 13 May 2013].

Van Baelen, W. (2006) Surround Sound Recording and Reproduction With Height. [Workshop] In: *AES 2006 Pro Audio Expo and Convention*, Paris, France, 20–23 May.

Van Baelen, W. (2013a) Update Auro-3D. *VDT Magazin*, [Online]. 1, 41–42. Available at: www.tonmeister.de/index.php?p=magazin/20131 [Accessed: 20 May 2018].

Van Baelen, W. (2013b) Multichannel Immersive Audio Formats for 3D Cinema and Home Theater. [Workshop] In: *AES 2013 International AES Convention*, Rome, Italy, 4–7 May.

Van Baelen, W. (2018) *Image Permission for PoMP2*. 19 May. [email].

Walton, T., Evans, M., Kirk, D. and Melchoir, F. (2016). A Subjective Comparison of Discrete Surround Sound and Soundbar Technology by Using Mixed Methods. In: *AES 140th Convention*. [Online] Paris, France: Audio Engineering Society. 1. Available at: www.aes.org/e-lib/browse.cfm?elib=18290 [Accessed: 4 May 2018].

Williams, M. (2004) *Microphone Arrays for Stereo and Multichannel Sound Recording*. Milano: Editrice Il Rostro.

Zielinsky, G. and Lawrence, R. (2018) *Proceeding*. 12 January. [email].

## FURTHER READING

For understanding many of the fundamentals that underpin multi-channel immersive audio:

Holman, T. (1999) *5.1 Surround Sound: Up and Running*. 1st ed. Burlington, MA: Focal Press.

For a deeper understanding of the guiding principles, and practical considerations for spatial audio reproduction:

Ambisonics, Rumsey, F. (2013) *Spatial Audio*. 1st ed. Burlington, MA: Focal Press.

Proven mixing approaches and studio fundamentals:

Senior, M. (2011) *Mixing Secrets for the Small Studio*. 1st ed. Burlington, MA: Focal Press.

# 10

## Producing 3-D Audio

Justin Paterson and Gareth Llewellyn

### INTRODUCTION

For the purposes of this chapter, 3-D audio can be defined as the reproduction of sound that is perceived as coming from all around us, including behind and above or even below. This is, of course, the manner in which we perceive the real world in daily life. The recreation of such perception straddles many disciplines, from recording to psychoacoustics, to reproduction itself, which Michael Gerzon (1973) referred to as periphony. While this chapter is centered on "production" (a multifaceted term in itself), it is also necessary to contextualize this with a number of references to these various disciplines.

Clearly, when working in 3-D, the most radical departure from stereo[1] or horizontal-surround production is that of source placement in the sound field, and so a perspective of this will be presented. Having said that, periphony brings other creative opportunities and hazards to the production process, and some of these will also be discussed. Further, the word *audio* generally refers to music but might be taken in a broader context to include dialogue and non-musical sounds that might accompany video.

Beyond academic research, little has been published on periphonic production at the time of writing (fall of 2017), and so this chapter also has a responsibility to introduce the topic in order to serve as a primer for the numerous texts that will doubtless be published in the future but also to draw the readers' attention to specific aspects of the periphonic arena so that they might seek out and read further specialist texts.

There are three main paths by which 3-D immersive audio can be constructed: channel-based, ambisonic, and object-based. These approaches can, of course, be combined in any given playback scenario. At some point each of these requires rendering to speakers or headphones for playback. Whilst the details of each system differ and a good understanding of their workings is important, there are a number of audio-production approaches which tend to be similar across the board, although the manner of "getting there" may require slightly different approaches.

## THE CONTEXT OF 3-D AUDIO

Periphonic reproduction has been the subject of research for a number of decades. Over this time, the degree of interest has ebbed and flowed, and various systems have been developed that facilitate so-called "immersive audio". It is worth first considering some of the context and various applications that employ 3-D audio.

In acousmatic music, composers have styled their music to be played through multichannel (often periphonic) speaker arrays. In "live" performance, a musician designated as the "diffuser" might dynamically spatialize the performance in a given array, moving particular instruments from speaker to speaker, and many such arrays have been configured in universities and performance venues. The term *engulfment* has been used in such circles to refer to representation of height.[2]

In the past five or six years, companies such as Dolby and Auro-3D have developed multi-speaker periphonic playback systems for cinema. Most recent Hollywood blockbusters have embraced this approach, and consequently, 3-D audio creation is now commonplace in post-production. Special effects and atmospheres can exploit the ability to be placed anywhere in a 3-D sound field around the listener, and this can enhance synchresis[3] and opens many possibilities for non-diegetic immersive audio. The listeners might perceive ambient sounds all around them, or experience dynamic effects, such as aircraft flying overhead that can disappear into the distance behind. The associated music composition/production is also beginning to exploit such playback.[4]

The "soundbar" is an increasingly popular add-on for large-screen televisions; it attempts to extend the perceived width of the sound sources beyond those of the television screen and might work both horizontally and vertically through a process termed wave-field synthesis – sometimes also bouncing sound off the ceiling in order to gain the perception of height.

Binaural audio can give the impression of front-back and up-down sound-source localization on headphones. It does this by incorporating natural phenomena, such as the difference in time that a sound takes to reach each ear and the physical shape of the pinnae, which together contribute to how the brain deduces the point of origin. However, because it is dependent on the listener's physiological shape, binaural performs with varying degrees of accuracy for given individuals. Binaural localization can be achieved either through special recording techniques or via artificial "synthesis". In March 2015, the Audio Engineering Society (AES) published the AES69–2015 standard that provides a framework to facilitate widespread adoption into creative content in handheld devices. This will continue to accelerate development and uptake of binaural technologies.

Virtual reality, augmented reality, and mixed reality are technically different, but for convenience, they will now be referred to collectively by the common acronym "VR". VR represents one of the biggest technological revolutions yet to be seen and has a revenue stream forecast to reach

$108 billion by 2021 (Digi-Capital, 2017). The attendant audio is generally binaural. Localization is greatly enhanced by the head tracking that is a feature of head-mounted displays (HMD), whereby the position of a sound appears fixed relative to the natural movement of the head. Whereas VR is currently dominated by "early-adopter" gaming applications (the gaming industry can be considered as residing "within" VR for the purposes of this text), its impact will become more ubiquitous as it expands toward productivity, education, and health. VR and its associated revenue have precipitated a step change in interest and momentum for 3-D audio.

Also in ascendance is 360° video. In March 2015, YouTube launched a dedicated 360° video channel, and Mobile VR (use of a smartphone with a low-budget HMD "holder") is driving its popularity. These videos typically require 3-D audio soundtracks.

Broadcasting is another industry that is increasingly experimenting with 3-D audio. Both the BBC and NHK in Japan have broadcast with a 22.2-channel system, and the BBC has increasingly created binaural content both with and without accompanying visuals, e.g., *Dr. Who* and *Radio 3*.

All of these require specialist equipment, and thus another sector that is rapidly expanding is that of 3-D audio tool developers, both hardware and software. Established international players, such as Dolby and Pioneer, are creating hardware, as are emergent start-ups, such as Smyth. One interesting example is OSSIC X headphones, which measured physiological characteristics of the listener's head and ears, feature spaced directional drivers to help create the sound field, and have inbuilt head-tracking. Unfortunately, this product was withdrawn from the market, although similar systems may yet evolve. Software is coming from a similar range of developers, from the international DTS:X protocol through to niche developers, such as plug-in manufacturer Blue Ripple Sound.

There is a great deal of academic research into the many areas associated with 3-D audio, spread across the humanities, science, and medicine. These include music composition and performance, sonic art, human-computer interface design, audio post-production, cognition/perception, architectural space-modeling, recording techniques, software design, creative applications, assistive technologies, and more. Publication is increasingly widespread, and in 2016, the AES held its first dedicated 3-D audio conference, and the Art of Record Production held its first such event in late 2017.

## SPATIAL PLACEMENT

We must first consider some of the issues associated with 3-D audio "placement" in reproduction. Our ability to identify a given audio source's point of origin in space – with a meaningful resolution, requires the right kind of information to be delivered to the ears, so that the brain is able to reconstruct an ostensibly correct (or at least intended) placement of that audio.

Creating a 3-D audio experience requires the presentation of usable spatial information to the ears so that this mental reconstruction can occur. The production of useful spatial information can only come via three mechanisms:

## • "Captured" From the Real World

The first is to try to capture some of the spatial cues that exist in real acoustic environments using two or more microphones and reproduce these cues over speakers or headphones. A single monophonic microphone recording may contain a number of clues as to the environmental acoustics of the space and the sound source being captured, but some of amount of "directionality" is needed if one wants to begin expressing spatial positioning from native audio recordings, and this requires two or more microphones. Some of the inherent spatial information can then be captured and later reproduced from the timing and intensity differences between the capsules at the moment of capture.

## • Synthesized Artificially

When it is not practical or desirable to capture real-world spatial information, one can model the acoustics that would be generated by a sound source in an acoustic environment synthetically. There are myriad ways of achieving this, including using spatial impulse responses, real-time room modeling, and synthetic HRTF modeling, some of which will be discussed later.

## • As a Function of the Playback System

Panning a sound within a 3-D speaker array naturally generates spatial information at the moment of playback that is not inherent in the source audio itself. By the nature of the physical relationships between the speakers, the room, and the listener, 3-D positioning may be induced from power-panning alone. Good localization may thereby be achieved in a speaker array only from these factors. This reproduction of spatial cues will potentially be partially or completely absent from headphone reproductions of the same material.

> The dynamic between these three fundamental approaches is at the center of all 3-D audio production. The intended playback system (or systems) that any given 3-D audio production is expected to play back on will fundamentally influence the approach that is required. There are potentially major differences between the aesthetics and approach for producing a large-format theatrical electronic-music track and an acoustic orchestral piece of music for a 360° video. The real challenge is where and when it is appropriate to use different capture and synthesis techniques and where and how transformation between formats should occur. So, to elaborate:

## Spatial Capture

Specialized recording techniques have been designed in order to capture audio in three dimensions. One seminal approach that is currently regaining popularity is ambisonics. This concept was developed in the 1970s by Gerzon and others, and an excellent collection of ambisonic resources has been compiled by Richard Elen (1998).

Ambisonics can function at various resolutions that determine the accuracy of the perceived source localization, and this determines the required number of channels.[5] These resolutions are categorized by "order". For instance, "first order" is a four-channel system. A very important aspect is that these channels are encoded/decoded, meaning that the signals derived from (say, four) microphone capsules are not the same as those transmitted – they are encoded into a different four for transmission, and these are not the same as those passed to four speakers for reproduction, which must first be decoded (also known as rendered) from those transmitted. Alternatively, decoding could be into other formats such as 13.1 or two channels for binaural playback on headphones.[6] This offers a great deal of flexibility for different capture, transmission, and reproduction configurations. Importantly, the transmitted signals – which are called B-format – are easily manipulated through computationally efficient multiplication by a scalar, and such operations can, for instance, rotate the sound field.

One reason for the resurgence in popularity of ambisonics is that such rotation is what typically happens in a VR HMD, where the sound field appears stationary independently of the movements of the head: an accelerometer/magnetometer/gyroscope combination generates such a scalar to manipulate the B-format signals before decoding. Such dynamic panning greatly resolves the front-back confusion typical of binaural playback and gives an overall much greater sense of realism. Such operations are typical of contemporary HMDs.

Higher-order ambisonics (HOA), sometimes called scene-based audio (SBA) (Shivappa et al., 2016), is when additional channels are incorporated to represent further "spherical harmonics",[7] and these give much better localization and a larger sweet spot as the order increases. The trade-off in this approach is the increased channel count, and in the case of seventh order, 64 channels[8] are needed to convey a single track of audio, and this puts considerable strain on any digital audio workstation (DAW) as the track count increases during music production. It should be remembered that beyond ambisonics, there are other ways to capture 3-D audio too, e.g., those of Michael Williams (2004).

## Synthesized

Where it is not possible, desirable, or practical to capture the spatial information that is present at a given recording situation, then spatialization cues can be synthesized after the recording event.

Spatialization of monophonic audio cues can take two forms – firstly simple panning through a periphonic system that reproduces HRTF effects, and second, through the simulation of room acoustics that aim to localize and place the audio event realistically in 3-D space. With regard to the latter, in a real-time game engine, this acoustic simulation may be largely procedural and can include reflections based upon room geometry, the occlusion of sounds behind physical objects, sound radiation patterns, and frequency roll-off over distance – all of which are dependent on the game-software code that is deployed and the processing limitations of the system.

With linear playback materials, all these effects can be generated offline by engineers using similar tools but with more direct access and control over the materials.

The addition of reverberation to sounds has been proven to increase a sense of externalization when listening on headphones, at the expense of localization accuracy (e.g., Begault, 1992). Real-time game engines, such as Unity and Unreal Engine, already have such built-in audio physics, and these can be further enhanced through a number of proprietary plugins, code or middleware specifically designed for more realistic synthesis (e.g., Steam Audio, Oculus Spatializer, G'Audio, and so on).

Synthetic spatialization of music elements will frequently take the form of individual monophonic sounds being placed in 3-D with a specific direction and distance from the listening position. This is particularly true for more "abstract" electronic sounds, which do not have the same restrictions of placement and acoustic realism demanded by "acoustic music", e.g., orchestral or acoustic folk. While this approach can be enough to generate the necessary "spread" of sounds, some judicious use of *spatial early reflections* is often helpful in placing these sounds satisfactorily.

Synthetic spatial early reflections are timed delays that try to emulate the timing *and direction* of the first reflections to reach the listening position before the reverberant field takes over. If these are not captured with some form of multidirectional microphone array at the time of the recording, it can be useful to use synthetic versions to achieve good placement. It is possible to create such spatial reflections with combinations of multiple delay units – if the engineer has the processing and patience necessary to set things up appropriately. Martin and King (2015) give an in-depth description of just such a system in the description of their 22.2 mix setup. There are a number of emergent bespoke systems that can also achieve this, as will be discussed. It is very important that any audio potentially destined for later synthetic spatialization must be recorded "clean, close, and tight" if believable results are to be achieved.

## Playback

Sound is literally spatialized as soon as it is played back through a periphonic system by definition of the physicality of the 3-D array, positioned in the array by amplitude panning.[9] The configuration of the speakers and the acoustic properties of the room all come into play as a holistic system to make sounds appear from different places around the listeners. Similarly, the configuration of the panning system also has an effect, which might follow one of a number of mathematical approaches, e.g., ambisonic panning, vector base amplitude panning (VBAP) (Pulkki, 1997), or distance base amplitude panning (DBAP) (Lossius et al., 2009), together with the nature of the co-ordinates that are used to describe placement. When transforming audio between systems, these issues need to be taken into consideration.

In a binaural version of the same mix, all of the effects of speaker placement, head morphology, and room acoustics are typically absent or grossly simplified (although modeling is rapidly developing). Thus, if translation between headphone- and speaker-based mixes is required, the approach must be considered with care to either maximize "natural compatibility" or allow for the various mix components to be optimally reassembled.

## OBJECTS AND CHANNELS

There are some fundamental differences between channel-based and object-based 3-D production. "Objects" are aspects of audio, including both PCM-type files and associated parameters that exist discretely, only to be reassembled at the point of playback, and this approach offers great flexibility around the modes and devices of reproduction (Pike et al., 2016). In 3-D work, a given audio file might be associated with a specific point in space and thus be linked to its spatial playback information. This approach is independent of the number of speakers or size of a given auditorium. It is employed in current 3-D cinema systems, such as Dolby Atmos, and is gaining increasing use in broadcast.

Channels (in this context not to be confused with the constituent audio streams of multi-channel tracks as earlier) refer to the routing of an audio source to a discrete speaker output. This paradigm is familiar in stereo[10] and has also been traditionally used with horizontal surround systems. Such "native" recording allows for the capture and reproduction of timing differences to be conveyed between mic-capsules and speakers. These differences are crucial for convincing 3-D reproduction, and such an approach allows some of the real spatial information that is present in the recording location to be encoded and transmitted over speakers (or binaurally rendered to headphones) at a later stage.

Conversely, one cannot record natively in any meaningful way for object-based workflows. The engineer is limited to reproducing notionally monophonic recordings for some kind of monophonic-object playback/rendering system. The objects can be placed in a periphonic system using amplitude panning, but a "believable" placement requires (at least) convincing early reflections since these are a major part of psychoacoustic source localization. Adequate tools for generating satisfactory reflections (and hence externalization) in object-based systems are still developing, since until recently, the CPU cost of generating "scalable" reflection/reverb systems often outstripped their commercial application, and the approach can adversely affect the balance and spectra of the mix in different rooms. De Sena et al. (2013, 2015) have made considerable progress in this regard with a computationally efficient reverb algorithm that uses a network of delay lines connected via scattering junctions, effectively emulating myriad paths that sound like propagation around a modeled room. Also, FB360 employs a system that models some early key reflections to both enhance binaural placement and also allow the subsequent application of "production reverbs" in the processing chain. Oculus and Google are currently developing near-field binaural rendering that takes account of the perceptual issues caused by head shadowing (Betbeder, 2017) and also the modeling of volumetric sounds that emanate from areas larger than a more typical point-source (Stirling, 2017; Google, 2017).

One current way of approaching this reflection/object problem is to allow reflections and reverb to be generated in (channel-based) beds, whilst the objects themselves are rendered as "dry" objects at playback (perhaps with monophonic early reflections/reverb "baked-in"). This may give a reasonable semblance of realistic positioning; however, the effectiveness of this approach is very dependent on the audio material, and there may be perceptual differences at different playback locations in the sound field. Equally, different kinds of audio material have varying degrees of dependence on the reproduction of these kinds of "realistic" reflections. For example, in orchestral material, both the recording environment itself and the positions of the instruments therein are crucial to the musicality of the piece, and so such music is heavily dependent on "authentic" spatial reproduction. More "abstract" materials (e.g., pop or electronic) may not be as dependent on realistic acoustic placement and so may translate equally well in an object-based environment – since the clarity and depth of each sound's position is inherently less fixed and thereby open to a less strict spatial interpretation.

There are also implications for using effects processing in channel- and object-based systems. Dynamic range compression (DRC), as usually used in standard stereo systems, starts to take on a more complex character when working spatially. Unlinked channel-based compressors create deformations of the spatial image, and there are few linked compressors that work beyond stereo or 5.1. Bryan Martin (2017) states that in 3-D, DRC begins to sound "canned" and tends toward making sound sources appear smaller rather than larger. Further, when working in pure stereo, the artifacts of compression might tend to become desirable yet subtle components, but these same artifacts become increasingly apparent and unpleasant in large 3-D formats.

It is not yet feasible to have a true object-based compressor (that is, one that will operate consistently at the playback stage), and control of dynamic range – such as is possible – can only really be made on the original monophonic sounds before they enter the panning system. Having said that, the MPEG-H 3-D audio-coding standard provides an enhanced concept for DRC that can adapt to different playback scenarios and listening conditions (Kuech et al., 2015). Given the increase in popularity for object-based approaches, it is likely that solutions will soon emerge that unify the object production chain. Aside from technical challenges, keeping within acceptable processing requirements can be an issue. Systems that offer hardware rendering, such as PSVR or Dolby Atmos, tend to avoid such concerns. Indeed, for theatrical applications, these issues are generally less problematic, since cinema sound tends not to rely on the *character* that compression effects bring to material, but for creative music production, the compromised nature of this familiar toolset tends to have greater implications.

Indeed, these issues apply to any traditionally insert-type effects that might affect objects spectromorphologically. It is also worth noting that because ambisonic audio is encoded throughout the production chain between the point of capture and reproduction, it is not generally possible to apply any kind of effects processing to it either, since such "distortion" will conflict with the delicate spherical harmonics, especially at higher orders. There are a few ambisonic plug-ins that offer EQ, delay, and reverb, but these tend to be limited to lower-order operation.

As a side note, it is worth noting that object-based production workflows are nothing new per se. At all times during a DAW-based mix, one is effectively working with objects. The fundamental difference is that during "normal" mixing, the rendering to speakers happens at the same moment as the mix proceeds, as opposed to an object-based mix that postpones that rendering to the moment of playback at the consumer end. An obvious implication in addition to this is that the normal kinds of "buss" processing which engineers are used to are no longer available when the mix is fragmented into its constituent objects, since the summing of the playback/production chain has been deferred to the moment of reproduction. This is why we currently see hybrid systems of objects and channels, and "pure" object-only systems are still in evolution. Modern game engines also follow this logic – and in fact in many ways have been leading such workflow architecture for some time – blending stereo and 5.1 tracks with real object-based sounds to achieve pleasing results.

## CAPTURE CONSIDERATIONS

When assessing the value of the channel/object parts of a given system, one must look at the material conditions of the recording in question. If one wants to capture something of the "acoustic" and the spatial relationships between elements as they were in the room at the time of recording (and have real coherence between speakers at playback), then some form of multi-microphone/channel-based system currently tends to be preferable to the object approach (at present). This state of affairs is likely to be the case only so long as synthetic acoustic-modeling tools lag behind the "real thing". As soon as the audio equivalent of photo-realistic visual effects is achievable, the flexibility advantages of an object approach may make it the dominant form.

It is worth considering the difference between working with spaced microphones and an ambisonic microphone (e.g., a Calrec Soundfield). Of the former, Williams's psychoacoustic research (2004) looks at the way in which multi-microphone systems create "coverage" between any given pair of capsules. Williams notes that the stereo recording angle (SRA), coverage, and angular distortion[11] are all dependent on a combination of the microphones' polar patterns, their relative angles, and their distance from each other, and this points to potential issues with ambisonic microphones that are frequently overlooked. An ambisonic microphone utilizes (theoretically) coincident capsules, and this means that it can only record intensity *difference*. Crucial timing information from the recording location is lost (and cannot be recovered). Further, the capsule types and angles imply large sections of SRA overlap in the horizontal plane once summed for playback, and this gives rise to angular distortion at regular intervals around the full circular field of capture (which becomes increasingly apparent outside of a very small sweet spot). The lack of timing information, coupled with the overlap of polar patterns, can lead to an unsatisfactory image that lacks the spaciousness and accuracy of a spaced-array recording (Williams, 1991).

Some of the issues of first-order ambisonic microphones have been improved with the recent introduction of higher-order ambisonic microphones. These greatly improve some of the angular distortion issues associated with first-order variants, but they can come at the cost of spectral-colouration issues, such as the low frequency roll-off associated with more directional cardioids or potentially high-frequency phase artifacts from their large number of capsules. As is so often the case, there is a trade-off to be made between spectral and spatial quality on the one hand and practicality and cost on the other, and the engineer must evaluate this for each project. New tools are emerging to support this, such as Hyunkook Lee's et al. (2017) Microphone Array Recording and Reproduction Simulator (MARRS) app, which predicts "the perceived positions of multiple sound sources for a given microphone configuration" and can also "automatically configure suitable microphone arrays for the user's desired spatial scene in reproduction".

In a VR workflow, there is often an assumption that ambisonic recording is "native" recording. Whilst this may be true enough for some kinds of single-perspective 360° video recordings, it can be less so for "true" VR that utilizes 6DOF[12] – where the listener's locative movement within a virtual space and its associated sound field requires a shift in sonic perspective. Spaced arrays and multi-microphone setups can be reproduced in VR if care is taken over their reproduction in the audio system of a game engine. Issues with phase tend to be dependent on the material being reproduced, the placement of the "array", and the particular qualities of the binaural decoder. Recordings with low levels of correlation in each channel tend to work best. Game engines also allow for other more sophisticated ways of placing and spatializing monophonic sounds, and ambisonics might be only one aspect of the greater sound mix, perhaps layered with binaurally synthesized elements and head-locked audio.[13] Further, coverage from multiple microphones can be blended into "scenes" using synthetic acoustic placement. Summing synthetically positioned spot-microphones and "spatial" ambisonic recordings can achieve very satisfactory results and becomes necessary to satisfy the requirements of balancing dry/close recordings with more distant or reverberant elements. This approach has been corroborated by Riaz et al. (2017) in a scientific setting, although the full results of that test are not published at the time of writing. Phase is likely to play an important part in the success of such "scene" combinations, and so the engineer might evaluate this and adjust it for optimal timbre and placement.

At this time, a common problem with HOA is that of channel count, since the host DAW perhaps needs to accommodate say 36 (for fifth-order ambisonic) channels for each "monophonic" track of audio, and this quickly becomes demanding in terms of CPU load and streaming bandwidth. Mixed-order ambisonics can offer a solution whereby different orders are combined, perhaps with first-order ambisonics for sounds like beds[14] without need of more precise localization and HOA for key sounds in need of accurate placement. This approach might also be implemented by providing decreased resolution in the vertical plane to which the ear is less sensitive; the ears sit on a horizontal plane and are therefore more attuned to it (Travis, 2009). However, in either case, care is needed with the type of decoder used:

> A "projection" decoder will have matrix coefficients that do not change for the first four spherical-harmonic components for 1st, 2nd and 3rd order. So simple summation in the equivalent of the "mix-buss" will work here, regardless of the layout. In contrast, a pseudo-inverse decoder won't behave this way and will have different matrix coefficients for the first four spherical-harmonic channels for each order. Thus the layout must be as symmetrical/even as possible, for this to give results similar to projection.
>
> (Kearney, 2017)

When working in linear environments like cinema (as opposed to real-time, for instance in VR or gaming, where spatialization requirements might change dynamically), there is scope to render high-quality acoustic spatial elements. For instance, it is possible to ameliorate the aforementioned CPU and bandwidth issues by "baking-in" higher-resolution reverbs into higher-order ambisonic beds. Another approach taken by the PSVR system uses an object-based approach, sending each voice with its associated azimuth and elevation information via middleware to bespoke audio libraries in the PSVR hardware for rendering along with mixed-order components. This gives excellent localization for the objects – which can also be moving, if tied to animated "emitters" associated with visual elements of a game or application – but loses the ability to perform more typical channel-based production processing en route.

## MIXING FOR BINAURAL

The pinnae, head, and torso modify sounds before they enter the ear canal (filtering via reflection effects) in unique ways that are individual to the listener and dependent on the morphology of *that* person. Whilst ITD (inter-aural time difference) and ILD (inter-aural level differences) are important for determining the position of sounds in the horizontal plane, spectral content, filtered largely by the pinnae, is mostly responsible for the perception of elevation or sounds in the median plane[15] (Wightman and Kistler, 1997), although this is slightly augmented by head and torso effects. Combining the preceding into filters whose response is given by such a head-related transfer function (HRTF)[16] allows a sound to be processed[17] by that filter in order that the brain perceives it as coming from a particular point in space. Such HRTFs need to be provided for every point in space that might be emulated, and the appropriate HRTF filtration applied to a source sound relative to a given localization placement. This is the mechanism of binaural reproduction in headphones and forms the basis for the binaural "synthesis" previously mentioned. Although HRTFs are person-specific, there have been a number of one-size-fits-all ones developed, e.g., Furse's (2015) "Amber", which exploits statistical averages to gain the best (compromised) optimization for the largest number of people. A good introductory treatment of binaural theory is given by Rumsey (2001).

The implications of the preceding for recording and mixing[18] in 3-D are important. The first is that binaural 3-D mixing on (for) headphones, using a non-personalized off-the-shelf HRTF will tend to lead to mixes which may or may not work as intended for others depending on the relative differences between the morphologies of 1) the mixer, 2) the HRTF model that was utilized, and 3) the end listener. Whilst some tools allow the engineer to select a preferred HRTF, although this will undoubtedly give a truer and more responsive experience to that individual, it is less likely to successfully translate to the largest proportion of end listeners. One option is to first mix to a favored HRTF, switch to one such as Amber, and then optimize the mix for that prior to final distribution. Clearly, such an approach is an idiosyncratic choice.

Another strategy is to mix as much as possible over a 3-D speaker array and use a conversion transform at the end of the mix process to complete the 3-D processing to a headphone format. This reduces the error distance between the mixer's morphology and the end user, since the engineer will not attempt to correct the HRTF model to suit his or her own peculiarities. If this approach is adopted, then care must be taken to ensure that the conversion tools are high quality, since there can be large differences in the quality of encoders/decoders that convert between speakers and headphones and vice versa. This approach can represent something of a leap of faith for the engineer, who simply has to accept the algorithmically created binaural mix. There is some hope that more personalized HRTF models will ameliorate some of these margins of error in the medium term, but it remains an issue at the point of writing.

There are numerous sets of ambisonic software tools available at the time of writing, many of which are freeware. Several also feature binaural decoders and can therefore monitor and render for headphones. Many of these tools tend to be first order, although the excellent ambiX suite (Kronlachner, 2014) goes up to seventh order.

A typical DAW workflow would be to insert an ambisonic panner (an encoder) onto a monaural audio track, and the panner would be able to send its multi-channel output to a buss, with the track's main output muted. The panner will convert the track to (say) four-channel for first order – which can now accommodate 3-D spatial information. A similar setup should be implemented in other tracks. The four-channel busses should be routed to another track of the same channel width, and a suitable binaural decoder should be inserted on this track. This will have four inputs and two outputs, the latter representing the headphone feed. The inserted panners can then be used to spatialize the various tracks, thus localizing monophonic sound. This basic setup can be extended by having a parallel destination track with an ambisonic reverb inserted on it and then fed into the main destination track prior to the decoder. To work at higher orders, panners and decoders designed for this can replace the preceding and the channel count be increased accordingly for all tracks and busses.

People will perceive the output of such a system in different ways to different degrees of the intended effect, a function of the HRTF issue but also variations in headphones, such as their physical structure[19] and more. It is generally possible to create definite localization to the rear, although sometimes this is just perceived as ultra-wide stereo. Elevation tends to be less successful. However, interesting mixes can be created with good separation and width. Of course, the engineer cannot know exactly how others will perceive it.

As mentioned, much of the current interest in binaural ambisonic audio is coming from VR HMDs, and crucially, these feature head tracking. As also mentioned, an ambisonic sound field can be rotated through simple multiplication by a scalar, and if head-tracking data is converted to such a scalar, then the sound field can be rotated in counter motion to the head to give the impression of sounds that are fixed in space and independent of the listener's head movement, as is of course the case with real life. Such head tracking is very effective at removing front-back confusions and further greatly enhances perception of elevation, even if that is only through the listener being able to "look up" at the audio and bring it into frontal auditory focus to affirm that it is up there. To monitor this in a DAW, a "rotator" plugin can be inserted before the decoder, and head-tracking data routed to the rotator's parameters.[20] To replicate the effect for the end user, an HMD needs to be supplied with B-format ambisonic multi-channel audio so that the rotation can be rendered "live".[21] At the time of writing, various head trackers are available that can be attached to normal headphones, bypassing the need for an HMD, although they typically have various compatibility issues, for instance being tied to a given manufacturer's software. Hedrot (Baskind, n.d.) is a cheap DIY head tracker that is widely compatible via OSC data.

The nature of HRTF-based spatial modeling means that filtering (particularly of upper frequencies) is readily discernable in many listening directions. This becomes much more apparent with dynamic head tracking – manifesting itself as something of a sweep, and so care must be exercised regarding what elements of a mix are panned where. If, for instance, a drum-kit stem is placed at the front, as the head turns, there will be noticeable attenuation of the cymbals, and this might be less preferable than the desired effect of spatialization. Head locking[22] of such parts provides a solution. Sounds with more arbitrary timbre are more tolerant of such panning since there is a less-polarized "right" sound, and so these might be better placed around the sound field. Good effect can be gained by fixing a key sound, such as bass, to the front, since the anchored low-frequency energy gives a strong impression of rotation regardless of what else is panned, but care must be exercised since rotating low frequencies can also induce nausea, and so judicious filtering – possibly splitting a given musical part into separately locatable tracks – might be necessary. Interesting effects can be readily created, for instance, with high-pass filtered reverb only apparent when looking up, which can give the impression of a high ceiling. Something that should be considered is that a unique musical journey can be created dependent on head position. The music might literally be presented differently dependent on the user's aural "field of view", analogous to visuals on a train journey where one might look in different directions each time the journey is taken. It is quite possible that head tracking will become a standard feature of many headphones, perhaps especially as augmented reality proliferates.

## SOME SPEAKER CONSIDERATIONS

It is commonly accepted that vertical perception is guided by spectral cues, whereby the higher the frequency, the higher the perceived height relative to a single source loudspeaker, and in fact, for a center speaker (on the median plane), frequencies below 1 kHz tend to be received as coming from lower than the speaker. This is supported by several scientific studies and is called the pitch-height effect (Cabrera and Tilley, 2003). Lee (2015) developed an upmixing[23] technique called perceptual band allocation (PBA) that could enhance the 3-D spatial impression by band-splitting horizontal-surround-recorded ambience and routing the bands to the higher and lower loudspeaker layers of an array (the upper frequencies to the upper layer). The PBA system outperformed an eight-channel 3-D recording in listening tests, and a center frequency of 1 kHz was proposed for this to work.[24] Lee (2016a, 2017a) also examined the degree of vertical image spread (VIS) – the overall perceived height of the sound field, this time with phantom images. This was also done via PBA, and he experimented with the control of the VIS on a per-band basis by mapping multiple frequency bands to speaker layers. Lee found that the vertical locations of most bands tended to be higher than the height of the physical speaker layer and that it was generally possible to control the VIS via different PBA schemes.

There is another psychoacoustic phenomenon related to the pitch-height effect – the "phantom-image elevation effect" (De Boer, 1947) – that also influences our perception of elevation. This is the observation that for perfectly coherent stereoscopic phantom-center images on speakers (e.g., a vocal, panned centrally between a pair, but not routed directly to any center speaker), different frequencies perceptually "map" themselves to different elevations in the median plane. It is a common approach in channel-based mixing with a speaker array to employ a conventional stereo source routed to a pair of speakers, which of course might typically form a phantom image in the center. In this context, Lee (2016b, 2017b) again explored the relationship between frequency and psychoacoustic perception of height using a variety of loudspeaker pair base-angles, including the conventional (stereo-like) 60° and 180° pairs that were orthogonal to the median plane – directly to either side of the head. For the 60° pair, broadband frequencies appeared to be elevated, and for the 180° setup, both octave bands of 500 Hz and 8 kHz and sounds with transient characteristics were the most prominently elevated, and the 1 kHz region was often perceived to be behind the listener for all base-angles. Further, as the base angle increases from 0° to 240°, the perceived phantom image is increasingly elevated. Overall, elevation was negligible (along with general directionality) for frequencies below 100 Hz. Although the higher frequency elevations align with Blauert's (1969) "directional bands theory", Lee hypothesizes that the sub-3 kHz effect is due to acoustic crosstalk from a pair of speakers being interpreted by the brain as reflections from the shoulders.

The application of such theory to mixing offers an enhanced understanding and extends the possibilities for spatial placement. When mixing to "single" speakers, EQ-ing to roll off above 1 kHz can extend the sound field to below that speaker, and in general, higher frequencies will be perceived as coming from progressively higher. Although the pitch-height theory pertains to the center speaker, this effect might momentarily extend to any speaker in the array since as the listener rotates his or her head to face it, it comes into the effective median plane. Sounds panned to the height layer of a speaker array that contain a certain amount of higher frequencies may be perceived as emanating from even higher, offering an extension of the sound field, and high-pass filtering might offer a further exaggeration of this effect.

Phantom centers present interesting opportunities. When working with 60° stereo pairs of speakers that create a coherent phantom center, most frequencies will elevate, and these might be balanced against any signal actually routed to the center speaker to create a vertical panorama on the median plane. Mid-range content (possibly band-pass filtered, since the 250 Hz octave elevated less) panned equally to a 180° pair will again elevate the phantom image, and this could be implemented with either a main speaker layer or indeed the height layer to gain greater sonic elevation; the same applies for 8 kHz. Increasing elevation might be generated by automating the routing to progressively wider pairs of speakers, but this should be done incrementally rather than progressively to avoid comb filtering and transient smearing in the process. The 1 kHz octave band has a "center of gravity" behind the listener, and thus panning and EQ can be arranged to exploit this. Conversely, it is useful to remember not to "fight" this phenomenon.

Also, where bespoke 3-D reverbs are not available, one can follow the PBA ideas and achieve good results by band splitting a reverb at, say, 1 kHz, and placing the hi-passed components in the height layer. This expands the vertical perception of the space and has fewer of the problems of using two reverbs to achieve the same result.

Thus, these theories present a basic premise for working with height in certain circumstances, although naturally any such spectral presentation is likely to be subservient to the greater tonality of the mix. It should be remembered that much of this applies principally to phantom-center images, and other components of a mix might easily compensate – or fight – spectrally and hence spatially. There is a copious body of literature on 3-D capture for the interested reader to explore; the archive of the Audio Engineering Society is a good place to start.

## MIXING: GENERAL APPROACHES

As one moves toward higher spatial resolution in 3-D, the intrinsic quality of the original component recordings / parts of the piece become increasingly laid bare, allowing a greater focus on the individual elements of a mix. These now have more space to be heard clearly, just as might be desired by three-point panning in stereo. That aside, in stereo, when all sounds come from just a pair of speakers, there is a lot of spectral and transient masking and blending that can greatly obfuscate many of the elements within that mix. This can, of course, be a very desirable aspect of balancing the competing elements of a complex music mix and may in fact be the reason that "mixing" might be required in the first place. As the speaker count (or binaural headphone resolution) increases, so do the opportunities to more precisely discern individual elements (assuming that both the total number of audio elements in the mix remains constant and also exploits the extra "space"). As this happens, the otherwise "hidden" qualities of the underlying elements are increasingly exposed – for better or for worse. This is doubly so when utilizing monophonic sounds in an object-based system, where background noise and recorded reflections can compromise the ability to repurpose the sound in 3-D space. Bryan Martin (2017) attests to all this and also emphasizes that edits must be much tighter than are often accepted in dense stereo work.

For these reasons, 3-D mixing invariably demands a number of new things of the mixer:

1.  There needs to be a greater amount of sonic material to "fill" the newfound space, and this can be approached in two ways: from a compositional/arrangement perspective or from a pure *mix* perspective. Clearly, compositionally, there is physically more *space* for the arrangement of elements in a 3-D space than there is in a stereo mix. Adding new layers of instrumentation (which need not represent harmonic extensions of the extant musical parts) is one approach to filling this space, although this approach overtly changes the character and arrangement of the original music. The same approach works more readily in film where ambience can more easily be added to and augmented into height/surround layers based on the content of the original stereo ambience. With music, this approach can lead to complexity and adornment that may be undesirable, and it is an approach that only tends to work with abstract (non-acoustic) performances. Of course, many composers will relish the opportunity to embrace such a medium and will tailor their arrangements in order to exploit the newfound possibilities. As 3-D delivery proliferates, such approaches will likely become accepted and commonplace.

Perhaps the more readily adopted approach is to use mix techniques to expand the sonic space of the original, in ways that do not compositionally change the piece in any radical manner. There are a number of approaches to achieving this kind of "spread" of the component elements. The first is to simply pan elements to their new positions in the 3-D sphere. The downside of this is that in their own right, each direction's audio can appear somewhat spartan in content, as the original busy stereo field is decomposed into individual elements – separated and exposed over the larger playback area.

The second is to add width and size to those elements. If the placement of a sound renders it too small in the overall sound stage, when, for instance, it is emanating only from a single speaker, then the mix engineer might wish to spread it over more than one unit. This can have a number of effects. If it is (say) evenly panned between two speakers that subtend around 60° to the listener,[25] then it will generate a new phantom center that will not necessarily sound any bigger than the original – just displaced, and biased panning will simply move the phantom's origin. Further, if transient material is present, then smearing can occur at certain points in the listening area, and this of course will be a function of the relative distance to each of the two speaker units.

If sounds are shared over two or more speaker channels, it is always a good idea to ensure there is some kind of decorrelation, so that the sound is not exactly the same in all speakers – this at least mitigates the phantom-center problem, although not the transient smearing. Sound sources in the real world are essentially decorrelated through reflections and asymmetries of the listening environment. Decorrelation within a mix environment can be achieved through slightly offsetting EQ settings, delay times, or pitch. Such EQ may not be desirable tonally, and strategic decisions might need to be taken to balance such tensions. Decorrelation can also be achieved by use of reverb systems and impulse response convolutions. Some commercial spatialization systems feature a parameter called "spread" or some such that can automatically change the size of a source in the sound field although their effect is not always as intuitively expected.

2.  Time-based effects (at least delays and echoes) can be most effective if adapted to spatially modulate over the front-back and median planes. There are also some excellent "3-D" reverbs that have come onto the market that can achieve great results, but stereo or mono algorithms may be used in multiple channels, so long as the settings are adjusted in each to ensure a satisfactory level of decorrelation, and of course interesting effects can be generated by spatially separating the "wet from the dry", to varying degrees.

    However, delays and phase shifts can also be used to place sounds in "the sphere", according to the precedence effect; there are situations where it is better not to rely solely on amplitude panning alone. The "ambience" around a source might need to change accordingly – particularly for moving sources – and to achieve this without overtly muddying the mix is a question of balance, taste, and technique. Smaller "room" settings with short early-reflection times and shorter reverb tails will often still work well if too much reverb is problematic for the greater mix.

Martin et al. (2015) makes a detailed examination of the use of syn-thetically generated early reflections in a 22.2 channel-based system and gives clear descriptions of the utilized timings. There is some dis-agreement over how well the precedence effect works in the vertical, but it is a viable option for placing sounds and the individual mixer will have to determine the impact of such use in their own mix. If translation to binaural might be later required, then the effectiveness of such panning mechanisms and indeed the reverbs must also be con-sidered in that context. Reverbs in particular can cause problems when "collapsed" from speakers into a binaural render.

3.  DRC in HOA has spatial and spectral artifacts that can be most unde-sirable, and such a process falls into the category already mentioned that will corrupt spherical harmonics, leading to a degradation of the spatialization. Also, if working with objects, it should be understood that they are by their nature hard to compress in any natural sounding fashion – just as sounds in the real world are not naturally compressed.

DRC and EQ in both 3-D speaker-based systems and 3-D binaural headphone mixes must be approached with more caution than in a stereo or even surround system. As the 3-D resolution increases, the artifice of heavy compression and EQ become increasingly revealed. Martin (2017) warns mixers who wish to embark on 3-D chan-nel-based mixing to "forget compression – it's the opposite of what you want". He also described the counter-intuitive experience of using stereophonic compression techniques in a 3-D-mix environment:

> Compression fails – because it makes it even smaller. You have this huge panorama to present, and compression just makes it even smaller – so it actually just fights you. People think compression is going to make it more "there" but it makes it smaller, when you want it to be bigger. Compres-sion is not your friend.

Use of DRC as a "glue to hold it all together" becomes increasingly problematic in 3-D. Coupled with the increasing bias toward creating a realistic (or at least believable) sound field around the listener that is engendered when working in higher resolution 3-D, traditional "ste-reo buss" techniques, such as buss compression feel increasingly out of place – not to mention technically unachievable at present. Richard Furse's Blue Ripple toolset (2017) has an interchange plugin which can transform an ambisonic mix to a 20-channel A-format that allows the use of standard stereophonic-type effects processing, followed by a re-en-code to HOA. Although it is preferable to remain in the ambisonic realm when mixing "ambisonically", Furse's approach allows for signal pro-cessing that is not currently possible with the currently available amb-isonic toolset. Perhaps the future of mixing in high-resolution 3-D will move toward increasing reliance on the highest quality reproduction of the musical elements themselves, within a dynamic and intelligent phys-ical environment with metadata-linked ends of the processing pipeline.

## CONCLUSION

Immersive audio indicates by its very name that some kind of presence in the performance is anticipated for the listener. This also implies that the overall effect of the sound elements should have a clarity and some kind of all-encompassing physical expression free from the limitations of a 60°-wide front-facing stereo track. It is not always obvious quite how detached a lot of stereo music reproduction is from sounds in the physical world, perhaps most greatly exemplified by classical music, and this leads to issues when extending and translating stereo techniques to 3-D, which invariably lean toward bringing sounds into a more open, physically encompassing expression in three dimensions.

There are a number of approaches to capturing and creating 3-D audio, and each has its merits and limitations. While stereo and horizontal-surround music production is an extremely well-established art form with a number of conventions and aspects of accepted good practice, many of these are subverted by the increased complexity, both perceptual and technical that nascent 3-D production entails.

As was the case with 5.1, it is unlikely that multi-speaker setups will routinely make it into the domestic living room with appropriate configuration, but as wave-field synthesis, transaural binaural reproduction, and distributed-mode loudspeakers evolve, so will home 3-D. However, the VR industry is expanding rapidly and bringing head-tracked audio into the mainstream expectation, and regular audio-only head-tracked headphones will soon follow in their droves. This will turn relatively banal binaural listening into something much more dynamic, and while the limitations of such systems will preclude their usage in lots of music, even small aspects of dynamic binaural reproduction will make it into a large proportion of mainstream popular music mixes. Although binaural is at the mercy of HRTF-matching at present, future solutions will overcome this and emancipate headphone listening further still.

3-D music production is evolving and will deserve increasing attention in order to develop the art form, and it is not possible to comprehensively cover the topic in a single chapter. Doubtless, much will be written in due course, and accordingly, we can look forward to developing increased understanding of the praxis. The creative possibilities are exciting – presenting one of the biggest single opportunities to forward music production in a long time. These possibilities await only innovative individuals with suitable tools to build a new sound world that will represent a step change in the creation and consumption of music.

## ACKNOWLEDGMENTS

The authors would like to thank the experts who graciously contributed interviews for this text: Hyunkook Lee and Bryan Martin. Special thanks go to Hyunkook Lee for the guidance around his own work.

## NOTES

1. Perhaps ironically in the context of this chapter, "stereo" comes from the ancient Greek *stereos*, which means solid – with reference to three-dimensionality, albeit with regard to the formation of words.
2. A body of research has formed around this, for example (Barrett, 2002) and (Lynch and Sazdov, 2011).
3. Synchresis is the psychological linking between what we might see and hear when such events occur simultaneously.
4. Media composers such as Joel Douek and Michael Price are notable for embracing 3-D approaches.
5. In fact, conventional stereophony is a subsystem of Ambisonics (Gerzon, 1985).
6. Jot et al. (1999) provide a useful text on encoding/rendering.
7. Spherical polar-coordinate solutions to the acoustic wave equation.
8. If the channel order = N (which defines the angular resolution of the spherical harmonics), the number of audio channels required for that order is given by $(N+1)^2$.
9. Although this also applies to any multi-speaker setup, be that stereo or horizontal planar surround.
10. E.g., in stereo: record an acoustic guitar with a single microphone and pan it to playback from a single speaker, or indeed centrally, when it is mapped to both speakers with equal amplitude and so on.
11. Williams used this term to describe how changes in the relative angles of microphone capsules might shift the perceived position of the instruments on the horizontal plane, once reproduced.
12. 6DOF: six degrees of freedom, which refers to movement in a 3-D-space: front/back, left/right, up/down, plus pitch, roll, and yaw of the head position. In other words, the user can navigate the space and might expect a subsequent shift in sound field both whilst moving and looking around. In contrast, 360° video (often presented as "VR") does not permit locative movement within its space.
13. A separate audio stem in the overall mix that does not respond to head tracking to provide elements such as narrative or beds.
14. A bed is an audio track that might be spatialized but is not dependent on precise localization of any of its content. It might typically be a sonic foundation on which to place other more localized sources.
15. Also known as the mid-sagittal plane, this is the plane that bisects the human body vertically through the naval. As such, strictly speaking, it can represent perception of sonic height but only when looking directly ahead.
16. The HRTF actually exists in the frequency domain. The temporal equivalent is known as head-related impulse response (HRIR).
17. Such processing is performed by convolution, a CPU-intense mathematical operation commonly used in reverbs whereby every sample in the source is multiplied by every sample in the filter in order to impose a spectral fingerprint on the source.
18. For both fixed two-channel and real-time multi-channel ambisonic that might also respond to head tracking.

19. HRTFs are linked to in-ear or over-ear designs.
20. Typically: pitch – analogous to nodding, yaw – looking side to side, and roll – moving the ear toward the shoulder.
21. Other systems are also possible, which might spatialize individual tracks in middleware or a game engine.
22. As also in 13, head locking is where the sound is independent of movement, just as with normal headphone listening. A separate buss that bypasses the rotator is required for such parts.
23. The process of converting one playback format to another of a greater number of channels.
24. In fact, it was shown that a cutoff frequency could be anywhere between 1 and 4 kHz for this effect to hold.
25. To maintain something close to stereo theory, although this only holds when facing forward toward the speakers and will break down for, say, a lateral placement.

## REFERENCES

Barrett N (2002) Spatio-Musical Composition Strategies. *Organised Sound* 7(3): 313–323. doi: 10.1017/S1355771802003114.
Baskind A (n.d.) Hedrot. Available from: https://abaskind.github.io/hedrot/ (accessed 11 July 2017).
Begault DR (1992) Perceptual Effects of Synthetic Reverberation on Three-Dimensional Audio Systems. *Journal of Audio Engineering Society* 40(11): 895–904. Available from: www.aes.org/e-lib/browse.cfm?elib=7027.
Betbeder L (2017) Near-Field 3D Audio Explained. Available from: https://developer.oculus.com/blog/near-field-3d-audio-explained (accessed 11 October 2017).
Blauert J (1969) Sound Localization in the Median Plane. *Acta Acustica United with Acustica* 22(4): 205–213.
Cabrera D and Tilley S (2003) Vertical Localization and Image Size Effects in Loudspeaker Reproduction. In: *Audio Engineering Society Conference: 24th International Conference: Multichannel Audio, the New Reality*. Available from: www.aes.org/e-lib/browse.cfm?elib=12269.
De Boer K (1947) A Remarkable Phenomenon With Stereophonic Sound Reproduction. *Philips Technical Review* 9(8).
De Sena E, Hacıhabiboğlu H and Cvetković Z (2013) Analysis and Design of Multichannel Systems for Perceptual Sound Field Reconstruction. *IEEE Transactions on Audio, Speech, and Language Processing* 21(8): 1653–1665.
De Sena E, Hacıhabiboğlu H, Cvetković Z, et al. (2015) Efficient Synthesis of Room Acoustics Via Scattering Delay Networks. *IEEE/ACM Transactions on Audio, Speech and Language Processing (TASLP)* 23(9): 1478–1492. Available from: http://dl.acm.org/citation.cfm?id=2824192.2824199 (accessed 3 October 2017).
Digi-Capital (2017) After Mixed Year, Mobile AR to Drive $108 Billion VR/AR Market by 2021. Available from: www.digi-capital.com/news/2017/01/after-mixed-year-mobile-ar-to-drive-108-billion-vrar-market-by-2021/ (accessed 16 July 2017).

Elen R (1998) The Ambisonic Motherlode. Available from: http://decoy.iki.fi/dsound/ambisonic/motherlode/index (accessed 30 August 2017).

Furse R (2015) Amber HRTF. *Blue Ripple Sound*. Available from: www.blueripplesound.com/hrtf-amber (accessed 7 November 2017).

Furse R (2017) 3.1 O3A B->A20 and O3A A20->B Converters. *O3AManipulators_UserGuide*, User Guide. Available from: www.blueripplesound.com/sites/default/files/O3AManipulators_UserGuide_v2.1.4.pdf.

Gerzon MA (1973) Periphony: With-Height Sound Reproduction. *Journal of the Audio Engineering Society* 21(1): 2–10. Available from: www.aes.org/e-lib/online/browse.cfm?elib=2012 (accessed 29 March 2017).

Gerzon MA (1985) Ambisonics in Multichannel Broadcasting and Video. *Journal of the Audio Engineering Society* 33(11): 859–871. Available from: www.aes.org/e-lib/browse.cfm?elib=4419 (accessed 29 March 2017).

Google (2017) Resonance Audio. *Google Developers*. Available from: https://developers.google.com/resonance-audio/ (accessed 7 November 2017).

Jot J-M, Larcher V and Pernaux J-M (1999) A Comparative Study of 3-D Audio Encoding and Rendering Techniques. In: *Audio Engineering Society*. Available from: www.aes.org/e-lib/browse.cfm?elib=8029 (accessed 29 March 2017).

Kearney G (2017) Ambisonic Question: E-mail.

Kronlachner M (2014) ambiX v0.2.7 – Ambisonic Plug-In Suite | *matthiaskronlachner.com*. Available from: www.matthiaskronlachner.com/?p=2015 (accessed 28 March 2017).

Kuech F, Kratschmer M, Neugebauer B, et al. (2015) Dynamic Range and Loudness Control in MPEG-H 3D Audio. In: *Audio Engineering Society Convention 139*. Available from: www.aes.org/e-lib/browse.cfm?elib=18021.

Lee H (2015) 2D-to-3D Ambience Upmixing Based on Perceptual Band Allocation. *Journal of Audio Engineering Society* 63(10): 811–821. Available from: www.aes.org/e-lib/browse.cfm?elib=18044.

Lee H (2016a) Perceptual Band Allocation (PBA) for the Rendering of Vertical Image Spread With a Vertical 2D Loudspeaker Array. *Journal of Audio Engineering Society* 64(12): 1003–1013. Available from: www.aes.org/e-lib/browse.cfm?elib=18534.

Lee H (2016b) Phantom Image Elevation Explained. In: *Audio Engineering Society Convention 141*. Available from: www.aes.org/e-lib/browse.cfm?elib=18468.

Lee H (2017a) Interview With Hyunkook Lee. Interviewed by Gareth Llewelyn for Producing 3D Audio.

Lee H (2017b) Sound Source and Loudspeaker Base Angle Dependency of Phantom Image Elevation Effect. *Journal of Audio Engineering Society* 65(9): 733–748. Available from: www.aes.org/e-lib/browse.cfm?elib=19203.

Lee H, Johnson D and Mironovs M (2017) An Interactive and Intelligent Tool for Microphone Array Design. In: *Audio Engineering Society Convention 143*. Available from: www.aes.org/e-lib/browse.cfm?elib=19338.

Lossius T, Balthazar P and de la Hogue T (2009) DBAP – Distance Based Amplitude Panning. In: *International Computer Music Conference (ICMC)*, Montreal.

Lynch H and Sazdov R (2011) An Investigation Into Compositional Techniques Utilized for the Three-Dimensional Spatialization of Electroacoustic Music. In *Proceedings of the Electroacoustic Music Studies Conference, Sforzando!*

*Electroacoustic Music Studies Conference*, New York. Available from: www. ems-network.org/spip.php?article328 (Accessed: 10 February 2018).

Martin B (2017) Interview with Bryan Martin. Interviewed by Gareth Llewelyn for Producing 3D Audio.

Martin B and King R (2015) Three Dimensional Spatial Techniques in 22.2 Multichannel Surround Sound for Popular Music Mixing. In: *Audio Engineering Society Convention 139*, Audio Engineering Society. Available from: www. aes.org/e-lib/online/browse.cfm?elib=17988 (accessed 1 July 2016).

Martin B, King R, Leonard B, et al. (2015) Immersive Content in Three Dimensional Recording Techniques for Single Instruments in Popular Music. In: *Audio Engineering Society Convention 138*, Audio Engineering Society. Available from: www.aes.org/e-lib/online/browse.cfm?elib=17675 (accessed 1 July 2016).

Pike C, Taylor R, Parnell T, et al. (2016) Object-Based 3D Audio Production for Virtual Reality Using the Audio Definition Model. In: *Audio Engineering Society*. Available from: www.aes.org/e-lib/browse.cfm?elib=18498 (accessed 21 March 2017).

Pulkki V (1997) Virtual Sound Source Positioning Using Vector Base Amplitude Panning. *Journal of Audio Engineering Society* 45(6): 456–466. Available from: www.aes.org/e-lib/browse.cfm?elib=7853.

Riaz H, Stiles M, Armstrong C, et al. (2017) Multichannel Microphone Array Recording for Popular Music Production in Virtual Reality. In: *Audio Engineering Society Convention 143*. Available from: www.aes.org/e-lib/browse. cfm?elib=19333.

Rumsey F (2001) *Spatial Audio*. Oxford; Boston, MA: Focal Press.

Shivappa S, Morrell M, Sen D, et al. (2016) Efficient, Compelling, and Immersive VR Audio Experience Using Scene Based Audio/Higher Order Ambisonics. In: *Audio Engineering Society*. Available from: www.aes.org/e-lib/browse. cfm?elib=18493 (accessed 22 May 2017).

Stirling P (2017) Volumetric Sounds. Available from: https://developer.oculus. com/blog/volumetric-sounds (accessed 11 October 2017).

Travis C (2009) "A New Mixed-Order Scheme for Ambisonic Signals" – Ambisonics. File. Available from: http://ambisonics.iem.at/symposium2009/proceedings/ ambisym09-travis-newmixedorder.pdf/@@download/file/AmbiSym09_Travis_ NewMixedOrder.pdf (accessed 10 October 2017).

Wightman FL and Kistler DJ (1997) Monaural Sound Localization Revisited. *The Journal of the Acoustical Society of America* 101(2): 1050–1063. Available from: https://doi.org/10.1121/1.418029.

Williams M (1991) Microphone Arrays for Natural Multiphony. In: *Audio Engineering Society Convention 91*. Available from: www.aes.org/e-lib/browse. cfm?elib=5559.

Williams M (2004) *Microphone Arrays for Stereo and Multichannel Sound Recordings*. Editrice Il Rostro. Milan, Italy.

# 11

## Popular Music Production in Laptop Studios

### Creative Workflows as Problem-Solving Within Ableton Live

## Thomas Brett

### INTRODUCTION

This article considers electronic music production techniques within Ableton Live, one of the most influential and widely used DAW (digital audio workstation) software programs of the past eighteen years. The chapter has three parts. Part one surveys key moments in the history of popular music production practices, provides an overview of DAWs and Live, and introduces the concepts of techno-musical systems, workflows, and a problem-solving perspective on creativity. Part two presents case studies on electronic dance music (EDM) producer-DJs who use Live. The first examines Diplo and Skrillex's collaboration on Justin Bieber's dance pop hit, "Where Are Ü Now?" as profiled in a *New York Times* video documentary. The second case study examines Kieran Hebden (aka Four Tet) remixing Michael Jackson's *Thriller* for the Beat This series on YouTube in which he creates an EDM track in ten minutes. I suggest that the workflows described and depicted in these examples are representative of how Live is used to compose through techniques for sampling, sound design, loop- and beat-making, and song arrangement. The workflows also show electronic music production as a recombinant craft whose elements musicians understand as modular and perpetually fungible. Part three turns to Dennis DeSantis's *Making Music: 74 Creative Strategies for Electronic Music Producers*, a pedagogical manual published by Ableton that maps techniques and aesthetics of digital music making. I show how some of its production predicaments and proposed solutions amplify the lessons of the Diplo-Skrillex and Hebden case studies.

### SOME KEY MOMENTS IN THE HISTORY OF POPULAR MUSIC PRODUCTION PRACTICES

The history of popular music production can be told as a series of stories about how technologies and ways of using them overlap with innovations in musical practice and style. For example, after the development of magnetic tape-based recording in Germany in the 1930s, the recording studio and its growing assemblage of equipment (e.g., microphones, mixing boards, effects, and other hardware) became the locus for producing music. In the early days of studio production, the goal was to faithfully capture live performances – whether it was Louis Armstrong's Hot Five playing "Stardust" in the 1930s or Elvis singing "That's All Right" in the 1950s. As technologies evolved and production incorporated part overdubbing, editing, and sound design, the performance-oriented conception of music gave way to *producing* music as a meticulously constructed artifact. From the intertwined histories of avant-garde and popular styles, some innovations stand out as foundational to the practices and sounds of contemporary electronic music production. In the 1940s, the French radio engineer Pierre Schaeffer, in his quest to create *musique concrète*, spliced together tape recordings to create "Chemin de fer", a composition built from train sounds. This piece showed the power of rhythmicized soundscapes about which Schaeffer later noted that "what the locked groove allows you to do is . . . make the instrument sound like another instrument" (Dilberto 1986). After multi-track tape recording arrived in the 1960s, producers began assembling music out of multiple tracks: George Martin used intricate tape collages on the Beatles' *St. Pepper's Lonely Club Hearts Band*, Phil Spector built his "wall of sound" through overdubbed parts, and Steve Reich looped multiple field recordings into hypnotic tape pieces. In the 1970s, Jamaican sound engineers deconstructed reggae songs using their mixing boards, highlighting drums and bass *riddims* and adding delay and echo effects to create "a layered yet decayed sound" (Sullivan 2014: 42), while disco and hip-hop DJs in New York used turntables to seamlessly mix together recombinant percussion breaks into a repeating groove flow (Katz 2012: 57). In 1979, Brian Eno summed up these production experiments by describing the recording studio as a "compositional tool" for producing music through an additive process by which to "paint" sounds onto the "canvas" of magnetic tape. Live performance had become just one of many components of record production. "As soon as something's on tape", Eno said, "it becomes a substance which is malleable and mutable and cuttable and reversible" (Eno 1979).

## MIDI and Early DAW Software

The composition processes of popular music production and the studio's configuration as a compositional tool expanded in the early 1980s with the development of MIDI (musical instrument digital interface) and digital music hardware. MIDI, an industry standard electronics communication protocol that allows synthesizers, drum machines, and sequencers to synchronize with one another, offered "an unprecedented degree of postperformance control" (Barry 2017: 158) and immediately changed pop's sound, as musicians could now incorporate programmed drum patterns and harmonic step sequences. This machine-generated sound defined new wave pop and what would soon become known as techno and house, as electronic dance music producers used MIDI-connected drum machines and synthesizers to build their tracks. By the mid-1980s, newly affordable digital samplers joined MIDI to suggest new production workflows. For example, hip-hop musicians used the E-mu Systems SP-12 and Akai MPC60 to create their breakbeats. Although electronic sampling practice extends back to Schaeffer's loops of train sounds, the capabilities of digital machines "made sonic material much more reusable, malleable and open to transformation" (Strachan 2017: 5). It was the sampler, more than any other digital technology, that spurred electronic music's pixelation into its myriad rhythmic and ambient styles.

Along with MIDI and digital hardware, the 1980s saw the emergence of first-generation computer music software programs, such as Hybrid Arts' ADAP Soundrack (for the Atari ST computer) and Mark of the Unicorn's Performer, which were based around the organization and display of MIDI (Brøvig-Hanssen and Danielsen 2016: 11). ADAP was the first commercial computer sampling software, offering a static graphical representation of an audio waveform – a "sound conceptually frozen in time" – that musicians today take for granted (Strachan 2017: 74). An enthusiastic review in 1986 described the software as "a rather clever way of getting the best musical use out of contemporary computer technology" by providing "an assortment of cut and paste-style editing tools which can rearrange passages of music or dialogue into any shape you can imagine" (Davies 1986). Within a few years, DAW software that integrated MIDI sequencing with audio recording had arrived. In 1989, Steinberg released Cubase for the Atari ST; in 1990, Mark of the Unicorn released Digital Performer; in 1991, Digidesign launched ProTools; and in 1992, Emagic created Logic. These programs were joined in the late 1990s by Image-Line's Fruity Loops (1997), Propellerhead's Reason (2000), and Ableton's Live (2001). Today, DAW software is a long-established and indispensable tool used by most electronic music producers.

Early DAW software looked visually similar, providing an arrangement page in which musical data are displayed on a linear, left-to-right time line, coupled with a mixing board display with track volume faders, effects sends, panning knobs, and other virtual controls. With their time lines, pop-up windows, database layouts, and drop-down menus (Strachan 2017: 77), as well as drag-drop, cut-paste, and numerous other audio-editing functions (Holmes 2008: 301), DAWs suggested new conceptions of music within "a malleable digital landscape" (Prior 2009: 87) in that their graphic representation of musical events configured musicians to think "of sound as an object rather than a stream" (Zagorski-Thomas 2017: 134). At the same time, DAWs encapsulated the history of music production practices by reproducing the functionality of the recording studio assemblage. Using a DAW, "MIDI composition, recording, editing and mixing could now be done in one environment" (Strachan 2017: 77), and at once, the electronic musician became a composer, performer, producer, and engineer (Moorefield 2010). By the 2000s, a generation of musicians "with little or no experience of the hardware studio [was] learning to make music in software environments" (Prior 2008: 924).

## Ableton Live

Addressing the needs of this new generation of multitasking music creators, Ableton Live was designed in 2001 by two German electronic musicians, Gerhard Behles and Robert Henke, who intended their software to seamlessly integrate composing, production, and live performance. Unlike Logic, Digital Performer, or Cubase, DAW software whose visual interfaces were modeled on the studio mixing console and other analogue hardware (Lagomarsino 2017), Live avoided skeuomorphic representations, conceived instead as a *sui generis* "technoscape" with a distinctive look and feel (Prior 2008: 923). Live was unique in two ways: it was the first software to automatically loop and beat match audio samples and MIDI sequences, and its Session View page enabled users to organize these musical data or *clips* into orderly vertical rows. In Live's minimalist design, the clip is the essential "unit of creativity" (Zagorski-Thomas 2017: 76) that can be looped individually as well as arranged into horizontal *scenes* of clips to be triggered collectively in beat-matched synchrony. As different as Live appeared, its functionality distilled key moments in electronic music's history outlined earlier; using the software, it's not difficult to create *musique concrète*–style sound collages, canonic loops, dub remixes, or hip-hop breakbeats. It's no surprise then, that Live has earned a loyal following among musicians working in both experimental and four-on-the-floor dance idioms.

Perhaps more than any other DAW software from the past two decades, Live changed how producers compose and helped shape the process and sound of electronic music.[1] In an era when EDM has developed through many sub-styles and producers build their tracks through sequences of scenes and clips, Live epitomizes a non-linear, modular, and recombinant musical aesthetic. Its sampling and beat-matching/time-stretching capabilities and Session View clip layout (which is also integrated into Ableton's hardware controller, Push) encourage musicians to experiment with audio and MIDI as interchangeable blocks of sound and build their tracks by juxtaposing brief repeating units. Live facilitates playing with music's recombinant potential, enabling musicians to turn a microsecond of audio into an ongoing groove or even, as Schaeffer predicted, a new instrument. Live's protean applications are reinforced by Ableton's online creative ecosystem, which includes video tutorials and interviews with musicians about their production techniques and Loop, an annual conference on music, creativity, and technology. Ableton's ecosystem suggests that the company views and markets its software as an essential component of the modern composer's creative toolkit.

## Techno-Musical Systems, Workflows, and Creative Strategies

In electronic music production, each producer's toolkit revolves around a techno-musical system with which they create. This system often consists of a connected collection of hardware and software technologies (Butler 2014: 93), such as synthesizers, a MIDI controller, effects devices, and DAW software. Borrowing a perspective from ecological psychology, a techno-musical system could be said to embody a set of *affordances* (Gibson 1979) which favor, without determining "particular kinds of creative processes" by which to use it (Prior 2009: 83). The musician navigates the affordances of his or her system using a *workflow*, a sequence of practical-technical moves.[2] Though each producer has different priorities pursued in different ways, workflows commonly include techniques for designing sounds, manipulating samples, and making beats. As I will show in my case studies, observing workflows is an effective way to understand creativity by mapping how electronic music producers make music.

Recent research on creativity in electronic music production and the musicology of record production (e.g., Strachan 2017; Zagorski-Thomas 2017) is increasingly attuned to the phenomenology of sounds as heard (Clifton 1983: 14), as well as the techniques producers use to create them. To some extent, technologies configure creativity. For example, Robert Strachan explores how DAWs favor particular production practices, identifying actions that have arisen in the context of sound visually represented as "distinct, temporally located blocks of musical information" (Strachan 2017: 99). These actions include drawing, cutting and pasting, looping, zooming, and undoing (ibid.: 99). Strachan connects these techniques for manipulating sound using a DAW with the development of "distinct aesthetic trajectories within electronic music" (ibid.: 103) which he elaborates upon through his notions of non-performance-based musicianship, aural thinking, and experimentation (ibid.: 122, 125). Building on this insightful work, I suggest a *problem-solving* perspective on electronic music making. Understanding workflows as forms of problem-solving shows musicians tinkering via trial and error to generate small mistakes that are rich in information, positioning them for discovering "something rather significant" (Taleb 2014: 236). Moreover, a problem-solving perspective shows musicians systematically mobilizing electronic music's enchanting aura, thereby dispelling some of the mystery of composing "in the box" of a computer and its DAW software. Evocatively referring to the digital musician as an alchemist, Glen Bach observes how "the laptop shaman can access worlds heretofore unknown" using digital software tools (Bach 2003: 5). This idea of production *alchemy* – transforming music through a seemingly magical process – in turn recalls the work of Alfred Gell, who describes art-making as a technology of enchantment (Gell 1992: 43) and the artist as "half-technician and half-mystagogue" (Gell 1992: 59). Gell's vivid descriptions aptly evoke the realm of electronic music production. Just as magical thinking in art-making "formalizes and codifies the structural features of technical activity" (Gell 1988: 7–8), producers structure and organize their creative alchemy by using techno-musical systems to solve problems raised by their tracks-in-progress.

## USING ABLETON LIVE: TWO CASE STUDIES

Online music instruction videos are windows onto electronic music production, and YouTube in particular has become a vast public platform for "peer-to-peer" education, where musicians of varied skill and accomplishment share videos of themselves at work (Miller 2012: 184). On YouTube, we see producers interacting with their techno-musical systems through problem-solving tinkering with sound, demonstrating the workflows they use to create tracks. Hundreds of videos feature Ableton Live, including tutorials produced by Ableton as well as others by amateur producers in their home studios. These videos often teach topics such as how to build a beat, how to recreate a specific sound (Brett 2015), or how to improvise using the software. This section considers two case studies to illustrate some production techniques within Ableton Live.

### Diplo, Skrillex, and Justin Bieber's "Where Are Ü Now?"

"Where Are Ü Now?" is a collaboration between EDM producers Diplo and Skrillex, with vocals from Canadian pop singer Justin Bieber. Released in 2105, the song reached the Billboard top 10 and earned a Grammy Award for Best Dance Recording. The catchy, hyper-synthetic dance pop is representative of the increasing collaboration between prominent DJs and pop vocalists – for example David Guetta and Sia's "Titanium", Coldplay and Avicii's "A Sky Full of Stars", and Swedish House Mafia and Usher's "Numb". These collaborations illustrate the deepening influence of electronic music techniques on the sound of popular music. "Where Are Ü Now?" was the subject of a *New York Times* video documentary produced by critic Jon Pareles, who interviewed Diplo, Skrillex, and Bieber. In the video, the musicians explain and show how they used Ableton Live to construct the song.[3]

"Where Are Ü Now?" is built upon a piano and an a cappella voice demo that was emailed to Diplo.[4] Diplo's first production step was creating a loop out of the piano chords and experimenting with Bieber's vocal track. "The first thing we made with the song was just me just literally playing with the vocals. . . . We took a lot of his vocals and did different things to it".[5] One of these things was creating vocal harmonies from Bieber's a cappella. Turning to the vocal audio file in Ableton Live's Session View, Diplo plays a clip of a lower bass harmony: "We added natural [sic] harmonies because we didn't have anything from [Bieber]. So we created our own harmonies". Diplo and Skrillex explain their rationale for their vocal experimentation. "Anyone can copy any synth now – you can find the synth sounds, you can find presets", says Diplo. "But if you manipulate vocals [to create a new sound] it's something really original". Skrillex, a former singer, elaborates on how "the voice and mouth and the whole skull structure is like . . . an organic synthesizer. You have your vocal cords that create a vibration that travels up through the rest of your face and hits the roof of your mouth and bounces off and depending how you open your mouth – those are just changing and shifting the frequencies".

Following this organic synthesizer idea, Skrillex, and Diplo degrade the sound of Bieber's voice by applying Live's audio effects (e.g., Dynamic Tube, Erosion, Overdrive, and Redux), which simulate the sound of low-fi analogue and digital filtering and distortion. "The thing about Ableton is that you can really almost destroy a sound so many times it loses quality", says Skrillex. "People are always trying to avoid digital distortion, but we always like using it". Diplo notes that he and Skrillex will destroy sounds in the course of their ongoing search for new timbres – "literally always trying to find something that you've never heard before". It was the process of experimenting with copying, re-pitching, and degrading sounds that led the producers to the song's signature "dolphin" melodic hook, whose sound evokes a violin mixed with a flute. Tinkering with a single Bieber phrase ("need you the most"), Diplo explains that the vocal melody is "pitched way up, distorted, bounced, re-bounced again so it sounds worse, almost" to create a half-human, half-machine timbre. Skrillex adds that they "took a little pattern and created [sic] it into a whole different sound, but it still has the elements of some human thing – like a warmth in the track". He compares sound designing in Live to a painter mixing pigments: "So much of the process until the eleventh hour is literally just getting colors. . . . You mix a little of this color and that color and you create a new shade that maybe you haven't seen before".

The Diplo-Skrillex documentary offers several insights onto creative workflows within Live. First, we see visual evidence of the producers having created numerous clips in Live's Session View page in order to experiment with variations on the song's parts and overall arrangement. In one demonstration (4:32–5:00), Skrillex plays through several song section variations that were not included in the final version. This is an example of how the producers consider "Where Are Ü Now?" to be what Skrillex terms "a work in process" – a phrase that remixes the conventional work-in-progress idea of production. Skrillex sums up the sound design tinkering he and Diplo used to assemble their song: "It was like processing stuff and finding the right sounds and then minimizing that at the end". A second insight of the video is that it shows how the producers' experimentation is aimed toward creating unusual sounds that can propel the music. Bieber's vocal track leads them to derive multiple harmonies, create stuttering rhythmic effects (which are not explained in the video), and recognize the potential of the "dolphin" timbre. "It was always the goal", says Diplo, "finding that one little thing which was the dolphin singing, the dolphin sound". A third insight of the video is how it reveals techniques of the producers' workflow. In a section where Diplo praises Skrillex as his favorite drum programmer (3:24–3:32), we see Skrillex tapping the letter keys on his laptop to trigger the song's log drum percussion sounds. Skrillex explains his drumming workflow: "when I do MIDI, I usually draw it in" rather than finger drum patterns on drum pads or a keyboard. A final insight resides in what is left unsaid: "Where Are Ü Now" was composed without the producers ever working with Bieber face to face in a studio. Constructed out of an emailed audio a capella, the song is a simulacrum of a performance.

## Kieran Hebdan's *Thriller* Remix

Among the more interesting electronic music instructional videos on You-Tube are those in which producers are given ten minutes to create a new track. In timandbarrytv's *Beat This* and FACT TV's *Against the Clock* series, for example, we see electronic music producers working in their home studios with DAWs, synthesizers, drum machines, and turntables. Ableton Live is often the centerpiece of these videos that demonstrate producers' workflows under a time constraint that accelerates their thinking as they create a track. By showing the transformation of one sound into another, these production workflows distill the creative process into its essential parts. One of the most compelling videos in the "Beat This" series features Kieran Hebdan (aka Four Tet), an electronic dance music veteran who emerged in the 1990s with a *folktronica* style that combines "fragile acoustic fragments, brutal beats and glitchy electronica" (Inglis 2003). The most striking aspect of Hebden's video, in which he remixes Michael Jackson's *Thriller* using samples from the original recording, is the simplicity of his techno-musical system: a PC laptop running Ableton Live, and a turntable.[6] Like Skrillex programming his beats, Hebden does all of his production work inside Live, without using a keyboard or any other MIDI controller.[7] This minimalist setup allows us to focus on the steps and decisions of Hebden's workflow, as we watch him finding and sampling sounds, creating sample instruments, editing audio, and arranging parts. The process guides his problem-solving as he shapes a piece in ten minutes.

The video begins with Hebden playing an LP of *Thriller*, casually moving his turntable needle along the songs on Side 1 and quickly recording a few seconds from each into Sound Forge, an audio editing program. At this early stage, Hebden does not fuss over the specifics of his sample choices; the goal is rather to grab sonic elements he can work with. Next, he imports his audio samples into Ableton Live's Drum Rack and Simpler, one of Live's sampling instruments. By loading samples into a Drum Rack and Simpler, Hebden creates numerous potential starting points for his remix's component parts. After this preliminary sound organization, he isolates a kick drum hit from "Billy Jean", trims the sample's length by adjusting Live's loop brace, and boosts its low-end frequencies using Live's equalization effect, EQ 8. With the tempo set around 130 bpm, Hebden mouse-click draws four MIDI notes into a Live clip to create a four-on-the-floor kick drum pattern to trigger the "Billy Jean" kick sound from the Drum Rack. On a second track, he follows a similar procedure for the hi hat and snare drum parts, drawing MIDI notes in off-beat patterns to trigger the percussion sounds. A snare hit from "Billy Jean", with its pitch transposed higher, becomes the hi hat, while the original snare sound on every fourth kick drum hit provides halftime feel backbeats. Within three and a half minutes, Hebden has a beat going.

Over this percussion bed Hebden creates an indeterminate-sounding track using a sample from Jackson's "Human Nature" with flanged electric guitar and layers of vocals. He chooses a half-second slice of the section, lowers its pitch, and adds side-chain compression (to create a "pump-ing" sound that ducks around the kick drum's hits) and EQ effects. With the effects on, Hebden continues nudging the sample loop brace left and right, in search of an even more compelling bit of audio (4:58–5:24), then mutes the other parts for a moment so he can better listen to his choice of re-pitched and side-chained guitar and vocal parts. Fourth and fifth tracks are generated in a way similar to the first two percussion tracks, with Hebden drawing MIDI notes in a syncopated rhythm pattern to trigger the gong sounds from the opening of "Beat It". With four minutes left, Hebden switches from Live's Session View to Arrangement View, roughly organizing his remix while continuing to tinker with its sounds – boosting the mid-range of the "Human Nature" sample and the kick drum track's low-end using EQ 8 and transposing a few notes of the syncopated "Beat It" rhythm one octave higher, to create variation. In the final two min-utes, Hebden shapes the remix's arrangement by positioning its parts in staggered entrances and creating DJ-style breakdown sections by muting regions so that the music unfolds in four-bar sections. When the time is up, the videographer asks Hebden if he will work on it further. "No, it's done", Hebden says, jokingly adding that he'll play the piece "in the club later this weekend". As compelling as Hebden's hard-edged dance track sounds, it was only an exercise in creative workflow against the clock.

Since Hebden's remix video was posted in January 2014, it has gathered over 400,000 views and hundreds of mostly enthusiastic comments. You-Tube viewers are impressed by the simplicity of Hebden's techno-musical system, noting how "he doesn't even touch the keyboard for shortcuts. Clicking for days" (Blair Maclennan) and that production is "not about the equipment it's about what you can do with that" (Lautaro Ciancio). Some knowledgeable viewers point out the sophistication of Hebden's remixing workflow, noticing that the "video is liked not because of the finished beat but because the sampling technique is unique and pretty genius. In fact it's one of the best techniques I've ever seen" (beatz04). Another viewer, Jer-emy Ellis, an electronic musician known for his advanced finger-drumming skills, praises Hebden's remix by jokingly measuring it against EDM pro-duction clichés: "Good god! He didn't squash the shit out of the mix, add any noise sweeps, and there are no snare drums rolls indicating when the moment of prime enjoyment should begin. It obviously sucks!" In a fol-low-up comment, Ellis clarifies what makes Hebden's workflow unique: "I think that this video is pure genius, especially in the concept of work-flow. Starting with the various midi notes already in place and then search-ing for bits of audio to fill them. . . . It's just so unusual and wonderful. Definitely not something I'd thought of before seeing this and I'm pretty decent at chopping up bits of music and replaying them".

Hebden remixing *Thriller* offers several lessons. First, it shows that when producers have a general idea of their workflow ahead of time (based on previous work), even a few minutes is sufficient to build a new track. Second, a limited set of sample material can provide a plethora of options: Hebden spends minimal time choosing samples from *Thriller* because he trusts that whatever he records will offer sufficient remix grist when run through his workflow mill. Third, the video shows Hebden incorporating creative constraints, specifically the technique of drawing MIDI note rhythms and then assigning his samples to them. Limiting creative options in this way helps Hebden swiftly generate new parts in the style of four-on-the-floor EDM. Finally, the video shows the simplicity of working "in the box" – using Ableton Live as one's entire techno-musical system. Hebden's workflow demonstrates the virtues of less is more by strategically using just a few of Live's virtual instruments and effects: a Drum Rack, a few Simpler instruments, compression, and EQ.[8]

## AMPLIFYING ELECTRONIC MUSIC PRODUCTION WORKFLOW PROBLEM-SOLVING: DENNIS DESANTIS'S *MAKING MUSIC*

The videos of Diplo-Skrillex and Hebden are illustrative of how established electronic music producers assemble tracks through a process of skilled problem-solving tinkering. Each in its own way, the videos are accomplished performances, as Diplo-Skrillex share (and perhaps retrospectively dramatize) elements of their methods and sections from their finished song, and Hebden coolly improvises his workflow under the scrutiny of a time constraint and video close-ups. Yet the producers' display of ease contrasts with the difficulties many amateur electronic musicians may encounter when trying to compose. These difficulties are in part due to the open-ended nature of DAWs and other music software tools: sound-wise, almost anything is possible, and therefore the question of where to begin and where to go with one's musical project can be intimidatingly wide open. While Diplo-Skrillex and Hebden have their reliable techniques for building tracks, many musicians typically face creative conundrums as they choose from the many workflows options that are possible when using DAW-based, techno-musical systems.

Creative conundrums are the subject of Dennis DeSantis's *Making Music: 74 Creative Strategies for Electronic Music Producers* (2015), an insightful book published by Ableton that maps the techniques and aesthetics of digital music making, specifically the applications of DAWs for composing through problem-solving. By interrogating the nature of DAW composing, DeSantis's "relentlessly pragmatic" text is relevant to thinking through the workflows of any style of electronic music making (Richards 2015). Through its enumeration of production moves and creative stances, the book models rigorous compositional thinking by offering and then analyzing examples arising from what DeSantis calls problems of beginning, progressing, and finishing a piece of music. In this way, *Making Music* explains how electronic music producers might apply creative strategies for systematically thinking through the many permutations latent in their materials.[9]

It is illuminating to read the Diplo-Skrillex and Hebden case studies through chapters from DeSantis's book to amplify some of their shared problem-solving strategies. A few chapters are particularly relevant. First, "Arbitrary Constraints" (DeSantis 2015: 42–45) considers the problem of DAWs offering so many options for shaping sound. One solution to this problem is to derive all of one's sounds from a single source to create "a 'narrow frame' of possibilities, and then act entirely within this frame" (43). This narrow frame strategy is used by Diplo-Skrillex to create new sounds from Bieber's voice, including additional vocal harmonies and the song's "dolphin" hook, while Hebden works under the constraints of a ten-minute time limit and using only sounds from Side A of *Thriller*. Second, DeSantis's chapters "Goal-less Exploration" (78–81) and "Creating Variations" (130–145) suggest exploring and developing musical material, either in a "kind of free, non-directed way" or by duplicating ideas and making "one meaningful change to each of the copies" (79, 131). These exploratory strategies lead Diplo-Skrillex to tinker with Bieber's voice by re-pitching it, re-sampling it, and adding effects to alter its timbre, while Hebden tries out different loop brace locations along his *Thriller* samples, auditioning which locations sound most compelling. Finally, in the chapters "Arranging as a Subtractive Process" (300–303), "Layering as Form" (316–319), and "Maximal Density" (282–284), DeSantis describes techniques of layering and removing musical parts "using a small amount of material subjected to ever-changing levels of textural density" to maintain musical interest (318). We see these textural strategies in play when Diplo-Skrillex share alternate iterations of their song, each of which has a different combination of beats, vocal variations, harmonies, and effects processing. Similarly, near the end of Hebden's video, he roughly shapes an arrangement by using subtractive methods to mute parts so that sounds enter the mix one at a time, creating drama by pacing the rate at which the mix builds toward its maximal state. In these ways, DeSantis's problem-solving approach identifies some of the most creatively effective electronic music production workflows.

## CONCLUSION: CREATIVITY AS TRANSFORMATION IN ELECTRONIC MUSIC PRODUCTION

This chapter has surveyed key moments in the history of electronic music production, considered two case studies on production techniques within Ableton Live, and examined these workflows through a pedagogical text about creative strategies for electronic musicians. The workflows described and depicted in the Diplo-Skrillex and Hebden examples are representative of how producers use Ableton Live to compose through sampling, sound design, loop- and beat-making, and song arrangement. I have offered a view of musical creativity as problem-solving tinkering, showing how producers use Live to achieve unforeseen musical ends. In my case studies, the creative process is most clearly articulated in the *transformations* undergone by sounds – those brief moments where Bieber's voice becomes an organic synthesizer timbre or when the opening of Michael Jackson's "Beat It" shape-shifts into a pulsating sound cloud. Both "Where Are Ü Now?" and the *Thriller* remix demonstrate how producers use Live to spark starting points, to inspire them to build their tracks from slivers of musical material. In this regard, the Diplo-Skrillex and Hebden workflows show the power of a recombinant view of music. By affording "different kinds of phenomenality, creativity and play" (Prior 2009: 95), Live frames music as a fluid process, its sounds always on the cusp of their next potential transformation. Skrillex captured the matter best by describing electronic music production as a work in *process* rather than a work in *progress*.

Finally, my case studies show how deeply Ableton Live's functionalities are integrated into musicians' production workflows insofar as their aural thinking and approaches to sonic experimentalism are to some extent a response to Live's considerable sound-shaping powers. As recent scholarship suggests, DAWs do more than *organize* action: "They are in themselves, actors within the process of creativity, helping to shape and direct particular ways of working, thinking and doing" (Strachan 2017: 104). The visual, conceptual, and sonic affordances of Live encourage workflows for manipulating sound as malleable material, while the software's clips and scenes layout invites thinking about music as a modular structure. Yet in making possible the techniques used by Diplo-Skrillex and Hebdan, Live's encapsulation of the recording studio assemblage also raises a question: Would the producers be creating their musics in the same way, with the same sounds, were they not using this software? In sum, insofar as music production can be approached analytically and is somewhat contingent upon the configuration and constraints of one's techno-musical system, this article has tried to show that *creativity happens in action* – in the moment-to-moment details of problem-solving as musicians choose this sound over that one by tinkering with timbres, beats, and form. Ableton Live's design and capabilities usher its users across music production's enchanted terrains by opening workflow possibilities whereby even the most unrelated sounds find ways to get along.

## NOTES

1. To cite just one index of Live's influence: DJ Shadow, a long-time user of AKAI MPC samplers (which he used to make his hip-hop instrumental, *Introducing*, in 1996), switched to Live in 2013, noting that the software "is the most intuitive music-making program I've used since the MPC" (Davis 2016).
2. Workflow can also be framed in terms of the "usage trajectories" (Tjora 2009) musicians construct within the implicit "scripts" (Akrich 1992) of their techno-musical systems. For example, Aksel Tjora applies Madeline Akrich's scripts work to show how users of the Roland MC-303 Groovebox "move between various usage modes or scripts, constructing their personal user trajectories" (Tjora 2009: 175). This usage can include idling, extra-musical experiences: "waiting-enthusiastically, playing-around-with-it, being-disappointed,

finding-advice-on-the-Internet, narrowing-to-one-advanced-use, and so on" (ibid.:163).

3. "Bieber, Diplo and Skrillex Make a Hit" (https://youtu.be/1mY5FNRh0h4).

4. The earliest iteration of the song, written by Bieber and Jason Boyd (aka Poo Bear), dates to 2009. See Hooten (2015).

5. At 3:08 in the video, a screenshot of Live shows a vocal audio effect on one track labeled "Treble Color" and on another track, Live's Ping Pong Delay effect.

6. "Beat This Episode 19 FOUR TET" (https://youtube/TUDsVxBtVlg). On Twitter in October 2017, Hebden shared a photo of his simple home studio along with a caption: "This is where I recorded and mixed the album and all the gear I used". The photo and caption quickly became a meme that inspired numerous humorous responses (Wilson 2017).

7. However, Hebden changes his musical system regularly. In a 2013 video for Red Bull Music Academy (Hebden 2013), he explains how he uses Ableton Live in conjunction with other hardware.

8. Hebden's less-is-more workflow may connect to his rationale for not knowing thoroughly the functionality of Ableton Live. As he tells students in a master-class, "I don't know and I don't want to know everything Ableton does because I don't want to sound like everyone else" (Hebden 2013).

9. The book recalls the self-help aspect of *Oblique Strategies*, a deck of cards created by Brian Eno and Peter Schmidt in 1975 intended to help musicians think differently about their work by approaching it in terms of problem-solving. One of the most famous Oblique Strategies is "Honor thy error as a hidden intention".

## REFERENCES

Akrich, M. (1992), "The De-Scription of Technical Objects," in W. Bijker and J. Law (eds.) *Shaping Technology – Building Society*, Cambridge: MA, MIT Press, pp. 205–224.

Bach, G. (2003), "The Extra-Digital Axis Mundi: Myth, Magic and Metaphor in Laptop Music," *Contemporary Music Review*, vol. 22, no. 4, pp. 3–9.

Barry, R. (2017), *The Music of the Future*, London: Repeater Books.

Brett, T. (2015), "Autechre and Electronic Music Fandom: Performing Knowledge Online Through Techno-Geek Discourses," *Popular Music and Society*, vol. 38, no. 1, pp. 7–24.

Brøvig-Hanssen, R. and Danielsen, A. (2016), *Digital Signatures: The Impact of Digitization on Popular Music Sound*, Cambridge, MA: MIT Press.

Butler, M. (2014), *Playing With Something That Runs: Technology, Improvisation, and Composition in DJ and Laptop Performance*, New York: Oxford University Press.

Clifton, T. (1983), *Music as Heard: A Study in Applied Phenomenology*, New Haven, CT: Yale University Press.

Davies, R. (1986), "Hybrid Arts ADAP: Computer Sampling System." Available from: www.muzines.co.uk/articles/hybrid-arts-adap/973.

Davis, J. (2016), "DJ Shadow on Key Influences Over the Years and the Gear He Just Can't Do Without." Available from: www.musicradar.com/news/tech/dj-shad ow-on-key-influences-over-the years-and-the-gear-he-just-cant-do-without-643390.

DeSantis, D. (2015), *Making Music: 74 Creative Strategies for Electronic Music Producers*, Berlin: Ableton AG.

Dilberto, J. (1986), "Pierre Schaeffer & Pierre Henry: Pioneers in Sampling." Available from: www.emusician.com/artists/1333/pierre-schaeffer--pierre/henry-pioneers-in-sampling/35127.

Eno, B. (1979), "The Studio as Compositional Tool." Available from: music.hyperreal.org/artists/brian_eno/interviews/downbeat79.htm.

Gell, A. (1992), "The Technology of Enchantment and the Enchantment of Technology," in J. Coote and A. Shelton (eds.) *Anthropology, Art and Aesthetics*, Oxford: Clarendon, pp. 40–66.

———. (1988), "Technology and Magic," *Anthropology Today*, vol. 4, no. 2, pp. 6–9.

Gibson, J. (1979), *The Ecological Approach to Visual Perception*, Boston, MA: Houghton Mifflin.

Hebden, K. (2013), "Studio Science: Four Set on His Live Set." Available from: https://youtu.be/9klvnLBF7vU.

Holmes, T. (2008), *Electronic and Experimental Music: Technology, Music, and Culture*, New York: Routledge.

Hooten, C. (2015), "Justin Bieber Wrote 'Where Are Ü Now' When He Was 15 and Here's the Video to Prove It." Available from: www.independent.co.uk/arts-entertainment/music/news/justin-bieber-wrote-where-are-u-now-when-h.com.come-was-15-and-here-s-the-video-to-prove-it-10506035.htm l.com.

Inglis, S. (2003), "Four Tet." Available from: www.soundonsound.com/people/four-tet.com.

Katz, M. (2012), *Groove Music: The Art and Culture of the Hip Hop DJ*. New York: Oxford University Press.

Lagomarsino, J. (2017), "Why Are There So Many Knobs in Garageband?" Available from: https://theoutline.com/post/2157/why-are-there-so-many-knobs-in-garage-band?

Miller, K. (2012), *Playing Along: Digital Games, YouTube, and Virtual performance*, New York: Oxford University Press.

Moorefield, V. (2010), *The Producer as Composer: Shaping the Sounds of Popular Music*, Cambridge, MA: MIT Press.

Prior, N. (2009), "Software Sequencers and Cyborg Singers: Popular Music in the Digital Hypermodern," *New Formations*, vol. 66, pp. 81–99.

———. (2008), "OK COMPUTER: Mobility, Software and the Laptop Musician," *Information, Communication & Society*, vol. 11, no. 7, pp. 912–932.

Richards, S. (2015), "Hit or Miss: Can You Learn Musical Originality From a Book?" Available from: www.google.com/amp/s/amp.theguardian.com/music/2015/apr/15/hexadic-cards-and-creative/strategies.

Strachan, R. (2017), *Sonic Technologies: Popular Music, Digital Culture and the Creative Process*, New York: Bloomsbury Academic.

Sullivan, P. (2014), *Remixology: Tracing the Dub Diaspora*, London: Reaktion Books.

Taleb, N. (2014), *Antifragile: Things That Gain from Disorder*, New York: Random House.

Tjora, A. (2009), "The Groove in the Box: A Technologically Mediated Inspiration in Electronic Dance Music," *Popular Music*, vol. 28, no. 2, pp. 161–177.

Wilson, S. (2017), "Four Tet's Studio Just Inspired the Month's Best Twitter Meme." Available from: www.google.com/amp/www.factmag.com/2017/10/27/.com four-tet-studio-meme/amp/.

Zagorski-Thomas, S. (2017), *The Musicology of Record Production*, Cambridge: Cambridge University Press.

# 12

# Real-Time, Remote, Interactive Recording Sessions

## Music Production Without Boundaries

Zack Moir, Paul Ferguson, and
Gareth Dylan Smith

## INTRODUCTION

Inhabitants of the 21st century live and work in a world in which many
aspects of daily interactions with other humans are no longer hampered
by physical distance. Indeed, scholars and practitioners in fields such as
music and communications are finding that there is much to be gained,
both epistemologically and experientially, through collaborating across
(often) vast distances. Relatively recent developments in Internet technology have "led to the advancement of new collaborative cultures which use
the network as a medium for exchanging creative materials in an electronic
form" (Renaud et al., 2007). The tremendous potential in such collaborative endeavors has been explored in depth by social scientists such as
Richard Sennett (2012) and by management researcher Peter Gloor (2006,
2017), whose groundbreaking work on collaborative innovation networks
(COINs) illuminates an exciting and dynamic mode of activity. Smith and
Gillett (2015) describe COINs in the domain of DIY scenes punk scene.

People in today's technologically advanced societies frequently engage
in almost instantaneous communication with friends and colleagues across
physical distance, often in remote locations, in the form of email, SMS,
social media, and various messaging apps. We have free-to-use software
(such as Skype and Google Hangouts) that allows us to conduct meetings
with people anywhere in the world. We employ cloud-based word processor and spreadsheet tools (such as Google Docs) and online file-sharing
repositories (such as Dropbox and Google Drive) to facilitate real-time
workflows that allow us to collaborative and interact productively with
people anywhere on the planet, provided they have an Internet connection. However convenient these systems may be for users, issues impact
the functionality of some of the technologies that enable us to work in
this way.[1] Specifically, we have to wait for files to upload or documents
to update, for example; or, in the case of video calls, we deal with imperfections in audio or video signals (or both). In many everyday scenarios,
such problems with the usability of these technologies rarely prove catastrophic because users do not typically engage in activities that require

actual real-time interaction. That is to say that a short delay in file transfer is unlikely significantly to hinder the collaborative editing of a text document, and a slight lag in a video call is probably not going to render a conversation incomprehensible or futile. However, when playing music over the Internet, such glitches can prove disastrous.

Recent studies (e.g., Ferguson, 2013, 2015) have shown that, given sufficient Internet connectivity and specific hardware resources, musicians and producers working in different parts of the world are able to collaborate and engage in recording sessions remotely and in real time. Pignato and Begany (2015: 121) borrow from Jordan's (2008) work to help explain how, by extending experience through technology, people's interactions have become increasingly "deterritorialized", which was certainly the case for the research team in the projects discussed in this chapter.[2] In keeping with this, it is also worth invoking here the notion of multilocality (Rodman, 1992), since the participants were clearly in plural, respective locations. Pignato and Begany (2015: 121) note how "multilocated experiences are decentered experiences", and we describe, in the following, how each participant musician experienced de-centering with a simultaneous feeling of conjoining in the newly curated "virtual" space.

Through low-latency networks and emerging cloud-based workflows (described in greater detail later) that enable musicians/producers to collaborate on the same project/session, real-time, remote music production will soon be a functional reality – and not just the outcome of an elaborate and perhaps esoteric research project – regardless of physical distance. From a commercial perspective, this is very attractive (particularly in the current economic eco-conscious climate), as it allows for projects to happen without the financial, temporal, logistical, and environmental implications of musicians having to travel or transport equipment to a single physical studio in a specific geographic location. The authors are currently working on a project of this nature in which the goal is to produce an album of original music using low-latency networks and cloud-based collaborations – to date, this project has reached two milestones. First, in November 2016, the team engaged in a highly successful real-time, interactive recording session in which musicians at Edinburgh Napier University played with a drummer at the Royal College of Music in London (separated by a distance of approximately 415 miles). Second, in July 2017, we achieved what we believe to be the world's first real-time transatlantic recording session, in which musicians at Edinburgh Napier University collaborated and recorded in real-time with a remote guitar player at Berklee College of Music, Boston, Massachusetts, USA.

Despite many developments in the use of network technologies in this emerging area, more work needs to be done to investigate the relationships between technical (i.e., hardware and software) and musical (i.e., musicians' experiences and communication) aspects when working in this way if it is going to be taken up more widely. This chapter is an exploration of the authors' experiences in this area and the ways in which our workflow influences our perspectives on music production. We continue by presenting a brief historical overview of attempts at networked performance and the current state of development in this area, before examining potential technical issues that may affect the quality and usability of the recorded material. Finally a discussion is presented, drawing on our experiences of engaging in the project and the ways in which these were impacted by various technical and interpersonal issues (including latency, quality/reliability of visuospatial information, musician/producer communication, and social interaction). We discuss how these affected the experiences of the musicians and producers participating and suggest some potential implications for meaningful, creative collaboration using the technologies and methods illustrated herein.

## HISTORICAL AND TECHNICAL CONTEXT

Every day, millions of people stream audio and video over the Internet, using services like Spotify and YouTube (Cayari, 2016; IFPI, 2017; Partti, 2012; Waldron, 2012, 2013). While "real-time transmission of audio data over the Internet has become relatively commonplace" (Cooperstock & Spackman, 2001), Internet streaming services are ". . . invariably characterized by relatively low-bandwidth, low quality, stereo sound" (Xu et al., 2000: 627) due to the somewhat limited bandwidth of the domestic and cellular networks used for such activity (Pignato & Begany, 2015). For most people who stream media recreationally, this type of transfer is sufficient and the reduction in audio fidelity (if even noticed at all) is an acceptable trade-off for the convenience afforded the user. However, for the growing number of audio engineers and musicians working in real-time remote interactive music performance, the rapid transfer of high-fidelity, multi-channel audio and video is imperative and cannot be achieved via standard networks.

The last 20 years or so have borne witness to the beginnings of several developments in multi-channel, high-fidelity audio streaming, and Bouillot and Cooperstock note that as of 2009, "current high-fidelity audio streaming activity covers a wide range of interactive applications between remote performers" (2009: 135).[3] While it is not the purpose of this chapter to give a detailed history of endeavors in this area, the following two pioneering examples illustrate developments in multi-channel transfer of audio. In 1999, researchers in the USA (Xu et al., 2000) conducted the first "concertcast" (Chafe et al., 2000: 2) in which multi-channel audio was streamed from McGill University, Montreal, Canada, to New York University, New York City, USA (a distance of approximately 370 miles), and they noted a delay of three seconds (due to compression processes, buffering, and so on). Cooperstock and Spackman (2001) report on a similar process in which an ensemble of musicians performed at McGill University, and 12 channels of uncompressed audio were sent to mix engineers at the University of Southern California, Los Angeles (around 2,800 miles). It should be noted that delays in transfer did not impact the success of these collaborations as delays are "not perceived when listening to or recording a concert provided the application is entirely unidirectional" (Chafe et al., 2000: 2); this is an important consideration for the context of the research discussed in this chapter.

## KNOWN ISSUES

Gabrielli and Squartini (2016) provide a comprehensive list of technical issues that impinge on the success of network music performance, citing the specific examples of "dropout" (21), "latency" (23), "audio blocking delay" (25), "algorithmic delay of audio compression" (27), "network delay" (29), "delay jitter" (32), "clock synchronization" (36), and "audio and video synchronization" (38). While several of these present significant obstacles for technicians and network engineers, the focus of the current chapter is on networked music production, so an analysis of some of these issues is beyond the scope of the chapter. Additionally, due to the technologies and workflows that we employed in the project discussed in this chapter, some of these issues were overcome or at least rendered less problematic for our purposes.

For the purposes of real-time, remote interactive music performance, two principal issues impact the performance of meaningfully recording high-quality audio. These are (1) dropout, which is the term used to describe the loss of audio packets (essentially small chunks of audio data) during transmission, and (2) latency, which can be described as the time lag caused by transmission of data in a system. Perceptually, dropout affects real-time performance and music production, as it introduces glitches in the audio signal. In the authors' experiences, these glitches have only a minor impact on our ability to perform interactively. However, they are catastrophic when trying to record streams of audio transmitted over networks, as the recorded files are effectively rendered useless for professional production purposes. Latency is a far more serious problem for interactive remote music performance. In computer networks, latency can be defined as the time taken between a source sending information and a host receiving it, thus we can conceive of latency as a delay or lag caused by the interaction of physical (i.e., hardware) and software issues. Literature suggests, as anyone with experience in this area might expect, that latency is one of the fundamental challenges for attempting to engage in real-time, interactive music making over a network.

As Chafe et al. (2004) note, the communicative power of ensemble performance "rests on exquisite temporal precision" (1), and while this may be true, such precision is not always easily achieved due to factors beyond the control of performers. Clearly, when musicians perform together, they depend on their ability to hear the sounds produced by themselves and others. As "musicians cannot be in the same exact location at the same time" (Bartlette et al., 2006: 49), whenever musicians perform together in the same physical space, there will always be some delay in the sound of one instrument reaching the ears of the other player(s). For example, musicians positioned on opposite sides of a stage experience a delay in hearing the sound of each other's instruments. The speed at which sound travels is approximately one foot per millisecond, so, for instance, "a double bass player listening to the triangle player on the other side of an orchestra will encounter . . . latency of around 80 to 100 ms" (Bartlette et al., 2006: 49). Bartlette et al. (ibid.) note that such latency can impact on (at least) three areas of musical interaction: (1) "pacing", relating to the tempo of a performance, (2) "regularity", which refers to the timing *within* parts, and (3) "coordination", which pertains to the timing *between* parts.

In live performance contexts, the issues just noted can be significant, but musicians develop strategies (often unconciously) for coping with the ways in which physical space (and separation of performers in the space) impacts interactive performance. Networked music making and performance is also subject to latency introduced by (a) the physical distance that the network covers and (b) the acquisition and conversion of audio and video at the transmitting end and subsequent decoding and output at the receiving end. As such, we must also consider aspects of the ways in which data are transferred. If we are engaging in the unidirectional transfer of data, such as the live streaming of a concert from one location to another, then, as noted, ensuing delays (regardless of length) are unproblematic. Under discussion in the current chapter, though, is *interactive* performance among remote musicians connected by a wide area network (WAN). When considering the lag in the transfer of audio between remote performers, network latency must also be taken into account, and owing to the need for musicians to be able to hear (and also, potentially, to see) and respond to *each other*, the cumulative "round trip" latency of the system becomes a crucial factor.

Delays, or lags, that can be attributed to latency in the context of real-time, interactive network performance can have a significant effect on the temporal accuracy of remote performers. For example, there is some agreement that there is a linear relationship between tempo decrease and delay-time increase (Chafe et al., 2004; Farner et al., 2009; Gurevich et al., 2004). Chafe et al. (2004: 1) state categorically that "rhythmic accuracy deteriorates as delay increases to a point beyond which performing together becomes impossible". Some researchers report that at very low latencies (below approximately 11 ms), tempo even increases for musicians connected via an audio-only network (Chafe et al., 2010; Chafe & Gurevich, 2004; Farner et al., 2009). A number of studies have investigated the effects of latency on temporal accuracy with the specific intention of trying to determine "ensemble performance thresholds", and there is some consensus that 25 ms is the threshold for latency, beyond which ensemble playing deteriorates, and "the groove-building-process cannot be realized by musicians anymore" (Carôt & Werner, 2007).[4, 5]

The studies mentioned here focus primarily on tasks in which the objective was for participants to maintain steady tempi, and that this is, arguably, not necessarily a particularly useful indication of the musical success of (a) remote interactions or (b) any recorded products of such sessions. As Gabrielli and Squartini (2016) note:

> Close to no reference is found in the literature regarding perceptual aspects related to the success of interaction, related to tempo, or the importance of latency for loose synchronized pieces, such as those found in the contemporary acoustic and electroacoustic classical music repertoire.
>
> (Gabrielli & Squartini, 2016: 24)

## BREAKING THE NON-REALTIME AUDIO/VIDEO STREAMING BARRIER

The LOLA (LOw LAtency audio video streaming) development project is a collaboration between Conservatorio Tartini and *Gruppo per l'Armonizzazione delle Reti della Ricerca* (GARR).[6] GARR runs the Italian National Research and Education Network (NREN), providing high-speed links between education and research establishments. The aim of the collaboration was to reduce the 200ms – 500ms latency incurred by video conferencing systems to the point that real-time performance became possible. The result is LOLA: highly optimized software coupled with fast acquisition cameras and monitors, resulting in a best-case end-to-end latency of less than 20ms. This order-of-magnitude improvement comes at a cost: LOLA makes the fundamental assumption that the collaborating sites are connected via a high-performance network[7] that exhibits low-latency and low-jitter (small variations in the time taken for packets of data to traverse the network) performance. By making this assumption, the LOLA design team were able to remove the large data buffers and the message repeat/error recovery mechanisms that are essential in conventional video conferencing systems running over the public Internet. GARR understands that in practice, no network is perfect and that some packets will be delayed and even lost, resulting in audible clicks in the audio together with visible video artifacts. Our strategies for dealing with this, are discussed below.

The first test of LOLA in the UK occurred in September 2012 between Edinburgh Napier University and JANET at the Arts and Humanities Streaming Workshop.[8] Since then, academics at Edinburgh Napier University (led by Paul Ferguson) have been conducting tests and demonstrating the extent to which networked audio and video can be successful in "bringing remote performers together". The following are examples of three projects (Ferguson, 2015). First is a real-time remote recording and mixing between Edinburgh and Prague (approximately 833 miles). The performance took place in Edinburgh and was recorded in a studio in the Academy of Performing Arts, Prague. The connection was via the JANET, GEANT, and CESNET NRENS, which gave an each-way network latency of 16ms. Second is a near real-time bagpipes and drums performance between

Edinburgh and Chicago (about 3,700 miles) in which LOLA was used to transmit a drum core in 5.1 surround sound. This experience shows that distances greater than 3,000 kilometers (corresponding to latencies exceeding the commonly agreed 30–50ms maximum) may be possible. Third is an ensemble masterclass between Edinburgh, Trieste (Italy), and London which was the first UK test of LOLA by a professional ensemble and composer, James Macmillan. Emphasis was placed on creating a more immersive audio and video representation of remote horn players in Trieste and London than had been possible without using LOLA.

While these projects were successful and instructive in developing understanding of the networking technologies described, we as a research team have become particularly interested in developing ways of working that allow remote, real-time, *interactive* music production. In the case of the examples noted here, some of the issues introduced by the inclusion of the networked aspect of the work are effectively nullified because of the context. For instance, in example 1, latency was not a problematic issue because the transfer was unidirectional. In examples 2 and 3, issues related to signal quality, such as audio clicks and dropouts were not catastrophic, as the rehearsal/workshop situations did not require pristine audio. However, in a professional recording environment, in which remote musicians are performing interactively, it is *imperative* that network issues such as latency and dropout do not adversely affect the performance of the musicians or degrade the quality of recorded audio.

## REAL-TIME INTERACTIVE RECORDING STRATEGIES

To date, two strategies have been tested by the research team to mitigate potential problems and allow us to record. The first was explored in a test between Edinburgh Napier University and the Royal College of Music in London. In Edinburgh, Paul Ferguson ran the session, which involved two musicians (Zack Moir playing saxophone and Ewan Gibson playing bass guitar) performing interactively with a remote drummer (Gareth Dylan Smith) located in a studio in London, over 400 hundred miles away. The studio engineers in Edinburgh and London each ran separate ProTools sessions to record pristine audio in their respective locations.

In order to reduce the network bandwidth required for this session, only the minimum required data were transferred via LOLA. The engineer at the Royal College of Music in London sent a stereo mix of the drums (despite making a full and comprehensive multi-microphone recording), plus talk-back, and the video feed. Similarly, in Edinburgh, only one channel of audio for each instrument, one channel of talk-back for each musician, and the video feed were sent. We experienced round-trip latency of only 11 ms, which did not hinder the musicians at all. There were a number of audio artifacts that were concerning to the engineers from a network and quality-control perspective, but these did not mar the recordings, as they only existed in the audio that was transferred via LOLA, not the audio that was recorded in each individual location. This meant that after the session, we had pristine audio in a ProTools session in Edinburgh, containing the saxophone and bass tracks, and a separate multi-channel recording of the drum tracks in London, leaving us with the straightforward (albeit frustrating and time-consuming) task of manually combining the sessions and re-aligning audio to compensate for the lag between sessions and the difference in start-points of the recordings due to the recording being activated manually by two engineers with no remote synchronization.

The second strategy developed by the research team is more innovative and, at the time of writing, is believed to be a "world first". The team postulated that LOLA's real-time audio and video capability could be combined with a cloud-based project session sharing technology called AVID Cloud Collaboration to address limitations in both, i.e., LOLA's audio robustness and AVID's lack of real-time collaboration. AVID Cloud Collaboration appeared in ProTools version 12.5 (announced in March 2016) and deviates from a conventional ProTools session by creating a shareable "project" on a remote AVID Server (with local caching) instead of placing the media and session document on local storage. The originator of the cloud-based project is then able to invite other ProTools users to collaborate and share the same media and time line. It is significant to note that AVID Cloud Collaboration is not real-time and no audio flows directly between the sites. Instead, any audio recorded is cached locally and simultaneously uploaded to the cloud. This uploaded audio is then downloaded by the remote ProTools system and only appears on the remote time line when recording stops. The team proposed that LOLA could be used as a low-latency "headphone cue" between the remote musicians to allow them to play together in real-time, albeit with occasional clicks in the audio coming from the remote site due to variations in the performance of the network. These errors do not affect the final product, since glitch-free recorded audio is automatically transferred in non-real-time through the AVID Cloud Collaboration mechanism.

The recording session took place in real time between Edinburgh Napier University and Berklee College of Music in Boston, USA. Again, this featured musicians in Edinburgh playing interactively in real time with a guitarist (Joe Bennett) situated over 3,000 miles away in Boston. The studio engineer in Edinburgh created a cloud-based ProTools project, utilizing AVID's Cloud Collaboration functionality, and invited the engineers in Berklee to join this project (see Figure 12.1). The drums were tracked in advance and loaded into the cloud-based project to function, essentially, as "backing tracks". This recording session proceeded in much the same way as the first, but any locally recorded "pristine" audio was automatically uploaded to the cloud and shared between the ProTools systems in Edinburgh and Boston, meaning that the engineers and musicians could see and, more importantly, hear the project being constructed in (almost) real time.[9] Although the audio was available to both teams and there was no need to combine two sessions, some repositioning of audio was required.

For the purposes of this chapter, we focus on two areas framing the various issues pertaining to this project that have been particularly interesting to the research team and instructive in the development of our workflow

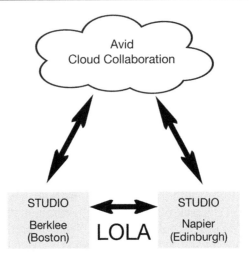

**Figure 12.1** LOLA with Avid Cloud Collaboration workflow diagram

and approach to networked music performance. While the technology and the innovative workflow that we employed enabled us to perform interactively and in real time, in the ways described earlier, it would be disingenuous to pretend that this was "just like any other session", for a number of key reasons, which are discussed in detail here. We focus on (a) the impact of technology and workflow (i.e., the way in which we used technology and how it impacted on our practice and ability to collaborate in this way) on our approach to music production and (b) the degree of immersivity and the impact on user experience and musical performance.

## THE IMPACT OF TECHNOLOGY AND WORKFLOW ON MUSIC PRODUCTION

The novelty and excitement of working in this innovative way meant that each member of the team was keenly aware of the fact that we were making music with remote musicians. From the perspective of the performers, working with technology as a vital mediator for the success of the music production practice and experience changed nothing fundamental about the approaches we took to collaborative music making. Many musicians with experience working in studios have become accustomed, during recording sessions, to being spatially separated (i.e., in different booths or rooms) for example in order to isolate signals and reduce sound-spill. Nonetheless, the experience was qualitatively different from that of traditional recording sessions in which the presence of a super-fast Internet connection is not central to the experience of interacting.

With regard to the immediate impact of technology and our workflow, a number of pressing issues needed to be handled in order for the session to even take place. These were most keenly felt by Paul Ferguson, the team's engineer/producer and networked-audio expert. He not only had to run the session as a recording engineer but also had to take responsibility for additional software implications introduced by LOLA and the associated network considerations. Working in this way fundamentally reframed the role of engineer/producer and meant that focus needed to be split between sonic/aesthetic issues and somewhat more mundane issues of monitoring and maintaining network connections. This basic (but important) finding would certainly suggest that situations like the sessions described in this chapter require an assistant engineer to deal purely with networking issues introduced.

It also became very clear that there is a need for a greater degree of synchronicity between studios when working in the manner described, in particular with regard to the importance of both studios starting their respective recordings at the same time and thus eradicating the need to realign audio after tracking to compensate for discrepancies in start times. It could be argued that this is a comparatively trivial issue, given the extraordinary benefits of working in this way. Nonetheless, since this repositioning of audio needs to happen in the session as soon as a take is complete before the team can even hear what was just recorded, it is not entirely quotidian, as it could profoundly impact the creative process and the ways in which musicians collaborate. While we found the types of issues introduced by the networked nature of the endeavor non-problematic for performance or recording (providing that both are well managed by the engineer), the issue of having to manually reconstruct sessions so musicians can hear playback needs to be overcome in order for the technology to become truly invisible and the process to feel as natural as possible for all involved. It is anticipated that a modified form of AVID Satellite Link could solve this problem; Satellite Link allows a master/'slave' relationship to be established between ProTools systems, with the master system controlling playback position and transport start/stop. The discovery mechanism used by Satellite Link is designed to function within a single network subnet[10] and therefore could not be used between Edinburgh and Boston in its current form.

## IMMERSIVITY AND INTERACTION

We can now confidently state, based on the experience of these highly successful tests, that when harnessed in the way described here, this technology and workflow can provide a reliable and effective way in which to engage in real-time interactive networked music making. They connect musicians in a virtual space in which the technology is practically invisible, so they can meaningfully improvise and respond to musical ideas and gestures, despite being in different locations and separated by vast physical distances. Regardless of how invisible the technology is from the musicians' point of view, this space is still "virtual" and, thus, activity of this nature involves remote musicians performing together in a way that is likely to be and to feel somewhat unnatural for most people. With this notion of virtuality in mind, we are using the term *immersivity* here to refer to the perceived degree of immersion or the extent to which the user (performing musicians, in this case) experiences the virtual space as "if it were real – producing a sense of presence, or 'being there' " (Bowman & McMahan, 2007: 36).

Stone-Davis (2011: 162) acknowledges that "the performance of music is an embodied event", one in which "the physical body becomes one with the instrument and with the sound produced". This is self-evident to experienced musicians and one of the fundamental reasons that people make music collaboratively in real time (Smith, 2017). This embodied sensibility was apparent in this online project; however, although the "sound produced" by each of our instruments was tangible and audible in our respective rooms, the "sound produced" by the band was somewhat disembodied. Again, in the context of contemporary recording studio practice, it is not uncommon for this to be the case. However, in such scenarios, musicians can (and frequently do) leave their booths to talk face to face with the other musicians or engineers. Clearly, this is not possible as remote musicians, situated hundreds or even thousands of miles apart, effectively meaning that the only channel of communication is through the same equipment used for recording, i.e., microphones and headphones. One psychological effect of working in this way is that, from time to time, the musicians become more aware of the headphones they are wearing and the fact that they are not only used for monitoring and playback but are vital in providing an aural and social connection to the virtual space that the musicians inhabit for the duration of the session. Gareth Dylan Smith, who played drums remotely in London (while the other musicians were in Edinburgh), for the first session, felt this powerfully and noticed that when he removed the "cans", he realized in his body and in the physical space of the recording studio, that he was physically and technologically disembodied from the sound produced by him, by the other individual musicians and by the band collaboratively.

Through our work in this area, it has become clear that one of the key factors we need to consider is the quality (or the qualities) of the immersive experience. We have experimented with the presentation of the video signal including displaying images on large-screen televisions, projecting directly onto the studio glass and projecting onto free-standing frosted glass panels to create a hologram effect. We have also experimented with the way in which the audio signal is presented including headphones, presentation in free-field via loudspeakers, and a combination of both. However, more research is needed in this area, and the next stage of our research will see us investigating the effect of immersivity as a crucial aspect of this way of working. Consideration of such issues is, we believe, not simply an exercise in the introspective exploration of the way we "feel" in this virtual space. Rather the degree of immersivity and the interaction and communication it facilitates are fundamental to our ability to produce music in this way, and it is a fundamentally important issue to consider for future endeavors in this area.

As we repeatedly performed (albeit with only ourselves and the engineers for an audience), we were reminded of Norman Denzin's observation, that performances "are the practices that allow for the construction of situated identities in specific sites" (2014: 31). This was at once very apparent, inasmuch as we were in recording studios in particular locations, but this project also problematized this notion, since we were keenly aware of realizing our identities, in and through the recordings, in a site that was somewhat less specific. This project has provided such a novel (and yet also very familiar) music-making experience that extant phenomenological and (meta-)physical explanations of real-time music making may need to be expanded in order to account for the ways that it feels and for both the qualitative and quantifiable extra space or "timespace" (Voegelin, 2010) that this musical project creates and inhabits. As Pignato and Begany (2015: 112) note, it is from collaborations such as this one that a paradigm shift can be seen to take place, as genuinely "new modes of music making emerge".

Pignato and Begany (2015) draw on Heidegger (1977), Nattiez (1990), and Aristotle (2014) to frame the notion of "creating with intentionality" (119) as taking place in poietic space. As pertains to the present study, they ask, "Where exactly is the poietic space in a distributed, multinational, distance collaboration?" They go on to write:

> It is our view that such collaborations at once stem from and encompass multiple poietic spaces. . . . These poietic spaces include the local contexts of each individual participant, the shared local contexts of co-located groups of participants, and macro collective poietic spaces, untethered to territory, that form in the virtual comings together. This third type of macro poietic space . . . exists because of and is dependent upon the technology mediating the collaborations.
>
> (Pignato & Begany, 2015: 120)

## CONSIDERATIONS FOR THE FUTURE

The authors strongly believe that real-time, remote, interactive recording sessions such as those described in this chapter are not simply a phenomenon that has occurred as a result of niche academic research projects. Rather, we see these as the early stages of developing systems and workflows leading to seeing this way of working become more commonplace or even desirable for musicians who wish to collaborate across large distances (especially) given the financial, organizational, and environmental implications of international travel. We do not wish to appear blazé about this remarkable situation, however, and we fully acknowledge that working in this way is far from the norm for a majority of readers. Sessions of this nature require equipment, expertise, and personnel that would not be part of traditional single-site recording sessions and thus are currently not practicable for everyone. Progress will require technological developments, primarily in the area of session synchronization, and refinements in the way in which we represent the virtual space. Given the capability of networks, the functionality of LOLA, and the workflow that we have adopted, we are confident that the traditional barriers to network performance no longer stand in the way of real-time, remote, interactive performance and recording.

## NOTES

1. Indeed, even a decade ago, such technology may have seemed unrealistic, and two decades ago, it may even have seemed more appropriate in the realm of science fiction.
2. However, as discussed in the following, there emerged a sense in which our praxis was not only *deterritorialized* but also *reterritorialized* in a new kind of space.
3. This has, at least in part, been facilitated by the development of high-speed research networks. As academics in the UK, we have access to the Janet Network (www.jisc.ac.uk/janet), which is a high-speed, high-bandwith, and very reliable network for the education and research community. Internet2 (www.internet2.edu/about-us/) provide similar connectivity for higher education institutions across the USA.
4. The experiences of this type of interaction and some of the "compromises" that the authors employed are discussed in detail in the following.
5. Kobayashi and Miyake (2003) found that musical interaction breaks down over 60ms.
6. LOLA Project Site. www.conservatorio.trieste.it/art/lola-project/old-lola-project-web-site/lola-low-latency-audio-visual-streaming-system.
7. To date, LOLA tests worldwide have connected using the NRENs in those respective countries.
8. This event was hosted by the Royal College of Music, in London.
9. File transfer time to and from the cloud is a necessary consideration when working in real-time.
10. It is common practice in large network installations to divide the network into smaller subnets, for example, music studios, post-production, newsroom.

## REFERENCES

Aristotle. 2014. The Poetics. In J. C. Douglas (ed.), *Aristotle Collection*. Translated by I. Bywater. Amazon Digital ed.
Bartlette, C., Hedlam, D., Bocko, M., and Velikic, G. 2006. Effect of Network Latency on Interactive Musical Performance. *Music Perception*, 24(1): 49–62.
Bouillot, N., and Cooperstock, J. R. (2009). Challenges and Performance of High-Fidelity Audio Streaming for Interactive Performances. *NIME*.
Bowman, D. A., and McMahan, R. P. (2007). Virtual Reality: How Much Immersion Is Enough. *Computer*, 40(7): 36–43.
Carôt, A., and Werner, C. (2007). Network Music Performance – Problems, Approaches and Perspectives. *Proceedings of the Music in the Global Village Conference, Budapest, Hungary, September 6–8.*

Cayari, C. 2016. Music Making on YouTube. In R. Mantie and G. D. Smith (eds.), *The Oxford Handbook of Music Making and Leisure*. New York: Oxford University Press, 467–488.

Chafe, C., Wilson, S., Leistikow, R., Chisholm, D., and Scavone, G. (2000). A Simplified Approach to High Quality Music and Sound Over IP. *Proceedings of the COST G-6 Conference on Digital Audio Effects (DAFX-00), Verona, Italy, December 7–9.*

Chafe, C., Gurevich, M., Leslie, G., and Tyan, S. (2004) Effect of Time Delay on Ensemble Accuracy. *Proceedings of the International Symposium on Musical Acoustics, Nara, Japan, March 31–April 3.*

Chafe C., and Gurevich, M. (2004) Network Time Delay and Ensemble Accuracy: Effects of Latency, Asymmetry. *117th Audio Engineering Society Convention*, Audio Engineering Society.

Chafe, C., Caceres, J. P., and Gurevich, M. (2010) Effect of Temporal Separation on Synchronization in Rhythmic Performance. *Perception*, 39(7): 982.

Cooperstock, J., and Spackman, S. (2001). The Recording Studio That Spanned a Continent. *Proceedings of the First International Conference on WEB Delivering of Music, Washington, DC, USA, IEEE Computer Society, November 21–23.*

Denzin, N. K. 2014. *Interpretive Autoethnography*, 3rd ed. Thousand Oaks, CA: SAGE.

Farner, S., Solvang, A., Sbo, A., and Svensson, P. U. (2009). Ensemble Hand-Clapping Experiments Under the Influence of Delay and Various Acoustic Environments. *Journal of the Audio Engineering Society*, 57(12): 1028–1041.

Ferguson, P. 2013. Using Low-Latency Net-Based Solutions to Extend the Audio and Video Capabilities of a Studio Complex. *134th Audio Engineering Society Convention*, Rome, Italy.

Ferguson, P. 2015. Real-Time Long-Distance Music Collaboration Using the Internet. In R. Hepworth-Sawyer, J. Hodgson, J. L. Paterson, and R. Toulson (eds.), *Innovation in Music II*. Shoreham-by-Sea, UK: Future Technology Press, 174–178.

Gabrielli, L., and Squartini, S. 2016. *Wireless Networked Music Performance*. Singapore: Springer.

Gloor, P. A. 2006. *Swarm Creativity: Competitive Advantage Through Collaborative Innovation Networks*. New York: Oxford University Press.

Gloor, P. A. 2017. *Swarm Leadership and the Collective Mind: Using Collaborative Innovation Networks to Build a Better Business*. New York: Emerald Publishing.

Gurevich, M., Chafe, C., Leslie, G., and Tyan, S. 2004. Simulation of Networked Ensemble Performance With Varying Time Delays: Characterization of Ensemble Accuracy. *Proceedings of the 2004 International Computer Music Conference, Miami, FL.*

Heidegger, M. 1977. *The Question of Technology*. New York: Harper.

IFPI. 2017. *Global Music Report 2017: Annual State of the Industry*. London: IFPI.

Jordan, B. 2008. Living a Distributed Life: Multilocality and Working at a Distance. *NAPA Bulletin*, 30: 28–55.

Kobayashi, Y., and Miyake, Y. 2003. Analysis of Network Ensemble Between Humans With Time Lag. *SICE 2003 Annual Conference*, August 2003, vol. 1, 1069–1074.

Nattiez, J. 1990. *Music and Discourse: Toward a Semiology of Music.* Translated by C. Abbate. Princeton, NJ: Princeton University Press.

Partti, H. 2012. *Learning From Cosmopolitan Digital Musicians: Identity, Musicianship, a and Changing Values in (In)formal Music Communities.* Helsinki: Sibelius Academy.

Pignato, J. M., and Begany, G. M. 2015. Deterritorialized, Multilocated and Distributed: Musical Space, Poietic Domains and Cognition in Distance Collaboration. *Journal of Music, Technology and Education* 8(2): 111–128.

Renaud, A. B., CarÔt, A., and Rebelo, P. 2007. Networked Music Performance: State of the Art. *AES 30th International Conference, Saariselkä, Finland, March 15–17.*

Rodman, M. C. 1992. Empowering Place: Multilocality and Multivocality. *American Anthropologist, New Series*, 94(3): 640–656.

Sennett, R. 2012. *Together: The Rituals, Pleasures and Politics of Cooperation.* New York: Penguin Books.

Smith, G. D. 2017. Embodied Experience of Rock Drumming. *Music and Practice*, 3. Available at: www.musicandpractice.org/volume-3/embodied-experience-rock-drumming/.

Smith, G. D., and Gillett, A. 2015. Creativities, Innovation and Networks in Garage Punk Rock: A Case Study of the Eruptörs. *Artivate: A Journal of Entrepreneurship in the Arts*, 4(1): 9–24.

Stone-Davis, F. 2011. *Musical Beauty: Negotiating the Boundary Between Subject and Object.* Eugene, OR: Cascade Books.

Voegelin, S. 2010. *Listening to Noise and Silence: Towards a Philosophy of Sound Art.* London: Continuum.

Waldron, J. 2012. Conceptual Frameworks, Theoretical Models and the Role of YouTube: Investigating Informal Music Learning and Teaching in Online Music Community. *Journal of Music, Technology and Education*, 4(2–3): 189–200.

Waldron, J. 2013. User-Generated Content, YouTube and Participatory Culture on the Web: Music Learning and Teaching in Two Contrasting Online Communities. *Music Education Research*, 15(3): 257–274.

Xu, A., Woszczyk, W., Settel, Z., Pennycook, B., Rowe, R., Galanter, P., Bary, J., Martin, G., Corey, J., and Cooperstock, J. 2000. Real Time Streaming of Multichannel Audio Data over Internet. *Journal of the Audio Engineering Society*, 48(7–8): 627–641.

# 13

# Tracking With Processing and Coloring as You Go

## Austin Moore

## INTRODUCTION

This chapter is a study of the use of signal processing during the tracking process. Microphone choice and placement have been well documented in recording and production literature, but the choices of microphone preamplifier, mixing console, dynamic range compressor, and equalizer have not been as thoroughly explored, specifically concerning their impact on sonic signatures (a notable exception being Howard Massey's *Behind the Glass* series and Great British Studios). Thus, the presented work fills a gap in knowledge by addressing how music producers use these elements of the recording chain with a particular focus on linear and non-linear distortion, a process I explore thoroughly in what follows.

Signal processing for tracking should not be confused with signal processing in mixing, although what's done in tracking affects decisions made during mixing. As Dylan Lauzon (2017: 115) argues, "microphone selection on some level is intrinsically a mix decision in that it's an irreversible sonic decision that profoundly affects how an element will behave in the final mix". Accordingly, the same argument can be made regarding the selection of other recording equipment. Preamplifiers, equalizers, compressors, and other signal processing devices will impart a sonic fingerprint on recorded audio that affects how it integrates into the final mix. Shaping audio to fit it into the mix is one reason producers use signal processing during tracking, of course, but there are other motives. Table 13.1 highlights the main reasons for, and against, the method. These motivations were drawn mainly from discussions arising during the interviews with producers I present later in this chapter.

## WHAT IS COLORATION?

As is evident in the interviews I present later in this chapter, it is common for producers to use processing to shape a track's "color", but what precisely constitutes "audio coloration"? In its broadest sense, coloration

**Table 13.1** Reasons for and Against Tracking With Processing

| Reasons to Track With Processing | Reasons Not to Track With Processing |
| --- | --- |
| To add character, texture, and color | Too risky / lack of experience |
| To correct a problem | No time to experiment with options / cost |
| To compensate for how tape changes the sound | Recording digitally so no need to compensate for how tape changes the sound |
| To build an aesthetic into the tracks to help the performer and make the mix easier | It's difficult to remove a bad decision at the mix |

can be thought of as any change to the sonic quality of an audio signal. Producers tend to consider coloration as the subtle, and sometimes not so subtle, changes in program material that manifest perceptually as variations in timbre. "Coloration" occurs when audio equipment alters features of the original program material, including (but not limited to) changes in the frequency response, dynamic envelope, and harmonic content, through the addition of harmonic components, a process which is more commonly called "distortion".

For clarity, it is important to note there are two forms of distortion, namely, (i) linear and (ii) non-linear. Linear distortion pertains to processes that, while affecting the time domain of an audio signal, do not add additional harmonic components (i.e., changes to the audio's frequency and phase response). Non-linear distortion, in contrast, occurs when a piece of equipment is stressed to work outside of its linear region, producing "clipping". All equipment has a maximum output voltage, and if an input signal exceeds this limit, the output is clipped. This transforms the signal in the time domain by removing any portions of audio which exceed the output limit, hence the name "clipping". Clipping occurs either as hard clipping, where the onset of distortion is abrupt, or as soft clipping, where the onset is gradual. In the frequency domain, the distortion process generates new harmonic components that were not present at input. Typically, these components are produced by harmonic distortion and, thus, are harmonically related to the input signal. Alternatively, they are created via intermodulation distortion (IMD), which creates sum and difference components that are not harmonically related to the input.

## EQUIPMENT AND COLORATION

Microphone preamplifiers (or mic pres) are an integral part of any tracking session. The mic pres job is to amplify low-level microphone signals as transparently as possible, but this has not stopped recording engineers from coaxing out coloration from these devices. Many professional producers will select mic pres because of their non-linear behavior and the coloration they impart when driven with gain. The sonic signature of a mic pre is significantly affected by the gain stages, the amplifier's class, the quality of its components, and its circuit topology. Neve mic pres, for example, obtain much of their character from the transformers used in the circuit, while API mic pres get much of their sonic identity from their API 2520 op-amps and transformers.

It should be added that perceptual differences between well-designed and maintained mic pres are small. Sam Inglis (2012a), writing for *Sound on Sound* magazine, conducted a comparison of mic pres by recording a Disklavier with three different microphones through six mic pres. He uploaded anonymized audio files for readers to rate and space to note their subjective impressions. The results revealed that the most expensive and revered mic pres were not rated any higher than inexpensive designs. The Mackie VLZ preamps, for example, were rated, on average, over three discrete tests, just as highly as a Neve 1073. Furthermore, the subjective descriptors one might expect to be ascribed to a particular mic pre, were not used. In a follow-up article, Inglis (2012b) posited that a possible reason that certain mic pres were not rated as one would expect was that the mic pres had not been driven as hard as some engineers might drive them in actual sessions. Another likely reason for the results Inglis (2012b) received is that his test only used a single track. It is a commonly held belief amongst audio engineers that tracking multiple sources through the same mic pre will add more color. This practice is called "stacking" by engineers. The theory states that a preamp's color will become more apparent as additional tracks are sent to it, the net result of coloration on many tracks being more audible than that heard on just a single track. To my knowledge, this has not been examined in a listening experiment, but the theory behind this claim has been disputed by Winer (2012: 58), who purports to have debunked the "stacking myth". Nonetheless, the use of different mic pres for coloration and non-linear processing is a standard working procedure for many producers and is explored in the interviews reproduced at the end of this chapter.

Mixing consoles were also designed to process audio material as transparently as possible, though engineers routinely use them to "color" tracks. An accepted process (budget allowing) is to select specific consoles for their sonic signatures, in fact. Producer Mike Poole said the following regarding his choice of mixing consoles:

> I auditioned a few consoles for mix down, including the big Neve at Ocean Way in Nashville, though that made the sound a little too warm and fuzzy. The recordings sounded great on the SSL at House of Blues: it had just the right amount of grit, along with clarity.
>
> (in Tingen 2010)

A typical working procedure is to use a colored console during tracking. A common choice for this work is a vintage Neve console for tracking and a less-colored console with sophisticated automation, such as an SSL, for mixing. Not all Neve or SSL consoles will sound the same, however, and many engineers claim a Neve needs to be of a particular vintage for it to generate the desired color. Similarly, consoles of the same model may not sound alike. As Kevin Mills, of Larrabee Studios, notes,

> [E]verybody knows that there are certain bad years for a car. For consoles, the same is definitely true. When you have a lot of SSLs, you find that there are certain years that, for whatever reason, just don't sound as good.
>
> (in Droney 2002)

As well as the console's intrinsic color, mix engineers may exploit its non-linear distortion by driving the channels and busses into clipping. When certain types of consoles begin to clip, they color the program material and affect the dynamic range of signal by gently shaping its envelope. Numerous professional rock and pop producers favor the sound of mixing consoles when signals are driven into clipping. Rock producer Spike Stent, for instance, is one such producer:

> You'll see red lights flashing everywhere when I'm overloading the EQ and things like that. The distortion and grit comes from the SSL compression and the clipping on the individual channels.
>
> (in Tingen 1999)

Dynamic range compression (DRC) is another, much-used process in all stages of music production which producers exploit to color signals. Traditionally, the process was implemented in tracking to control the dynamic range of program material and to minimize the potential for overloading recording devices. However, over time, DRC found use more as a creative tool and less as a preventative measure. In a professional recording environment, it is common for engineers to have access to several different DRC units, each with their own unique sonic signatures. Michael Brauer, in an interview with Paul Tingen (2008), notes that "eighty percent of my compressors are used strictly for tone". Similarly, Tom Lord Alge explains that "each one of my 40 compressors has a different sound, and I generally use them as an effect" (in Tingen 2000). Classic tracking compressors include the Teletronix LA2A (now the Universal Audio LA2A), the Fairchild 660/670 compressor, the Urei 1176 (now the Universal Audio 1176), and the dbx160. All of these compressors (and many others) have unique coloration and behavioral traits that affect their sonic signatures, and this in turn guides the decision-making process concerning their suitability for specific sound sources during tracking.[1]

Alongside DRC processors, equalizers (EQs) are also employed during tracking to sculpt the audio material's frequency content. As with compressors, there are many different types of EQ units available. An EQ's sonic signature is affected by its equalization curve, whether it features an active or passive design, its user interface, and how it distorts signal. During tracking, producers will work in broad strokes. Thus, an EQ with wide curves, gentle filters, and pleasing coloration is often a default choice. One of the most revered tracking EQ units is the Pultec, a passive design with flattering curves and coloration provided by the valves and transformers in its circuit design. Other popular choices include Neve and API EQs, all of which have transformers in their design and typically offer broad musical-sounding curves and a limited selection of frequencies.[2]

Recording to magnetic tape is well known to impart audio coloration during tracking. However, the producers who were interviewed for this chapter stated it was not a common recording medium, mainly because of the high cost of tape. Thus, tape recording is not discussed in the following interviews.

## THE INTERVIEWS

Professional producers were interviewed to garner their opinions on tracking with processing and, in particular, the issue of "color". The following discussion thus focuses on mic preamplifiers, compressors, mixing consoles, and equalizers, and these components can be thought of collectively as "the recording chain". Microphones are briefly considered, but they were not the focus of discussion. The producers who participated in the interviews are Ken Scott, Tony Platt, Phil Harding, Joe Kearns, and James Mottershead. The producers cover a wide age range and mainly work in rock and pop styles with Joe Kearns and Phil Harding having a particular background in pop genres. Like Devine and Hodgson (2017), the author includes these interviews because he believes that it is of great importance to include industry professionals in academic discourse. The responses from the producers are highly illuminating; there are many similarities in approach but also some largely expected, and edifying, variations of opinion present. The following is a transcript of the questions and a selection of responses from the producers.

## Talk Me Through a Typical Recording Chain for Vocals

*Joe Kearns:*   For pop stuff, it's usually a cleaner mic. I have moved away from the tube stuff. Normally an older u87 into a Neve pre and then into a Tube Tech CL1B Compressor. Still more than happy with an 1176 or an LA2A, but the tube tech has more detail and sounds fuller and cleaner. With pop vocals, I normally don't have much time to experiment, I maybe only have them for two or three hours, so I need to get the stuff quick and clean. I don't want any harmonic distortion or anything that will be a pain when it comes to mixing. That would probably be my go-to vocal chain. If you are working with a band, you might have more time; you might be tracking them for weeks. I might do something similar, perhaps with an SM7 or an RE20 as an option mic for something more aggressive sounding, but even then, I often go back to my go-to setup and add the processing later. Alternatively, I might use an 1176 or something that I can smash a bit harder. You can smash the Tubetech too, but it's a bit gentler and nicer. There's a bit more aggression with the 1176. Occasionally with the rock stuff, I will use a Distressor because you can really squash them and add in some of the distortion, but it's a bit risky doing that so I might split the signal and do a live parallel compression.

*James Mottershead:* Vocals will always get compressed, and EQ'd a bit on the way in. Ideally the mic pre will be an old Neve, 1081 or a 1073, the usual suspects. Occasionally, a Focusrite or an Avalon, mic and singer depending. The compressor is generally an 1176, which I prefer for tracking. Everything sounds good through an 1176, revision F ideally. I think there has only been one or two singers where I didn't go for an 1176 when I was tracking their vocals, in that case I'd go for a Distressor. The EQ is a Pultec; they sound great on vocals. That would be my "go to" starting point.

*Phil Harding:* Nowadays, it could be as simple as a mic pre straight into the recorder. I never used to believe in that, but it's something I have grown into. I'll do the coloration on monitoring within Protools or Logic. That would be a compressor setup, then an EQ, some basic reverb and delay, and I would have them set up before the vocalist arrives. During the analogue days, the big difference is I would have been adding the EQ and compression on the way to tape. For me, as a long-term engineer starting back in the '70s, there's a difference. It took me a while getting used to not using compression on the way in because as an analogue engineer, you would rely on the compression and limiting to not drive you into tape saturation, you're always looking to get good level onto tape to get a good signal-to-noise ratio.

*Ken Scott:* I would set up several mics and see which works out best, whatever vocal mic ends up being best for the artist, the one that sounds the best, that's the one I would use. It would go into whatever ever board I am using and into Protools or whatever. I would have a limiter in there and depending on what I was after, it might be a Fairchild because it has its own unique sound and that's the only limiter that gives that sound. Otherwise, it's generally a Universal Audio (1176) or Teletronix (LA2A), and I prefer to use hardware rather than software.

*Tony Platt:* I like to amass as many microphones as a I can find in the studio, and I set up the likely contenders and get them to sing a verse and a chorus through each of the microphones with no treatment on it whatsoever. In conjunction with the singer, I choose the one we feel is representing their voice the best of all and I'll use the same mic pre for this. Then with whatever the best microphone was, I will try out different mic pres and see which of them works out the best, and after that, I will try maybe three or four compressors on the chain, having chosen the mic pre, and we'll decide between us which one is giving the singer the right feeling in the headphones and the right feedback from their delivery. . . . All mic pres have different personalities; some are better at reproducing high frequencies, some are better at low frequencies, and some mic pres don't work well on high output microphones. There are certain mic pres that are better on female voices than male voices and vice versa, similarly, with compressors. I don't like using the 1176 on voices because it adds sibilance, especially on male voices, but the other day, I used one on a female voice, and it was absolutely superb.

## Talk Me Through a Typical Recording Chain for Bass Guitar

*Joe Kearns:*     I would: normally take a DI and one mic. For the mic I am actually going more for condensers lately, something like a U87 or the cardioid DPA. Sometimes with a small diaphragm condenser, you can get a nice crisp top end and the DI can fill in the gaps, the sub stuff. I have changed in that I used to think to record bass you needed a mic for bass like a D12, but I like the top end from the condensers. One of the things with bass is actually getting a smaller amp and operating it at a lower level. When you really turn up an amp, it can become a big distorted mess, and you find all the issues with the room when you do that. The best is the old port a flex Ampeg B15s. Back to the recording chain, I am probably going to say Neve pre for everything, but if there is a Neve pre available, I will use that and maybe a dbx160 because they are quite full sounding and you can tickle them. I wouldn't smash the bass too much on the way in. I might also use the 1176, but I am not looking to change the sound too much, just even it out.

*James Mottershead:*     I'll split the bass guitar up a bit when I am recording. There will be a dedicated chain to get the low end, which will be a DI, and that will go through a Neve and then a Vari Mu or Opto style compressor. I will then have a bass amp and that will have a large diaphragm condenser on it, such as a FET 47 or an AKG 414, that will go through a Neve pre, but if I want a bit more bite from it, I will go through an API.

*Ken Scott:*     Into whatever board and for compression, well, the Fairchild is for a very specific vocal sound; generally speaking, I wouldn't be using that, so it would be something like an LA2A.

*Tony Platt:*     The style of music influences the chain. Certainly, if I am going to record rock music, then I will lean towards Neve or that type of mic pre. I like it in studios where there is a whole host of different mic pres because I will choose the mic pre to suit the instrument that I am recording. The musical genre and the instrument I am recording is always going to have an influence on my choice of mic pres. As far as microphones are concerned, I tend to push myself away from using the same combinations all the time, but you end up having preferred places where you go. If I compress the bass, it's generally to enhance the low frequencies a little bit, make them punchier.

## Talk Me Through a Typical Recording Chain for Drum Rooms

*Joe Kearns:*  That's the thing I will properly crush. So a 1073 but if they are available, the old TG desks, which they have at British Grove and Abbey Road, and the compressors on those, I don't know anything that sounds better on room mics. When the meter is showing about -9 to -12 dB of gain reduction, they just sound amazing. I can't remember who it was told me why you compress room mics, but they said when you go out and stand in a room with a drummer, it sounds massive and amazing, and then you come back in and listen to your room mics and it sounds a bit nice. But this engineer believes there is a natural compression in the ear when things are too loud and maybe it distorts or it compresses naturally and that's why when you compress room mics, it sounds much more like what you would expect. If you can't get the TG desk, then a Neve into an 1176 because you can do the all buttons thing and destroy it.

*James Mottershead:*  In addition to a pair of mics that capture a stereo picture of the drums, I'll always have an excitement mic that could be anything from a 57 to whatever, something interesting, a bit trashy. That will go into a preamp I want to distort nicely. You're looking at an older Neve or a trident preamp; they are quite good for that. Then some abusive form of compressor like a Valley People Dynamite or the 1176 with all buttons in. That mic will work well in a larger drum mix where you need to turn it up to 11. If it's something with a more minimal drum mic approach, where the room mic is going to be providing a lot of the picture of the kit, the microphone choice will be wildly different, looking at something a bit more classic, a u47, u67, u87. That will go to a little more gentle. Preamp- and compressor-wise, it will be more broadband sounding. An API, a Neve, or a Trident is good, and the compressor will not be that aggressive sounding, so something like an LA2A, just taking a little bit off.

*Phil Harding:*  I will tend to immediately put those room mics into heavy compression to the point of distortion.

*Ken Scott:*  It depends on the room. With most rooms, it's not even worth it anyway. I would use whatever mic they have two of that I'm not already using and no limiter, just straight into the board.

*Tony Platt:*    You're always going to need a mic pre that has a good transient response and a lot of headroom, so APIs are particularly good under those circumstances. I wouldn't record that heavy compression, 1176 style. From time to time, I have put it on the monitor side so we are hearing something in the area it might end up, but with digital recordings, it's hard to hear how much you are going to require until you have the instrument in the mix.

## How Important to You Is It a Preamp Imparts Color?

*Joe Kearns:*    With modern pop, it's insanely bright, and if you add lots of plugins, you can find yourself down a hole if it's already recorded with coloration. That's the reason I don't color vocals on the way in with pop. It's a bit of an unromantic reason, but it's a logistical reason. There aren't that many second chances. But if I have an appropriate project, I will do it on anything that is rock or indy. The 1073 is a go to, but if it's a really poppy thing, then I might select something else because I know they add color, so I almost avoid mic pres when I know they will add color. Neves are particularly notorious for it. The mic pre coloration is very important, but I think I still find it to be a subtle rather than an overwhelming thing. I don't know if some of it is a placebo in my head, because on certain things, I think it needs to have a Neve on it, while on others, I think I want it clean, whereas in reality, if you don't work a Neve that hard, it's extremely clean.

*James Mottershead:*    The preamp color is very important, I think. If you are tracking drums, the color of an API preamp is brilliant, just because they are really fast and transient. The thing with preamps is to have uniform preamps for whatever you are doing. So, let's say I am tracking drums. I like going to a studio that has a quality console in there so I can use the preamps in the console for the drums, just so I have this nice color and it ties everything together. That's not to say the overheads might go to another preamp somewhere. I tracked some drums up in RAK studio 2, which has got the old API in there, and everything through that sounded brilliant. The room sounds really good as well, but everything working together through that console sounds great; the color is the same. The other thing as well is you can't go wrong recording anything with a Neve. I've done a load of drums recently up at Assault and Battery that have a Neve, and all the drums going through that sound really good because it just ties everything together.

| Ken Scott: | Not really, everyone is different, but you work with it. The important things in any studio as far as I am concerned are the monitors. If they are good, you can have the shittiest pre in the world, but you can still make it sound great as you know what you're listening to. |
|---|---|
| Phil Harding: | I am happy using a transparent mic pre and getting my coloration elsewhere, in other words mic choice, compressor, and EQ during the mix stage. I was quite happy with the mic pres on the Yamaha desks, which to me were quite characterless and transparent. |
| Tony Platt: | I would be choosing the mic pre because of the coloration it was likely to bring to the sound. I don't think anything is particularly transparent. |

## Do You Find the Color Changes When the Preamp Is Driven Harder?

| Joe Kearns: | The more you drive it to distortion, the grittier it sounds, but then again, some of that could be a placebo. Some of the ones that are favored for being colored at low levels are ridiculously clean, so it definitely seems to have more effect the more you push it. I remember Glyn Johns telling me he always runs the output on about -4 or -5 on a Neve and then pushes the input gain up so he is actually working the distortion on purpose. This is obviously a trick, but I never do it because it's too risky, but he thinks it's the only way you get that nice crisp drum sound. He's adding the harmonic distortion on purpose. It's quite interesting how it's really only an amplifier, but you can change the sound dramatically. |
|---|---|
| James Mottershead: | The coloration definitely changes. The more you drive the preamp, the more you distort it. It will start to compress it at a certain point. If you have an input and output on the pre, you can drive the input a little more. It's more obvious on some than others. If you go to the Neve VR consoles, it's a bit harder to tell with those; you can hear it compress a bit, but driving it isn't always a thing. The older design of the Neve 1083 or 1073 circuit, then the distortion comes into it a bit more and the distortion of those is really nice; it gives it this nice warm feeling. |
| Ken Scott: | Generally speaking, I don't want to distort. Just get a decent level and go in. |

## When Tracking, What Do You Use a Compressor For?

| Joe Kearns: | It's the presence it adds; I don't know how to describe it, but it's the "upfrontness". It can make a difference on |
|---|---|

how the vocal comes across. It makes you connect better with the vocal. I use the dbx for bass and drums a lot or Distressors.

*James Mottershead:*   When I am tracking, I try and use whatever is in front of me to put a color in. I use everything in front of me for color; it's just one big pallet. With compressors, you can use the release time to make things sound fuller, slow it down. A valve-based compressor can be used to find a sweet spot where it just about breaks up. . . . If you have a drummer who can hit properly, you don't need to be using the compressor for its normal function. If he can hit, his internal tone will be consistent, then you can go with a compressor instead of EQing it to get a bit more body of the drum. . . . If you have a distorted guitar, you don't need to compress it because it's already compressed, but if you really have an edgy, distorted, or fuzzy sound, an opto can help smooth it all together. It's like a tonal thing.

*Ken Scott:*   We always used the Fairchild on vocals; with Lennon, we were often hitting it so hard we were distorting it. That's what he liked. That's what I use the Fairchild for, that really compressed sound. . . . I can't describe the sound of it; it just sounds different from any other limiter.

*Tony Platt:*   In the days of analogue, I would use tape to do the compression, certainly on guitars. I have now found that sometimes I am using compressors a little bit more when I am recording, but I tend not to do that because I have so many more choices when I get to the mix. I use compressors when I am recording now, only if they are necessary. For a vocal, I use light compression to enable the singer to work the microphone in the way I would like them to.

## When Tracking, Do You Have a Preference for a Mixing Console?

*Joe Kearns:*   Now I would have loads of preamps and into Protools, but in a perfect world, I would have a console. It's more fun to track through. One of my favorites is the old API in RAK. It's got that glue thing. Plug a mic into a preamp in that console, and it sounds good. One of the reasons the EQs sound good on these consoles is because the curve is really broad. It's really easy. It's very musical. It's really smooth and nice.

   (As a sub-question, Joe was asked by the author how he would define "glue".)

   It's how the elements are squished together. Sometimes when recording, it's a battle, things sound separate, but some desks and tape can glue things together. There is something that makes it so much easier to do. It's something I can feel while I am working on the recording. If you are tracking and get your balance, everything feels very natural with little effort. There is something you get from tape and real consoles; it feels easier, but this could be to do with using the faders.

*James Mottershead:* If the choice is available, either an API or a Neve is ideal. The Neve 1081 or 1073 modules are always going to be fantastic as are the API consoles; even the new API's are really cool. If there is a console of that ilk, I will use it. Even if I am sat in front of an SSL, there will always be a rack of Neves or something similar. I use those for the key areas and then go to the SSL for their remainder. There is a really cool studio that has an old E series, the preamps sound really good; they kind of sound like how you'd like all SSLs to sound. But this is all just aesthetic. There's room for those clean, super broadband preamps the SSL does.

*Phil Harding:* If the budget was no object at the recording stage, I would go for the Neve, preferably a classic one over a current one, but either way, they are still great desks with great mic pres and EQ. My mixing preference, and this is partly because I mainly work in the pop genre, is the SSL, and that is partly because of the flexibility of the SSL in mixing and because of its sound and the years I spent on them. In the Strongroom studios, they had a Euphonic, and for pop, straight ahead boy band pop, I did grow to quite like the Euphonic because it was completely transparent. I would know if Ian and I had done all the coloration over in our production suite, in other words, we recorded everything onto tape and had used the Neve mic pres and compressors, we'd arrive at the Euphonic room knowing that the coloration and character was there. Then we were just going for a balance.

*Ken Scott:* I do have preferences, but it's purely mental rather than anything else. I've always hated Neve. In England, there were two major desks for a time, Trident A Range and Neve. I was in the Trident camp so I hated the Neve, and mentally, I have never got over that. I've worked on a whole lot of Neves over time, so I know it's just up here in my head. What you need to do is get the sound to the speakers and hear it. I worked at Trident studios, and whilst I was there, the management decided to build their own boards, and out of this came the Trident A Range. Malcom Toft was the guy building it, and he'd put something together, let's say the EQ, and bring it down to us, and we'd try them on sessions. We would then get back to him, and say we like this, but we don't like that, and he molded it, he molded the design and sound to how we liked it. I loved those boards; there were only thirteen of them made, and you'd go to each board and it sounds totally different. One of the things with that particular board was if you put it on a scope, it was absolutely shitty; you would never build a board that would give those readings. But it was one of the most musical boards I have ever worked on and that was because it wasn't technicians working on it, yes, there were, but they gave it to recording engineers to get their views on it.

*Tony Platt:*    If I was to have a perfect console, one-third of it would be Neve, one-third of it would be Rupert Neve, and there would be some API and Trident in there. They all have their place. If I were to be building a recording studio, I would be using some 500 racks and building it up like that.

## How Does Tracking with Processing Affect the Production Process?

*Joe Kearns:*    The more you do on the way in, the less you have to do later on. If you have tracked it well and added lots of color to the sound, then you find yourself doing a lot less later on; you don't need to do anymore production because you've kind of done it, but there's always a line to walk because if you do it too much, you can't really undo it. Decisions about which compressor and how much you compress it are in the same realm as deciding which guitar amp and pedal to use . . . I guess I want to commit more nowadays because I want to have less channels and keeping everything simple is my main motto as it were.

*James Mottershead:*    I am all for coloring as you go. You're building a picture as you're going. The coloration starts with the band stage, with the snare, the bass amp, the pedal, even what guitar you're using, even down to strings, that to me is where the coloration starts. Everything for me afterwards is just turbo charging that image. Your EQ and preamp choice is just as important as deciding between which Marshall. My old mentor said, mixing should just be a formality. A great mix starts with great multi-tracks. . . . If something is not right at the recording process, you should fix it there. If you are recording it really safely and allowing yourself an exit strategy, you will never have moments where the bands' ears prick up and they get inspired.

*Ken Scott:*    First and foremost, that's what I am used to doing because I started off on four tracks and we had to do that, I am 100 percent used to doing it. Number two, I like to keep the mix going, have it close to the mix all the way along the line as I can, for both the artist and also myself . . . the less you make decisions, the more you have to put down and it's harder to mix it in the end. Quite often, I don't even have to use EQ when I mix.

| | |
|---|---|
| *Phil Harding:* | You're making those early commitments, with confidence, and it helps your monitor balance process. You have got to remember that pre DAWs, you couldn't leave up your previous monitor balance. . . . With a budget with no limits, I would probably go back to processing during the recording stage, but with budgets being tight these days, my general view is its just quicker to save it to the mix. In the meantime, I just monitor what is required to give an indication to everyone how that mix is going to sound. |
| *Tony Platt:* | The mix starts for me during the pre-production period. By choosing the orchestration and the arrangement, you are creating a particular point in time and that mix continues the whole way through. When you are recording, you are gathering the components for the mix, you're actually quite a long way up the production process, and then you are able to enhance all that in the final mix. The overall picture is something I will have a conversation with the artist about, so we are creating the mix as we go. |

A final interview was conducted with Alessandro Boschi, who works in electronics and audio engineering. He also develops console libraries for the multi-effect plug-in Nebula, which makes use of Vectorial Volterra Kernels Technology to capture sophisticated impulse responses of hardware equipment. During the development process of his libraries, Alessandro makes numerous measurements, calibrates the equipment to a high standard, and conducts intensive critical listening. Consequently, he is familiar with the sonic signature of many classic mixing consoles. This interview aims to get a better understanding of the sonic signature of modern and vintage consoles from Neve, SSL, and API. The following is a transcript of the questions asked to Alessandro and his responses:

## How Would You Describe the Tone/Sonic Signature of a Neve 80 Series and a Neve 88RS?

The main characteristic of these consoles is that they soften the transient by giving a sense of roundness and improving the density of the mix with full-bodied lows and silky highs. These characteristics are more evident on the 80 Series. The Neve 80 Series was designed by Rupert Neve and built by Neve Electronics, while the Modern 88RS was designed by Robin Porter and built by AMS-Neve. In the 80 Series, the sound is airy, fuller, big, larger than life, colored, and punchy. Most of the sound is due to the Marinair transformers, which produce sub-harmonics, and some components, such as tantalum capacitors and the Motorola 2N3055 transistor working in Class A in the line amp section. The 88RS sounds clean and hi-fi, natural with some punch but less fat compared to the 80 Series. It has an exceptional sound quality, surround capability, and incredible routing and digitally controlled automation.

## How Would You Describe the Tone/Sonic Signature of a Vintage API Console and a Modern 1608?

API consoles all sound natural and well defined. The transients are clear and present, especially in the low end, which also has fantastic clarity. In the whole context, the mids are a little more forward. Both new and old consoles are quite similar in sound with a lot of character; it's clear, punchy, and tight with a solid bass. The vintage console is a bigger and warmer sound compared to the modern 1608. Both share the same technology and the differences are mainly in the modern electronic components, which tend to sound more thin and sterile.

## How Would You Describe the Tone/Sonic Signature of an SSL E and SSL G Desk?

The SSL sound is more electronic, but it retains analogue roundness with its characteristic in-your-face sound. The 4000E console has more crunch compared to the cleaner (but only a bit) 4000G console. Both are detailed, full, and punchy consoles. What I like more is the 242 EQ, the black one.

As can be seen, Alessandro notes there are some differences in the perceptual qualities of these consoles that vary between the brands, models, and ages of the components. Interested readers can listen to mixes run through a number of his console emulations at www.alessandroboschi.eu/html/alexb/consoles.htm.

## CONCLUSIONS

This chapter has elucidated the ideation producers follow when it comes to processing in tracking, and it explores how coloration is introduced to audio material. The interviews revealed that the producers used microphone preamplifiers, equalizers, and compressors to not only fix technical issues but also to impart specific colors and to simplify the mixing process. Additionally, it can be posited that tracking with processing allows the artist and producer to hear the aesthetic of the production as it is being developed, which in turn affects the technical and artistic decisions made by the artist and producer, potentially creating a more holistic production process.

Patterson (2017) notes that the roles of the producer and the mix engineer often overlap, and the notion of committing while recording is perhaps one such example of where the lines between the two are blurred. Moreover, Patterson (2017) argues that one disadvantage of the producer taking an active role in elements of the mixing process is that the "aesthetic becomes entirely constrained by the subjectivity, stylistic preferences and limitations of the individual" (79). While this is a valid concern, a counter-argument is that a producer, particularly working in rock and pop genres, who is merely putting a microphone in front of a source and recording it without error is not producing at all. This point is discussed in an interview between producers Warren Huart and Chris Lord Alge (2015), where Alge states,

> I wish the engineers that are making records now would focus more on making it sound like a record from the get go. Everyone's too safe . . . a pro goes, hey I've got a cool mic, I've some cool EQ, we're going to build something in, I'm going to create something.

Huart, expanding on Alge's statement, argues that a mixer engineer's job is to take what he or she (as producer), or another producer, has created during tracking and extend it "to the next level", not to work from a blank sonic canvas. Therefore, the almost infinite amount of possibilities which arise at mix down and the lower budgets and time constraints, have encouraged some producers to "play it safe" by not committing during tracking, allowing themselves an exit strategy later on in the production process. The danger with this conservative approach, however, is that it eliminates components of the production process that establish it as an artistic craft, rather than as a purely technical procedure.

While processing during tracking is an effective holistic production strategy to adopt, it should not be overlooked that there is a fine line between giving extreme consideration to the subtle differences in equipment and what Ken Scott calls "minutiae". Patterson (2017: 83) notes that "gear envy" is common with regard to analogue outboard equipment, and Nelson (2017: 122) observes that producers "can become side tracked by the latest piece of gear, plugin or update. They sometimes appear to be some holy carrot hung before us. It's so temping to endlessly research all the facts and figures, including the gear used on so many great records, and for me the result is I am buried in referential foundation". Consequently, gear obsession, coupled with a pernickety attention to detail, may result in judgments which are not made in the best interest of the production's aesthetic, or worse still, creative inertia.

## NOTES

1. See the previous work by the author for more detail on this area.
2. See George Massenberg's pioneering 1972 Audio Engineering Society (AES) paper for more information on EQ design and history.

## BIBLIOGRAPHY

Devine A and Hodgson J (2017) Mixing In/and Modern Electronic Music Production. In: Hepworth-Sawyer R and Hodgson J (eds), *Mixing Music, Perspectives on Music Production*, New York; London: Routledge, pp. 153–169.

Droney M (2002) The Agony and the Ecstasy of Choosing a Console. Available from: www.mixonline.com/news/facilities/agony-and-ecstasy-choosing-console/366640.

Inglis S (2012a) Pick a Preamp. Available from: www.soundonsound.com/re
    views/pick-preamp.
Inglis S (2012b) Preamp Post-Mortem. Available from: www.soundonsound.com/
    reviews/preamp-post-mortem.
Lauzon D (2017) Pre-Production in Mixing: Mixing in Pre-Production. In: Hepworth-
    Sawyer R and Hodgson J (eds), *Mixing Music, Perspectives on Music Produc-
    tion*, New York; London: Routledge, pp. 114–121.
Massenburg G (1972) Parametric Equalization. In: *Audio Engineering Society
    Convention 42*. Available from: www.aes.org/e-lib/browse.cfm?elib=16171.
Massey H (2000) *Behind the Glass: Top Record Producers Tell How They Craft
    the Hits*, San Francisco, Berkeley, CA, Milwaukee, WI: Miller Freeman Books;
    Distributed to the book trade in the U.S. and Canada by Publishers Group West;
    Distributed to the music trade in the U.S. and Canada by Hal Leonard Pub.
Massey H (2015) *The Great British Recording Studios*, Milwaukee, WI: Hal
    Leonard Books, an Imprint of Hal Leonard Corporation.
Moore, Austin (2016) *The motivation behind the use of Dynamic Range Compres-
    sion (DRC) In Music Production and an analysis of its sonic signatures*. In: Art
    Of Record Production 2016 Conference, Aalborg, Denmark, December 2–4
    2016, Aalborg University, Denmark. (Unpublished).
Moore A, Till R and Wakefield J (2016) An Investigation Into the Sonic Sig-
    nature of Three Classic Dynamic Range Compressors. In: *Audio Engineer-
    ing Society Convention 140*. Available from: www.aes.org/e-lib/browse.cfm?
    elib=18270.
Moore A and Wakefield J (2017) An Investigation Into the Relationship Between
    the Subjective Descriptor Aggressive and the Universal Audio of the 1176 FET
    Compressor. In: *Audio Engineering Society Convention 142*. Available from:
    www.aes.org/e-lib/browse.cfm?elib=18625.
Nelson D (2017) Between the Speakers: Discussion on Mixing. In: Hepworth-
    Sawyer R and Hodgson J (eds), *Mixing Music, Perspectives on Music Produc-
    tion*, New York; London: Routledge, pp. 122–139.
Patterson J (2017) Mixing in the Box. In: Hepworth-Sawyer R and Hodgson J
    (eds), *Mixing Music, Perspectives on Music Production*, New York; London:
    Routledge, pp. 77–93.
Produce Like a Pro (2015) Chris Lord-Alge: Studio Tour & Interview – Warren
    Huart: Produce Like a Pro. Available from: www.youtube.com/watch?v=81MKs
    U1CIAk.
Tingen P (1999) Spike Stent: The Work of a Top Flight Mixer. Available from:
    www.soundonsound.com/sos/jan99/articles/spike366.htm.
Tingen P (2000) Tom Lord-Alge: From Manson to Hanson. Available from: www.
    soundonsound.com/sos/apr00/articles/tomlord.htm.
Tingen P (2008) Secrets of the Mix Engineers: Michael Brauer. Available
    from: www.soundonsound.com/techniques/secrets-mix-engineers-michael-
    brauer.
Tingen P (2010) Robert Plant 'Angel Dance'. Available from: www.soundon
    sound.com/people/robert-plant-angel-dance.
Winer E (2012) *The Audio Expert: Everything You Need to Know About Audio*.
    Waltham, MA: Focal Press.

## BIOGRAPHICAL NOTE

Dr Austin Moore is a senior lecturer and course leader of sound engineering and music production courses at the University of Huddersfield, UK. He recently completed a PhD in music production and technology, which investigated the use of non-linearity in music production with a focus on dynamic range compression and the 1176 compressor. His research interests include music production sonic signatures, the perceptual effects of dynamic range compression, and semantic audio. He has a background in the music industry and spent many years producing and remixing various forms of electronic dance music under numerous artist names. Also, he is an experienced recording engineer and has worked in several professional recording studios.

# Part Three

# Concepts of Production

# 14

# Working With Sound in the DAW

## Towards a New Materiality of the Audio-Object

## Stace Constantinou

## INTRODUCTION

Contemporary music production occurs at the junction of a series of relations between creator, software, and hardware. The combining of computer software and hardware used for digital music-making contained within a relatively small physical space gives rise to the Digital Audio Workstation (DAW). In recent years, the software element in this series of relations has increasingly become the central element, so much so that today it tends to be regarded as the DAW itself. Music production software does not, however, constitute a neutral field for free creative exploration. Rather, it includes a set of assumptions about the ways in which the creative user will work (Marrington, 2011). Such a set of assumptions arise within the user interface, the onscreen visualization of the ideas that underpin the software's workings, with each DAW having its own particular set of music production features that inevitably channel the efforts of creative users in particular directions. And in a competitive market, the DAW's user experience becomes an increasingly important factor, as companies strive to maximize profits by delivering software that is simple for the customer to use.

A significant aspect of contemporary music production concerns the manner in which recorded sounds are employed as material for creative endeavor. Audio recordings are often given visual onscreen objects that users edit and manipulate. These visual objects then transmute into what I call the "audio-object", as a result of the alignment of the visual and tactile interactions between user and equipment (mouse, tracker pad, MIDI controller, and so on). This alignment also encompasses the seemingly aural response of the audio-object, as the cursor or playline is seen, onscreen, moving over it. It may consequentially feel like this audio-object has an onscreen material presence, even though it does not. Instead, the onscreen object simply represents the materiality of the sound, which exists as an array of binary markings stored on a hard drive.

My purpose in this chapter is to reflect on the nature of the audio-object and the manner of its employment, within the DAW, providing specific reference to electronic music production practice. In particular, I wish to consider the way in which the relationships between sound and object may impact upon electronic music producers' thinking about, and working with, audio as a material for creative use. This chapter draws on recorded interview material I collected between February 2017 and October 2018, from a select group of electronic music producers currently working in the electronic music scene. These electronic music producers are Manni Dheensa aka Manni Dee; Ewa Justka Thimitris Kalligas aka Kalli; Shelley Parker: Alan Stones and Jesse Kuye aka Jesse Tijn. I have included salient passages from the aforementioned recorded interviews that underpin the trajectory of this chapter. As part of my conclusion, I suggest new ways of working with digital sound, using a refined series of interactions between the creative user, visualized objects, and the intrinsic creative potentiality of those onscreen "audio-objects". In essence, what I propose is a new theoretical model for working with the audio-object within a speculative DAW, one that combines a free-flow creativity with touchscreen devices.

## TECHNOLOGY AND MUSICAL CREATIVITY: SOUND AS MATERIAL

Technology shapes the way in which music making occurs. For example, a singer's relationship to the vocal mechanism is markedly different to the pianist's relationship with the keyboard. The very nature of a piano, or voice, necessarily means that a certain kind of idiomatic writing for that instrument can occur. If instrument technology shapes the way in which this idiomatic music making, and ideation, occurs, or can at least be said to have a minimal impact on the final creative result, there is reason to suppose that the computer is no different. As Hans-Joachim Braun explains:

> Technology has always been inseparable from the development of music. But in the twentieth century, a rapid acceleration took place: a new "machine music" came into existence, electronic musical instruments were developed and composers often turned into sound researchers.
>
> (Braun, 2002): 9

The early basis of the computer in the mathematical and technical milieu of the scientist, consisting of a QWERTY keyboard, visual display unit (VDU), central processing unit (CPU), working memory (RAM), and storage (hard drive), initially meant that musicians needed to possess both technological skill and a keen sense of musicality. At the end of the 20th century, as the size of computers shrank, commercial manufacturing costs fell, and their power increased, the personal computer (PC) became a general commodity, in a consumer market, that could be used for a number of different purposes, including music making. Limitations in RAM size, processor speed, and storage meant, however, that the use of audio within

PC music systems was more problematic than the use of MIDI, the latter requiring less data consumption than that of digital recording. As the common musician did not own a computer powerful enough to run multiple tracks of digital audio with simultaneous recording and playback, digital recording typically occurred away from the CPU, on hardware digital tape systems, such as the Alesis ADAT (Théberge, 1997: 246) or the Tascam DA-38 instead. Digital recording and production systems were also available as a type of keyboard sequencer, such as the "most expensive commercial digital synthesizers like the Fairlight [CMI Series II] and Synclavier (each of which was originally developed as a studio machine)" (Braun, 2002: 55).

Initially, recording multiple tracks of digital audio required a considerable financial investment in studio space, mixing console, recording, and other physical gear (Bartlett, 2001: 4–5). Multi-track digital-audio tape machines, such as the Alesis ADAT, for example, were specifically manufactured for the affordable end of the home studio music market, as, according to the manufacturer,

> 1991 proved to be the ground-breaking year for Alesis, with the introduction of the ADAT Multi-Channel Digital Tape Recorder. Before ADAT, a studio would have to invest $50,000 in order to afford a multi-track digital recorder. With ADAT, the price tag came down to $4,000, essentially allowing every home and project studio to afford digital recording. This new technology allowed any artist or musician to record studio quality recordings.
>
> (Alesis, 2018)

In time, such outboard devices as the ADAT became obsolete as computing power increased and the cost of recording directly onto a hard-drive fell. Importantly, computers also offered hardware that allowed for a non-linear digital recording, meaning that musicians became able to afford high-quality digital recording equipment, as well as utilize its onscreen user-friendly (or friendlier) software. Initially prohibitive to the ordinary musician in cost, by 1998, innovations such soundcards were an important part of the computer audio market and were being included as "free" items within the cost of a computer. A computer soundcard, the "Sound Blaster Live! EMU10K1 APU, 2M Transistors", was reputedly

> a game-changer. A new PCI audio chip was introduced in the market, which powered the Sound Blaster Live! The EMU10K1 APU, with two million transistors and unprecedented audio processing power and performance of 335 MIPS, blew away the competition. Another new audio platform, Environmental Audio eXtensions EAX application programmer's interface API was introduced in the market. Established as the new audio standard under Windows, the EAX API was made available in the industry for free.
>
> (Soundblaster.com)

Today, the availability of affordable PCs, with their continuously increasing processing power and storage capacity, means that for many recording jobs, all that is required is a computer, an external audio interface, headphones, and microphone(s). A bigger recording space and multi-track device may still be needed for larger-scale drum recording, or ensemble sessions, but not for most overdubbing requirements, or mixing requirements, and likewise not for electronic music production. As a result, as Marrington (2016: 52) writes, "the Digital Audio Workstation (DAW) has established itself as the predominant technology for music creation and production".

Add to this the availability of vast sound libraries (pre-made material in the form of musical snippets ready for looping or editing), and it becomes clearly questionable whether one even needs to be able to play a musical instrument in the conventional sense. The PC *is* the instrument, so it is just a question of choosing the right sounds, as producer and DJ Jesse Kuye aka Jesse Tijn, in an interview with me, explains,

> Find sounds that don't need to be worked on from the beginning. If you hear a sound that you don't like a part of it, just don't use it, wait 'til you find a sound you like all of. And then you can bend it into what ever you like.
>
> (Kuye, 2017)

It is important to note the interchangeability of different terms that refer to the audio-object. Jesse Tijn, for example, uses the term "sound" in the citation I reproduced here. Elsewhere, DJ and producer, Shelley Parker, uses the term "samples"; see the following text from the interview with me. Parker says,

> For the last three years, most of the sounds/music I make is from samples. I love having a load of recorded material and then tucking into it, working out how I'm going to use the samples.
>
> (Parker, 2018)

My aim here is to locate a general, common, or umbrella term, like "audio-object", to describe a variety of simultaneous practices, whereby digital recordings are the main material used in creative music production – for example, use of samples, loops, recordings, regions, "found sounds", sound objects, and so on.

One way to begin understanding the audio-object is in terms of its intrinsic material properties. Different file formats are used as audio recordings creating digital material objects that exist on hard drives, as the result of different methods of encoding acoustical data, for example the standard Mac audio file format is AIFF (audio interchange file format), which is not data compressed, and can be compared to the MP3 (MPEG Level-1 Layer-3), which does compress the data, typically to around one-tenth of its original size (Bartlett, 2002). Arguably these different formats imbue the sound with different audio qualities. Producers may feel that there is a unique or specific sound quality that arises as a result of using a particular file format.

Electronic music producers can use digital file format features as a perceived or actual aesthetic medium for the creation of a particular type of electronic music. Andrew Burke, for example, explains, in his article *Trademark Ribbons of Gold: Format, Memory, and the Music of VHS Head*, how the producer VHS Head has a preference for audio material taken from VHS tapes (Burke, 2015 355):

> Comprised in large part of samples drawn from a collection of 1980s videocassettes layered over frenetic and fractured beats, the music of VHS Head points to the way in which memory and technology intersect. Occupying the space where glitchy electronica meets hypnogogic pop and futurist soul the tracks on VHS Head's debut full-length album, Trademark Ribbons of Gold (2010), complete a trajectory from VCR to the mp3.
>
> (Burke, 2015: 355)

Shelley Parker, too, draws inspiration partly from a combination of different media, including cassette tape:

> I've got a few hundred tapes, minidiscs, DATs, and digital recordings. 1996 is when I first started recording. I did lots of recordings in 1997 of an air vent in Kings Cross. I used to listen to a lot of music on tape, so I like that sound of C30s. There would always be that hiss but I liked it.
>
> (Parker, 2018)

The relative sound quality of an individual audio-object may also be important in order to produce a particular kind of desired end result, as electronic musician Thimitris Kalligas aka Kalli explained to me in an interview in 2017:

> I've always had to like work on my samples and see what problems they have with them and make them more accessible for myself. . . . If it's a high quality sample I can always mess it up. But there's more potential to mess it up. . . . If I'm using kick drums I'll look for something thuddy, something very organic sounding. I'm not really a big fan of using 808s, 909 drum samples . . . I put it through a lot of effect . . . chaining, and warping, which helps a lot.
>
> (Kalligas, 2017)

## THE EVOLVING AESTHETIC OF THE AUDIO-OBJECT

Making music from pre-recorded sounds, that is, using "samples", is a commonly recognized tenet of electronic music. The acceptability of this practice in the wider sphere of musical aesthetics was achieved only gradually, however. In the first part of the 20th century, for example, Pierre Schaeffer theorized about and worked with sound recordings to make musical compositions, a style he initially called "concrete music", and, later, "acousmatic" music (2012). Schaeffer used the term "sound object" to describe his samples, which he saw as a genuinely new material for music composition. He then detailed his sound experiments, as well as a series of compositional techniques in a number of diary entries entitled "In Search of a Concrete Music" (2012). Later, he developed and presented a more detailed work titled "Traité des objets musicaux" (1966) "Treatise on Musical Objects" (2017), which ultimately moved a group of composers to, broadly speaking, work in the field of acousmatic music. One of the themes running through Schaeffer's texts is the notion that sound composition may be considered in some sense akin to the science of acoustics. This view remains relevant even today, as can be seen in the comments of electronic musical instrument maker and performer Ewa Justka who says, in an interview with me:

> Sound has a form, a specific form. If you have an oscilloscope and plug that oscilloscope to the sound system you can see specific, I mean you can see the sound wave . . . sound is just a signal, it's a voltage signal so you can translate it to anything to vibration, to light, erm you can plug it into an oscilloscope and see actual sound waves, it's not really, yeah, it can be anything.
>
> (Justka, 2017)

Interestingly, Justka occupies a relatively uncommon space, insofar as she works without using computers in her live performance. Pre-performance, she prefers making her own setup of DIY electronic musical instruments, which she then "performs" live on stage. She continues, "It's more about the process of making things. My work is about making things" (Justka, 2017). In Justka's case, then, the primary object is the electronic music making machine itself, with the onscreen representation of the sound, the oscilloscope's image being of secondary importance. It is worth noting perhaps that Justka's live show includes lights that pulsate in time with the beat.

Another important aspect of working with audio-objects in the DAW concerns the application of digital signal processes (DSPs) to transform the sonic characteristics of sound. Such processes are often akin to practices familiar from the analogue era of production – for example, the transformation of the sound of an electric guitar using distortion or the modification of a vocal part using echo, delay, and reverberation. This allows for the creative manipulation of audio-objects into new and potentially novel types of sound. DSP transformation of audio-objects is ingrained in the process of electronic music production, as Kalli explains,

> Most of my music is manipulated sounds. Probably one hundred per cent of it is manipulated samples . . . turning things into what

they shouldn't be, or just trying to escape . . . say I've just turned the
sound of a train going past and I can manipulate that into anything.
(Kalligas, 2017)

Recreating hardware equipment (such as a mixing console for example)
in music production software that simulates the operations and sonic characteristics of earlier forms of technology means that pre-existing physical
modes of operation (moving faders up and down say) continue but in the
onscreen paradigm of the computer user interface. This method, of recreating hardware equipment in software, is described as "skeuomorphism" by
Adam Bell, Ethan Hein, and Jarrod Ratcliffe in their joint article "Beyond
Skeuomorphism: The Evolution of Music Production Software User
Interface Metaphors." Working with a physical mixing console involves
kinetically moving faders and dials whose positions are relatively fixed,
ergonomically. In contrast, the experience of interacting with DAW software relies on mouse, or touchpad kinesthetic movements that relate touch
to the onscreen visualization and then the resulting aural outcome. Skeuomorphism is therefore an attempt by software developers to reproduce in
the computer a digital form of an earlier hardware production practice/s
associated with older technologies (Bell et al., 2015).

The way in which audio-objects are contextualized within the broader
environment of the DAW is also of significance. For example, the common DAW-based paradigm for the visualization of audio-objects is the
linear sequencer that charts time from left to right (see Logic X, Cubase,
and Sonar, for example). This defaults to an onscreen grid, with vertical
lines denoting the passing of clock/ metronomic musical time, the latter
defaulting to a 4/4 time signature. Both audio- and MIDI-objects (both
termed regions by the software manufacturers) are then easily visually
aligned to this grid. A stack of audio- and MIDI-channels are used, most
commonly, with each channel being aligned to a separate sound, so it can
be more easily isolated during mix down. Quite commonly, perhaps, this is
achieved using the "mix window", an onscreen reminiscent of a hardware
mixing console. Transformations of the sound then typically happen as
a result of inserts being applied to each channel-strip with more control
being applied using automation of its parameters. Such inserts typically
assign DSPs, such as equalization (EQ), followed by compression, and
reverberation perhaps. Each insert may be turned on or off, as well as have
their parameters altered in real-time, a process known as automation. This
combines, therefore, the linear paradigm of the digital tape recorder with
the left-to-right music reading practice of standard notation.

A DAW such as Ableton Live, in contrast, is distinctive in that it allows
for the pre-selection and uploading of multiple audio-objects onto a single channel-strip. In addition, each loop can be easily turned on or off at
will by the electronic music producer, or DJ, in real-time. This difference
in the spatial hierarchy of audio-object utilization gives Ableton a live
performance functionality, a key feature of its appeal. Traditional time-
based paradigms are still present, however, in that the audio-objects are,

by default, set to trigger according to a seemingly ever-present, rigidly metronomic, 4/4 beat. Each sample is also automatically shortened or lengthened so it fits in with the beat, unless explicitly programmed to do otherwise.

DAWs, such as Logic X and Ableton, both carry vestiges of analogue ways of working (e.g., the channel-strip) combined with a, not essential but nonetheless telling, grid-time 4/4 beat. New ways of working with sound may come about by accident, or they may be sought. To seek a new way of working, one may employ either theory to rejuvenate the thinking process or embark on an experimental method, hoping that serendipity will strike. One could employ both methods perhaps, as Schaeffer did.

## THEORIZING THE AUDIO-OBJECT

In his book *Traité des objets musicaux*, Schaeffer considers the "sound object" (the analogue forerunner of today's digital audio-object) in these terms:

> This unit of sound [sound-object] is the equivalent to a unit of breath or articulation, a unit of instrumental gesture. The sound object is therefore an acoustic action and intention of listening.[1]
>
> (Schaeffer, 2002: 271)

Schaeffer was working with analogue, magnetic tape, as the means by which to record and store his sound objects, thus his experience of sound-as-object will have been similar but subtly different to that of contemporary producers using audio-objects within the DAW. In a DAW, as already mentioned, a way of imagining an audio-object is as a time-domain (oscilloscope) visualization. This emphasizes the amplitude of sound waves as they transform through various compression and rarefication states. This visualization becomes an onscreen object, that the listener links to the actual sound emanating through the speakers see Figure 14.1 below.

Such time-domain visualizations are based in modes of thinking and working that may be thought of as one of two ways of considering the audio-object. The first is to think of hearing as a passive organic machine absorbing sound – essentially the ear-brain mechanism whose salient associated features might be summarized as follows:

- The acoustical properties of sound
- The physical attributes of the ear
- The means by which the ear relays acoustical energy to the brain to produce the sensation of hearing
- The brain science of the impact of hearing on brain function (and form)
- Psychological phenomena that can be observed and measured.

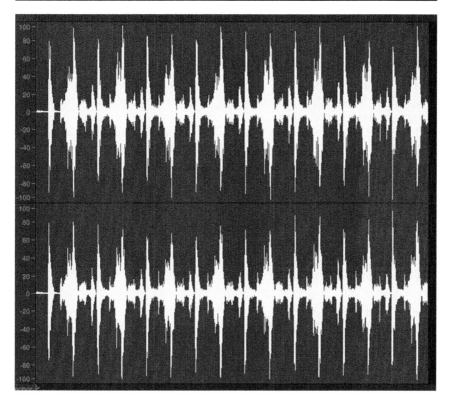

**Figure 14.1** Time-domain visual representation of an audio-object

These features are empirical because they can be studied, proven, or refuted using the scientific method. The second way is to think of listening as an activity that we do *to* our experience of hearing. Listening in these terms may be summarized as including the following features:

- Listening is something we do *to* sound.
- Listening is learned culturally – think of triadic harmony, melodic shape and instrumental arranging.
- Listening occurs in different cultural contexts including the concert hall, at home, in the car, using mobile devices / on the move, and so on.
- Different practices of listening have been developed especially from within the music fraternity – playing instruments requires specific ways of listening that will be different to those of the piano tuner, recording engineer, mix down engineer or mastering engineer.
- Philosophies and theories of listening have been developed that are then used as models for listening in practice.

These cultural and philosophical theories and practices of listening, and music making, are imbued therefore with social vestiges that, in the West, have a rich tradition dating back to the writings of Plato (c. 500 BC).

Unfortunately, the first method (empiricism) does not really assist us in understanding the complexity of art forms, nor suggest why we should care. The successful scientific method tends to focus acutely on either the natural phenomena of sound itself or the effects of sound on the sensuous faculty of hearing. Also, knowledge of hearing alone lacks an adequate account of how the sound is actually experienced by us, as conscious beings, in the form of music.

It is significant that in the "how to" literature for recording and music production, much thinking on sound remains of a rule-of-thumb kind, intertwining the empirical with the cultural. Take the term "muddiness", for example, which is used most often at the mix stage of the music production process. The term isn't scientific, because what it means for a mix to sound muddy is subject to difference of opinion, but the solutions may include precise remedies in the form of frequency and amplitude measurements and adjustments.

Bartlett and Bartlett, for example, list four possible causes of "muddiness" and offer several solutions:

Muddiness (Lacks Clarity)
If your sound is muddy because it lacks clarity, try these steps:

- Consider using fewer instruments in the musical arrangement.
- Equalize instruments differently so that their spectra don't overlap.
- Try less reverberation.
- Using equalizers, boost the presence range of instruments that lack clarity. Or cut 1 to 2 dB around 300 Hz.
- In a reverb unit, add about 30 to 100 msec predelay.

(Bartlett, 2002: 409–410)

Interestingly, such rule-of-thumb recommendations intertwine cultural and empirical measures, as if they both occupy the same space. And returning to our electronic producers, sometimes their language suggests awareness of the properties of sound in the empirical sense. As Manni Dee comments,

You do anything really, I think er what I look for in sounds is just a kind of frequency a resonance of a frequency, erm and it's completely malleable from there, I can do what ever I want with it, like a piece of dough just stretching it out and flipping it around, you do anything, it's great. So, yeah, I can impose the qualities I want on the sound even if they don't exist inherently in the sound.

(Dheensa, 2017)

In this instance, for Dee it is clear that the primary objective is to acquire audio material of a certain sonic character that is ultimately pliant – perhaps even neutral. There is an affinity with Schaeffer here, who made concrete music by recording many different sounds, spending months carefully listening to each, experimenting with different ways to transform their sonic characteristics, and then combining this recorded material to form musical compositions. If you are unfamiliar with Schaeffer's compositions, listen to "Etude aux chemins der fer" (Railway Study).

Schaeffer realized that sound recording stimulated in the listener a visual imagery (of trains in the case of the "Railway Study"). Such visualization, he worried, would distract the listener from the musical properties of sound. So to solve this perceived problem, he turned to thinking about the subject-object relationship. He hoped to achieve a way of making concrete music with no visual references whatsoever. A complete discussion of this topic is beyond the scope of this chapter, but the following summary attempts to highlight its main features with reference to Schaeffer:

- Subject (a) apprehends Object (z)
- Objects (n) are apprehended using sense perception (ears, eyes, touch, smell, and so on), and through careful study of its data, we gain knowledge of said object/s
- Knowledge becomes empirical through repeat test conditions that allow us to check the outcome is always the same, or consistently the same
- If we attempt to derive knowledge from thought alone, there's the potential that the subsequent understanding is erroneous. Empiricism helps us to confirm the validity of our thinking by giving our ideas over to experience.

Schaeffer attempted to create a theory of composing with sound that had an empirical element by conceiving of a "transcendental" mode of listening whereby the sound object is purified by the removal of its visual reference (Schaeffer, 2002: 268), the idea being that listeners are thus unencumbered by any visual reference that may have resulted had the sound not been purified, leaving them free to concentrate on the innate music character of the "objet sonore" (ibid). Such a listening situation has an analogy to the way in which the disciples of Pythagoras received their master's sage words from behind a curtain or screen, so as to leave his teaching unsullied by visual references that may have been transmitted through physical gestures that were caught by the eye. The disciples of Pythagoras were reputedly known as "Acousmatikoi" – hence the term acousmatic music (Constantinou, 2009). Acousmatic music works, therefore, because multiple subjects (n) listen to an object (z) using a transcendental listening (see Figure 14.2).

This listening situation is empirical, Schaeffer argues, because it utilizes an intersubjectivity of multiple subjects (n) that exist as a community who

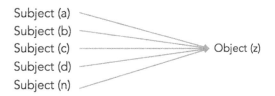

**Figure. 14.2** Pierre Schaeffer's transcendental listening

collectively agree on the purity of sound object/s (z) within the acousmatic situation. Schaeffer likens working creatively with sound to sculpture (Schaeffer, 2012: 14) and argues that finding sound objects is like walking along a beach looking for seashells: "the seashell enthusiast takes up the object and this object says something to him" (Schaeffer, 2012: 148).

## AN INTUITIVE APPROACH TO WORKING WITH AUDIO

Pop producers unhindered by the theoretical purity advocated by Schaeffer and his acousmatic followers have instead developed intuitive aesthetic approaches to the utilization of the audio-object. The Beatles song "Honey Pie" (1968) from the album *The Beatles* (1968), includes a section of spoken text by Paul McCartney saying, "Now she's hit the big-time". This spoken element is set in sonic relief to the rest of the production and is given a "music hall" sound through processing that references older recordings. To achieve the required sound quality the spoken passage was "heavily limited, chopping off the signals at both ends of the frequency range, and superimposing the sound of a scratchy old phonograph, to make the end product like a vocal from a very early and worn 78 rpm record" (Lewisohn, 1990: 159). Such a treatment offers the listener a sense of historical distance within the aesthetic framework of the song (Clarke, 2007: 56).

Another example from the commercial music domain can be heard in Björk's song "Scatterheart" (2000) from her album *Dancer in the Dark* (of the same year). Two vinyl scratch sounds can be heard at the beginning of the track, and gradually these morph into the groove of the main beat (Clarke, 2007: 56). Such uses of the audio-object for the reasons described might be regarded as essentially un-acousmatic because they rely on the listener having a reference point.

The highly competitive nature of the DAW market has resulted in a race toward an increasingly user-friendly software experience, aimed at the general user. Software such as Logic Pro, for example, has the potential to be used by many types of musician, not just the commercially minded and successful electronic music producer (Paterson, 2016: 82). Nonetheless, a handicraft, or process, aspect of making something remains, as composer and sound designer Alan Stones explains, "There's a kind of goal you're, generally aiming for, but it's about process as well. It's about what emerges as you make it . . . the process is definitely very important" (Stones, 2017). But whilst taking this into account, an important part of electronic music production is, for some, to have ready access to a plethora of readymade recordings that can be shaped into the desired sound. Such a readymade resource need not be fixed, however: it can completely change, as Jesse Tijn, who twice lost his hard drive, explains,

> it was good anyway because when I got new samples it changed how my sound was and stuff like that, I was happy about it in a way . . .

it's just like one of those weird things that I've accepted in my mind yeah, it's almost like I'm renting them, or borrowing the samples and then, when the time's right my hard drive will die again [laughs].

(Kuye, J., 2017)

Whilst contemporary DAWs and practices of electronic music production may vary between the purity of the acousmatic and the grittiness of techno, what generally unites this plethora is the use of the audio as a visualized onscreen object. Many DAWs are multipurpose tools designed for use in a wide variety of music-making practices that reference and direct a creative modus operandi rooted in an earlier era of analogue equipment. The analogue era was one that utilized hardware tools as well as developed production methods based on the conditions that such equipment levied on the overall practice and musical outcome. Up-to-date DAW design does not need to be limited to those conditions imposed on the production of music by earlier forms of hardware equipment. So rather than limit our creative potential to a skeuomorphic metaphor, why not instead conceive of an audio-object production software whose features are determined by the essence of the audio-object itself.

## TOWARD A NEW DAW-BASED PARADIGM CENTERED ON THE AUDIO-OBJECT

An audio-object centered DAW, or DAOW (Digital Audio-Object Workstation), would emphasize the audio-object, moving away from skeuomorphic models toward paradigms that feature the creative and exploratory manipulation of sound, as a material for artistic pursuit in and of itself. The following is therefore a speculative model for the possible realization of such a DAOW:

1. Ergonomics: the preferred technology would be the portable touch-screen device.
2. Neutrality: audio-objects will be visually represented (VAO, meaning visualized audio-objects) on a blank background unencumbered by paradigms of linearity, verticality, channel-strips, grid-time, and so on. Clicking and holding on the screen would cause a directory window to appear, allowing access to the sound library/ies. A user will be able to build up a composite sound made up of the variously playing audio-objects and the sum of the resulting sonic manipulations.
3. Workflow: the extent to which layers of sound can be shaped will be by both an intuitive tactile (using the touchscreen device) instantaneously auditioned, and responsive to the agency of the imaginative response of the artist. The manner of visualization of the audio-object, its "content", will remain the waveform but boxed within a thin line for visual clarity. There will be an emphasis on quick and easy layering of multiple DSPs. Each DSP added to VAOs will incrementally

change its appearance, operating as a visual guide to the object's trans-formation. Each DSP layer can be muted and/or automated. A (pre- or post-fader) bus-like DSP layer can be added, which will mean that a new auxiliary sprouting object will appear. Onto this object, more DSP layers may be added, thus providing users with the functionality of auxiliary sends but without the need for a visual skeuomorphism referring to the engineering practices of mixing consoles. Instead, multiple objects can be opened and organized onscreen. DSPs may be applied to a single or to multiple VAOs.

4.  Time: because there is no obvious visual reference to the time line, DAOW will run at their existing speeds. But the speed of each audio-object can be time-stretched; the resultant object may be man-ually or automatically synchronized to a tempo in beats per minute (BPM) or stretched at multiple points and completely mangled. To time-stretch the VAO, simply click, hold, and drag to the desired dura-tion. When tempos are used, a number box will show the exact BPM, allowing for synchronization of audio-object to beat, as necessary. And multiple VAOs can be synchronized either to a single BPM or to multiple tempi to form cross- or polyrhythms. One VAO can be synchronized to the same BPM as another simply by linking the two and designating each as either a lead or a follow. This will also work whereby, say VAO (A) is made to play in the same time as VAO (B). Or ten VAOs (B-K) all follow the tempo of audio-object (A). These follow and lead settings can be automated, so that they change at any moment, opening up the enticing prospect of a dynamic in-time shift in sonic texture from one moment to the next.

5.  Spatial relationships: should two VAOs touch, the sound will change at the point where they overlap. This idea comes from the visual arts, where by two or more colors are mixed to yield a new hue. In our new DAOW, however, the angle by which the VAOs overlap will have an effect on the resultant sound. By turning the audio in such a way as the rear of the object goes over the front means that front and rear swap place, so the VAO is now the opposite way round; in this case the sound will reverse. But what if you twist the VAO so that the rear goes under the front? This is a different action and will therefore transform the sound in a different way. And what if the VAO is twisted so that rather than being reversed, the rear now forms the top of the object? In this case, perhaps every other sample could be reversed, producing a semi-reversed sound. And if we accept this, then it becomes possible to image any gradation of sound in between, given enough processing power.

## CONCLUSION

This chapter has considered the ways in which particular DAW-situated paradigms can have a significant impact upon the way in which music is

conceptualized and produced. I have noted, for example, that in designing and making DAWs, software companies leverage a combination of
empirical knowledge and cultural vestiges, as seen in their foregrounding
of skeuomorphic features of visual design and the inclusion of DSP algorithms which model past concepts of audio processing. Within this, I have
focused on the conception of the audio-object in the context of electronic
music production, which is often represented, during the production process, as a predominantly onscreen graphic, an item to be placed in the
arrange window and synchronized to the beat.

I have suggested that working with sound recordings to produce music
involves using both empirical acoustics, as well as an understanding of
cultural norms, which are in practice not necessarily easily distinguishable.
To support this, I have drawn attention to the work of Pierre Schaeffer,
who used both theory and practice, by using sound objects to make first a
concrete music and then an acousmatic music. The latter relies on the consent of a community faithful to the intersubjective cause of a pure acousmatic sound, seemingly clinically isolated from any visual reference it may
potentially inculcate in the listener. I have suggested that contemporary
electronic music producers put much stock in the audio-object as a material. And whilst they need not necessarily draw creative energy directly
from Schaefferian ideals, such artists have developed the means of concep-
tualizing audio as material, whether in terms of its sonic aesthetics and/or in
combination with the agency of a DAWs' particular sonic modus operandi.

By firstly focussing on the audio-object and then secondly forming
working objectives, I have suggested that it may be prudent to devise new
creative modes of its engagement. The hypothetical DAOW, discussed
earlier, is designed so as to illustrate what might result from a model based
on emphasizing the audio-object as the central area from which creative
energy and processes emanate. It, of course, remains to be seen how, or
whether, such DAOW ideas will develop in actuality. Perhaps, rather than
veering in the direction of innovation and novel approaches, DAWs will
continue to homogenize around a skeuomorphic nostalgia. Or it may tran-
spire that the audio-object itself will cease to be thought of as a material
used in the production of electronic music.

## NOTE

1. Schaeffer, P., trans Constantinou, S. (2015) PhD thesis, Kingston University,
   p. 271. (Constantinou, S. (2015) *Processes of Creative* Patterning: *A Compo-
   sitional Approach*. Unpublished PhD Thesis, Kingston University, Kingston
   Upon Thames, UK.)
   Interviews: Audio Recordings transcribed.
     Dheensa, M. (2017) Interview with Manni Dheensa aka Manni Dee.
   Interviewed by Stace Constantinou for *"Working with Sound in the
   DAW: Towards a New Materiality of the Audio-Object"* 4 March 2017,
   London.

Justka, E. (2017) Interview with Ewa Justka. Interviewed by Stace Constantinou for *"Working with Sound in the DAW: Towards a New Materiality of the Audio-Object"* 20 April, London.

Kalligas, T. (2017) Interview with Thimitris Kalligas aka Kalli. Interviewed by Stace Constantinou for *"Working with Sound in the DAW: Towards a New Materiality of the Audio-Object"* 12 March 2017, London.

Kuye, J. (2017) Interview with Jesse Kuye aka Jesse Tijn. Interviewed by Stace Constantinou for *"Working with Sound in the DAW: Towards a New Materiality of the Audio-Object"* 20 April 2017, London.

Stones, A. (2017) Interview with Alan Stones. Interviewed by Stace Constantinou for *"Working with Sound in the DAW: Towards a New Materiality of the Audio-Object"* 9 March 2017, London.

Interview: Email:

Sender: Parker, S. Recipient: Constantinou, S. (2018) 'Shelley quotes re-edit', 26 October 2018.

## REFERENCES

### Books and Articles

Angus, J. and Howard, D. (2006) *Acoustics and Psychoacoustics*. Oxford: Focal Press.

Bartlett, B. and Bartlett, J. (2002) *Practical Recording Techniques*. London: Focal Press.

Bell, A., Hein, E. and Ratcliffe, J. (2015) Beyond Skeuomorphism: The Evolution of Music Production Software User Interface Metaphors. *Journal on the Art of Record Production*, (9).

Braun, H-J. (2002). *Music and Technology in the Twentieth Century*. Baltimore, MD: The Johns Hopkins University Press.

Buick, P. and Lennard, V. (1995) *Music Technology Reference Book*. Tonbridge: PC Publishing.

Burke, A. (2015). Trademark Ribbons of Gold: Format, Memory, and the Music of VHS Head. *Popular Music and Society, 38*(3), 355–371.

Clarke, E. F. (2007). The Impact of Recording on Listening. *Twentieth Century Music*, 4(1), 47–70.

Constantinou, S. (2009) *A Statistical Analysis of the Form of the Piece Etudes aux Chemins de Fer by Pierre Schaeffer*, www.orema.dmu.ac.uk/analyses/stace-constantinous-étude-aux-chemins-de-fer-analysis (30/7/2017).

Kane, B. (2014) *Sound Unseen: Acousmatic Sound in Theory and Practice*. New York: Oxford University Press.

Kant, I., trans. Meiklejohn, J. M. D. (2003) *Critique of Pure Reason*. New York: Dover Publications.

Lewisohn, M. (1990). *The Beatles: Recording Sessions: The Official Story of the Abbey Road years 1962-1970*. New York: Harmony Books.

Marrington, M. (2011). Experiencing Musical Composition in the DAW: The Software Interface as Mediator of the Musical Idea. *Journal on the Art of Record Production*, 5.

Marrington, M. (2016) Paradigms of Music Software Interface Design and Musical Creativity. In R. Hepworth-Sawyer, J. Hodgson, J. L. Paterson and R. Toulson eds, *Innovation in Music II*. Shoreham-by-sea: Future Technology Press, 2016, 52–63.

Marrington, M. (2017) Mixing Metaphors: Aesthetics, Mediation and the Rhetoric of Sound Mixing. In R. Hepworth-Sawyer and J. Hodgson ed., *Perspectives on Music Production: Mixing Music*, first edition. Abingdon: Routledge imprint of the Taylor and Francis Group.

McIntyre, P. (2012) Rethinking Creativity: Record Production and the Systems Model. In S. Frith ed., *The Art of Record Production*. Abingdon: Routledge imprint of the Taylor and Francis Group.

Paterson, J. (2017) "Mixing in the Box" Hepworth-Sawyer, R,. and Hodgson, J., (eds.) *Perspectives on Music Production Series: Mixing Music*. New York: Routledge, pp. 77–93.

Schaeffer, P. (1966) *Traité des objets musicaux*. Nouvelle édition. Paris: Editions du Seuil.

Schaeffer, P., translated by North, C., and Dack J. (2017) *Treatise on Musical Objects*. Berkeley, CA: University of California Press.

Schaeffer, P., translated by North, C., and Dack, J. (2012) *In Search of a Concrete Music*. London: University of California.

Théberge, P. (1997) *Any Music You Can Imagine: Making Music/Consuming Technology*. Middletown, CT: Wesleyan University Press.

## Web

The Alesis Story (2018), available at: www.alesis.com/company (accessed 11 March 2018).

About Sound Blaster (2018), available at: www.soundblaster.com/about/ (accessed 11 March 2018)

## Discography

*The Beatles*, 'Honey Pie', The White Album [mp3 download]. Amazon Media EU.á r. l, 1968.

Bjork, 'Scatterheart', Dancer in the Dark [CD] France: Zentropa Entertainment, 2000. TPLP151CD.

Dee, M., 'B2. Adorable Disorder', Throbs of Discontent (TPT073). Perc Trax, London 2017.

Parker, S., 'Fixed Connections', Sleeper Line [CD]. London, Entr'acte, 2013.

Schaeffer, P., 'Etude aux Chemins de Fer', Ohm+: the early gurus of electronic music [CD]. UK: Ellipsis Arts, 1948. CD3691.

VHS Head. 'Trademark Ribbons of Gold'. Skam Records (SKALD025) 2010.

# 15

# Coproduction

## Towards a Typology of Collaborative Practice in Music Production

## Robert Wilsmore

### INTRO: THE DOMINANCE OF THE ONE

The record producer has long been represented by the image of the lone figure of a man sat at a large mixing desk bestrewn with knobs and faders. All that has been captured from the virtuosic performers, creative artists, and genius songwriters is now at his fingertips with faders moving up or down at the inspired will of this solitary guru. The tracks laid down by the artists are just the raw materials to which this great sculptor of sound will roll up his sleeves, light a cigarette, and mold the music into the fixed-for-all-eternity artwork that is the recording.

It is a romantic and seductive image (albeit the "his" sticks out like a sore thumb to us now), and we do not need to credit him with the angst and expressive rawness of the artists who make the sounds; the producer is the one who saves the artist from their failings and cuts the rough diamond into the shape that gives the sound its aesthetic and increased monetary value. George Martin was not the first Beatle but the fifth (the "quintessential" Beatle), and this not only recognizes his importance as being part of the music but places him as that calm, in-control, but nearly out-of-sight mastermind behind the sound of the records. And this is no illusion, as far as any truth can be true (which is a tough call in this day and age), we rightly revere Martin and so many others who have sat behind the desk on the other side of the glass and pushed and pulled the faders and routed this auxiliary through that effect. There were and are many great controllers of sound who can lay genuine claim to this singular title of "the producer".

This reality is further made evident in the literature of producers. The reading lists for our students of music production are likely to include classic works such as Howard Massey's *Behind the Glass* (2000), with more than 30 chapters, each of which is titled only by the producer's name, it features interviews with producers such as Brian Wilson, George Martin, and Nile Rodgers. And I make no criticism, as for any budding producer here are the horses' mouths telling how they made the great works that we know and love, and beyond that, Massey brings out the human in these people. They are not gods born with superhuman powers capable of doing things we will never be able to achieve, not at all. As the front cover says, this is "how they craft the hits", for indeed it is a craft, and the producer does not always appear *deus ex machina* at the end to make the problems go away and bring a glorious final resolution to it all. I say "not always" because there is probably enough evidence to suggest that at times they are literally the god from, or with, the machine. And perhaps it is the machine that gets in the way of our understanding of producers and production.

Anthony Savona's *Console Confessions* (2005), from the same Back-beat publications as *Behind the Glass*, although different in approach to Massey's book, still has lists of those great names attached to the chapters: Les Paul, Phil Ramone, Herbie Hancock, and so on, and, like Massey, the book has those iconic images of the man and his machine, the console. In both these books, the bearded and bespectacled Ramone looks directly into the camera lens with the roomful of gear as the background land-scape, like a music studio take on a Reynolds or Gainsborough portrait but with knobs instead of trees. In another picture, Herbie Hancock relaxes, hands behind his head, smiling at the camera off-center to the left of shot so that the many faders and lights are visible to the right. In similar posi-tion, in Massey's book, sat relaxed with arms behind his head but with even more faders than Hancock, Humberto Gatica smiles (more cheekily than Hancock), and the shot is on a diagonal that maximizes the landscape aspect's capability of fitting in as much of the desk as possible across the diagonal from bottom left to top right. The console is of course an impres-sive instrument, and we are amazed that this relaxed and in-control expert knows what all these buttons and faders do. Most people would probably be clueless as to what to do if they came face to face with such a beast. It looks like rocket science, and these guys are creative engineers capable of getting a spacecraft to the moon and back. Yet they sit there and smile as if it's all in a day's work for them – which of course it is. The images are incredibly seductive, and yet they are also, in one sense at least, genuine. The literature and the recordings testify to their ability to use the equip-ment as the tools of their trade, so why not have Jack Douglas lovingly caressing the faders or Frank Filipetti sat at the console hands folded with an unlit cigar? Is it any different to picturing Jimi Hendrix with a Fender Stratocaster or Jimmy Page with a Gibson Les Paul? The answer is "yes, it is different" (and in one sense at least, disingenuous); hence, there is a mythology at the route of the producer's art that has been oversimplified in the iconography. To say that Jimi Hendrix played the guitar on "Purple Haze" is probably about as accurate as we might get with an attribution of what the artist did. He played the guitar, and we can hear it. The contri-bution is easily identifiable. We can make out the guitar quite clearly and attribute it (quite clearly) to one man. The guitarist is the person who plays the guitar; the producer is the person who produces. The obvious discrep-ancy is that where playing the guitar is clearly understood as the activ-ity taking place, the act of producing, as we know, is multifaceted with no clear boundaries as to what it includes and excludes, and because we (the fans, the interested) want clarity, then the desk becomes the catch-all placeholder signifying activity well beyond what the desk itself is capable of. Hence it plays into the fixity of signification that we associate with the guitar, but in reality, try as the image might, it lacks the guitar's distinct denotation and by attempting to represent something that has no such definition, it enters into the realms of dubious connotations and myth.

The iconography of the man at the desk as described here is a rough generalization, one that we will instantly recognize whilst at the same time we may have a nagging doubt that the music producer extends far beyond this simple, yet dominant, representation. The ideology of the producer (but not the reality) is in some ways akin to the notion of the classical composer. They too are, or were, almost exclusively male and singular in their authorship; whereas we can quickly go straight to Lennon and McCartney when asked to give an example of a songwriting duo, we are likely to struggle when asked to do so for classical composers. That symphony co-written by Beethoven and Mozart never happened, and that opera by Verdi and Puccini never happened either. Even the idea sounds odd, but why should that be? Songwriters collaborate all the time, yet somehow "composers" don't, or at least they only collaborate in complementary mode with other disciplines, such as with poets and librettists. Even now, one only has to take a look at new commissions for the BBC proms, and one will be hard pushed to find much else beyond the singular composer. The BBC website (BBC, 2017) lists the 2017 commissions, which include Julian Anderson, Mizzy Mizzoli, and Laurent Durupt among the composers of new works. But in contrast, production is not without its great teams. In the 80s, we might well have been used to the names Trevor Horn, Giorgio Moroder, and Quincy Jones, but then another name on everyone's lips at that time was SAW (Stock Aitken Waterman) with whom we associated the great 80s names of Kylie Minogue, Rick Astley, and Bananarama among so many others with whom they developed ways of turning out hits that 30 years on are still loved, admired, and respected. And maybe this has something to do with the team spirit and the collaborative processes out of which they were forged. A producer at PWL studios (the "factory floor" of SAW), Phil Harding, writes openly about the creativity of collaboration with his colleague Ian Curnow, writing with wonderful generosity that "There's something unique about working partnerships where you get to the point where you know what each other are doing and thinking without verbal communication. In other words, a telepathic understanding where you get on with creating what you feel is right" (Harding, 2010: 129).

The term *production* is immediately thrown wide when we speak of Stock, Aitken, Waterman, or of Motown before that. Hepworth and Golding in *What Is Music Production?* (2011) note three general stages of production as "capture, arrangement and performance" and each of these has many sub-categories as well. We have long accepted that the production-line involves stages that might otherwise be named differently, such as composition, performance, engineering, arranging, or even inspiring (definitely "inspiring", getting the best from one's performer), and it is hard to give it a definition least to say that usually there is a "recorded" element in there somewhere. It is easy for us to understand the idea of one person as the artist or as the producer and of the individual who has a role within the team. But what is *joint* authorship? What is coproduction?

Rather than attempting an ontology, the aim here (and the aim of the book on coproduction that is to follow in this series of publications on *Perspectives on Music Production* that Chris Johnson and I are coauthoring) is to find the different ways in which coproduction occurs. It would be nice to find that moment, one clear example, where we can with absolute certainty say "there is a singular joint authored moment", not a moment where we can identify individual contributions or where we cannot identify individual contributions – only because we cannot see the history of its production, but because this moment is genuinely a result of, dare I say it, "when two become one". In *Behind the Glass*, Massey transcribes this conversation with Brian Wilson:

*[Massey]*   Recently, though, you've been working with coproducers as opposed to being the sole producer. Is there a reason for that?

*[Wilson]*   It's because I needed to have the springboard of ideas between people. Because I'd run out of ideas – I had writer's and producer's block.
(Massy, 2000: 43)

There are many reasons for coproduction – the sparking of ideas, the complementarity of skills, the desire to work together, the potential for greater success of the whole through team effort. And there are many forms that coproduction can take from two producers knowingly working together to a whole world that works together on a singular project (either in denial or without realizing it). Even the individual may have to acknowledge that they are also others in themselves, those internal voices of one's influences. The aim here is to forefront joint authorship in producing music, production that is coproduction, that is multi-authored in all its manifestations from the agreed collaboration of two producers working closely together to the many and varied contributions to a project by contributors who have no such contract or contact with those making the work. Even solitary gurus cannot override all the decisions that have been made leading up to the point that has presented the material before them; production has already taken place, and they have to work with that as part of their product. Producer, songwriter (and collaborator) Chris Johnson and I have begun researching coproduction that explores these types, and this chapter attempts to set out some of the initial territory of that larger study and begin to put into types the various modes of coproduction that we will explore in greater detail later. Of course, the caveat is that these categories and types may change as the research progresses, but by way of introduction, the types presented here form a key part of our research methodology. With this, we note that these characteristics and types are not, as yet, case studies of actual production teams. Putting names and faces to the types will have to wait.

## TYPE OVERVIEW

Only part of the study will focus on those groups of producers who work together to produce their product in the traditional sense, that is, the team that is attributed to the finished recording of the song. Although that study deserves greater research in itself, to begin with, we want to throw the net as wide as possible and find not only the cutting edges but the extreme middle, a place that was sadly neglected in a modernist last century. We are under no illusion that the main interest will be in the teams of producers and how they work together, the actual people coproducing actual hits, but in this first attempt, we want to zoom out to view as much of the territory as possible, even if that means covering ground that some will find uninteresting and possibly unconvincing. This is not then a study just of those collections of individuals who form production teams, albeit that category and the types within it are part of what is outlined here, and we recognize that most interest is likely to lie in that area rather than in fanciful philosophical speculations and assertions (the outlying territories). There are different modes of collaborative practice that we can bring to this aspect of the study, in particular those identified by Vygotsky influenced collaborative theorist Vera John-Steiner, but we are also excited to venture into less conventional areas of coproduction and take a look at the individual as a collective in him- or herself, as well as the individual's place within a collective of actual others. Then we observe that music in general, and popular music in particular, is an ongoing coproduction by the many millions of us who are involved in producing this ongoing singular thing (popular music) that we have called "The Song of a Thousand Songs", which is something of an "all the world's a stage" approach to pop music collaboration. And we are aware that not everyone in coproduction is a willing collaborator and acknowledge that plagiarism is also a form of joint production. If collaboration can only be called such if all participants are willing and in agreement, if consent is a predicate of collaboration, then we are already at fault and you need read no further; this is a localized calibration for locally calibrated people (there's nothing for you here). We also wonder where it could all end up; this is in the realm of philosophical speculation, but we will look to a future point where production may have ceased because all has been collectively produced. Not a Hegelian "end of art", where art's function has ceased but where there is actually nothing left to produce. We will even attempt to do that ourselves. We will, without any doubt, fail, but perhaps we can put the wheels in motion and in doing so we might well spark a few thoughts and ruffle a few feathers. Good, it will have been worth the attempt if we do. The feathers of authorship and ownership in music production could do with a good ruffling.

To put these larger categories into manageable types, we list the top-level categories of coproduction as:

1.  Group coproduction (collaboration between individuals)
2.  Internal coproduction (the self as many)

3. Coproduction without consent (denial or unknowing collaboration)
4. Deproduction (the collective disappearance of production)

Each of these will be introduced separately, mostly with further subdivisions within categories. Some of these types are drawn from observation of the real world, of an engagement with the act of production and coproduction as well as the observation of them, but some of the models are also set up *a priori* so that they can be tested against real situations later. Hence the types will shift from the blindingly obvious to the utterly ludicrous (by which I mean "playful" in its exploration of ideas rather than demanding universal validity). Ultimately, it might turn out that coproduction is nothing more than a discipline-specific study of teamwork. If that is the case, it will still be an interesting area to study but not a paradigm shift into some exciting new realm of music production studies. That remains to be seen.

## Group Coproduction

Emerging from a Vygotskian social science approach to joint authorship, Vera John-Steiner in her seminal book *Creative Collaboration* (John-Steiner, 2000) identifies four "patterns of collaboration" that she carefully caveats as being on a fluid spectrum, and these move from the closest of collaborations to the widest and most open form (and vice versa). They are labeled as Integrative, Family, Complementary, and Distributed. These have associated roles, values, and working methods, and they form a useful model for understanding different types of coproduction. John-Steiner's model, neatly depicted as a circle in order to avoid a hierarchical taxonomy, starts from the widest heading and goes to the most integrated; however, we will work outward only because the most open collaboration (distributed) leads to later types, that of "coproduction without consent" and "deproduction".

## Integrative Coproduction

This is the mode in which we might best hope to find that purest of coproduced moments. The integration of producers at this point is such that their understanding has moved beyond the need for discussion, beyond the need to divide roles, in fact beyond any recognition as to who contributed what to the point where there is no difference between the operation of the one and the operation of the many. This moment may not actually exist outside of theory; even in a hypothetical example, it seems hard to see how there could be this pure moment. One might imagine two producers with hands on the same fader making adjustments almost as if receiving spiritual guidance on a Ouija board, but even then (ghostly externality aside), we are drawn to the separate decisions that provided the pressures on the fader and hence are thrown back into knowing that there is separation. Perhaps Integrative coproduction is in *proximity* of this pure moment rather than achieving it. John-Steiner's descriptions of this type are of "braided roles", "visionary commitment", and "transformative co-construction" (2000: 197), the "braiding" confirming that separate strands still exist, making this model far more discoverable than the pure moment might be. That said, Keith Sawyer, in his research on "group genius", points to moments of group decisions that are exactly this, that is, where the group finds a solution to a problem but none of the individuals within the group are aware of that solution. Following a narrowly avoided shipping disaster in 1985 on the USS *Palau*, researcher Ed Hutchins analyzed the responses and actions of those involved in averting the disaster and concluded that "the solution was clearly discovered by the organization itself [the collective group] before it was discovered by any of the participants" (Sawyer, 2007: 28). So there may be hope of finding similar moments of coproduction that are outside that of any individual control or contribution.

## Familial Coproduction

Trust and common vision are central in this type. Coproducers will be highly "familiar" with each other and will have become used to each other's ways of working and cross over roles easily. John-Steiner's term is "family", but we have used "familial" here to emphasize that the mode is "typical of a family" just to avoid any misunderstanding that shared genetics is essential. In this mode, any division of labor might change, and expertise becomes dynamic. This is a "pre-Integrative" model where there are still identifiable contributions but much cross over. Here, one might find coproducers feeding in and out ideas and activity in all stages of capture, arrangement, and performance, comfortable enough to know that one has permission (indeed encouragement) to do so from the "family". One producer has gone into the live room to move the kick drum microphone that the other set up without asking because he or she knows he or she can act upon his or her own informed decision to do so and that consent is implicit.

## Complementary Coproduction

Perhaps the most common form of coproduction is the mode in which roles are clearly identified and executed as such. In this way, one can assemble a team of experts trusted to do their activity best, and it will benefit the whole. One producer might be able to EQ the singer's voice better than the other, but the other has inspired the singer to give the best performance in the studio and so on. Separation of roles is key here but so is an "overlapping of values" (John-Steiner, 2000: 197) so that the final product works as one unified thing.

The artists themselves are perhaps the most frequent coproducers, and most likely, this operation will fall into the "complementary coproduction" type, that is where the division of roles is clear and complementary, but there is a shared vision (if it is to be successful) and the artists will often be with the "producer" validating their decisions or trying out ideas with him or her in the control room. Complementarity is also probably the best understood of collaborative types. Where John-Steiner uses "complementary", interdisciplinary collaborators have used the Deleuzian terms "striated" and "stratified" (see Alix, Dobson, and Wilsmore, 2010). Joe Bennett, in songwriting collaboration, uses the term "demarcation" (Bennett, 2011), and business teamwork theorists have this model worked out in great detail, in particular the roles identified by Meredith Belbin that produce successful teams (see for example Belbin.com).

## Distributed Coproduction

In terms of normal studio practice, this would seem to be a somewhat anarchic working model, as its methods are "spontaneous and responsive" with roles that are "informal and voluntary" and might even lead to sabotage (as we will see in the "weasel words" intervention later). This model suggests a rather more open approach that might see the production group widen to anyone who might want to "have a go". Hence, this is not likely to be a typical model in terms of song production in a studio. We will need to look wider to see this model in operation. Artists sometimes put out stems for fans to mix and produce as they will, and this fits this model well. Moving out even further, we might view the artist who coproduces by using a sample of a previous work, with or without consent. Further out still, we may see the whole of popular music as one production that everyone contributes to, and indeed that notion will be explored in more detail under the discussion of "coproduction without consent", which aligns with John-Steiner's role description of "informal and voluntary".

## Internal Coproduction

As with the patterns of collaboration, the types identified here will often overlap, or they may even be considered as the same model but viewed from a different angle. And so it is with internal coproduction. Notwithstanding one's ability to argue with oneself, we might also recognize that our decisions that ultimately manifest themselves in terms of what comes out of the studio monitors will include thoughts of those who *have* influenced us (past) and thoughts of those who *will* listen to us (future). Unlike the composer who walks a dangerous legal tightrope when stringing together a sequence of notes, the producer has been freer to take on influences of the past with less fear, though even that territory is now being claimed and fenced off. The Monopoly board of music production is expanding, and one has to be ever more careful as to which square one lands on (there is no longer a "free parking" space or a "get out of jail" card in this game). Within the self, one might have to identify one's influences *and* one's audience (past and future others). The discussion relates to artists of any type and any discipline with regard to decisions one might identify as one's authentic self and then those decisions that second-guess what the other wants (the audience). Heidegger notes that the inauthentic self includes the other as our "*they-self*" in that "The Self of everyday Dasein is the *they-self*, which we distinguish from *authentic Self* – that is, from the Self which has taken hold of in its own way" (Heidegger, 1962: 167). For Deleuze and Guattari, the "I" does not disappear but rather becomes much reduced. As they write with regard to their collaboration on *A Thousand Plateaus*, they want "to reach the point, not where one no longer says I, but the point where it is no longer of any importance whether one says I" (Deleuze and Guattari, 1987: 3–4). Not only are they collaborating with each other; they also acknowledge that as individuals, they are already a collective of others.

We can try out two models in this category, one that is before one's production and one that considers the "after" of one's production.

## Present, Past Internal Coproduction

The influence of existing production comes to bear on one's own production. To produce with producers of the past, one does so presumably without their consent (which also means this sits in another category), but if one is to stand on the shoulders of a giant, does one need to ask the giant's permission to do so? Although one could, it is not every case that one *has to* return to one's collaborator for permission after making a contribution. Perhaps this type does not need a new name; we can call it "influence". It is simply that of the self as constructed from the many, but it gives an opportunity to think of our engagement with the past as a collaborative venture. The past is tried and tested; hence, we might go straight for that Neumann U87 for the vocals and an AKG D12 for the kick drum because it has worked *before* for others. Or perhaps we can be inspired by Martin's discussions with Lennon and subsequent decisions that gave rise to the extraordinary "Strawberry Fields" recording, where the desire to achieve something new required a problem-solving approach to production – the past telling the future to try out new ideas. In this particular example, Lennon liked the first part of one studio recording and the second part of another and wanted to combine them. George Martin pointed out that they were in difference keys and had different tempi, to which Lennon replied, "Yeah, but you can do something about it, I know. You can fix it, George" (Martin, 1979: 200). And of course he did (*deus ex machina*).

## Present, Future Internal Coproduction

The future of one's production lies with the listener. In this mode, we are familiar with what they will like and give it to them, or we give them something we know they will like but that they don't know yet (a disruptive model in the manner of Henry Ford or Steve Jobs). Either way, although the second notion seems more creative, both involve the consideration of the judgment of the other that is yet to come. The wanting to succeed, the need to be wanted, the need for money, and so on drive an inclusion of the future other in the production decision. Do we have to tune that vocal? Whom are you doing that for? Is it because we believe the audience demands it? Actually, another "other" here are the artists themselves of course to whom (one version of) the producer must be subservient or simply be one voice in the crowd, but this model best sits in the complementary coproduction model. Richard Burgess's "type C" producer, the collaborator, has similar properties to this (see Burgess, 2002). Here we are in a dual position of basing future audience demand on past audience behavior or basing future demand on our intuition that goes against past audience behavior (they haven't behaved like this before because they've never heard it this way before). The latter model has been the dominant model of creativity, certainly in modernist times, but we are less convinced that it works now. It is time to retreat from the cutting edge (have we not noticed that it has become blunt?) and move instead toward the middle (toward James Blunt) for that is where the creativity of this age really lies.

## Coproduction Without Consent

Here we will outline two modes of coproduction without consent, although we will also have to say up front that there is also a "knowing" in this category as well as an unknowing; however, as a general heading, it will work for the moment. The first type is that of the distributed model of collaboration, as discussed earlier, but in this case, it has been widened out to encompass the whole world, so that everyone who produces "popular" music is a collaborator on one singular song, which we have called "The Song of a Thousand Songs". The second type in this category is that of denial or unknowing collaboration. In this type, it might indeed be "knowing coproduction" (but the person denies this knowledge), where an artist has deliberately taken the work of another and used it in their own creation, though more often than not, this category will not be aware that it is coproducing but rather will be asserting originality and authorship. This category has clear connections with that of the "internal coproduction" model described earlier, as it draws externality into an internal creative process.

## The Song of a Thousand Songs

This first type is actually where this whole project started, having its roots in a paper given at the International Festival for Artistic Innovation at Leeds College of Music in March 2014, which was called "The Song of a Thousand Songs: Popular Music as a Distributed Model of Collaboration" (IFAI, 2014), and bit by bit, the theory has expanded to encompass the many other modes that are now beginning to be explored. The notion of joint production, unknown (or un-thought about) collaboration where everyone in the world is continually adding to this singular song that is called "The Song of a Thousand Songs". Theoretically, we (it is now "we", that is, Chris and I) have found that the articulation of this concept works best as a combination of the social science "distributed" model of collaboration by John-Steiner and of the postmodernist ever-connecting rhizome of Deleuze and Guattari (1987). In this way, the extravagant and poetic philosophy of postmodernism is held to account by a grounded social science approach or at least by a large dose of pragmatism. And they are not mutually exclusive; in fact, the multiplicity of the rhizome is simply another version of distributed collaboration, decentered and de-ontologized in order to point out the singularity and connectedness of things rather than the separations which may indeed not be there at all. We put it in our last chapter in this series on the ontology of the mix, "The mix is, the mix is not" (Wilsmore and Johnson, 2017) that the separation of songs in our everyday world is nothing more than a prescribed division, where these separations (ends of songs) seem to have nothing at all to do with sound and as such should not be thought of as endings. We wrote that:

> If songs are separated largely by non-musical signifiers (the composers, the song titles, etc.) we can "nullify endings and beginnings" [Deleuze and Guattari, 1987: 25] for these starts and finishes are nothing more than segmentation caused by the effect of imposing non-audio signifiers onto audio. When we do this, we cease to operate within a representational system. There are no longer identifiable ones of songs – the removal of the artificial beginnings and ends has shown that they are actually all joined together. In fact, it is not correct to say that they are joined at all, once we have removed the sticky labels marked "beginning" and "end" we see underneath that there is nothing but continuity.
>
> (Wilsmore and Johnson, 2017: 196)

In one respect, the model is far-fetched and a long way from our everyday understanding that individual songs are composed by individual artists or bands, but then it is also a familiar notion that music has a life of its own and that "rock 'n' roll ain't gonna die", even if that is only as a metaphor rather than a belief in music as a living thing. And the main focus here is not to indulge in transcendental materialism (although we will indulge, it's just that it is not the main focus) but rather to point out the grounded observation that endings and separations of songs are indeed mostly about something other than the sound itself, and in doing so, we hope to clarify what endings are in the "real" world (our everyday phenomenology) and how this might affect considerations of what to produce and what coproduction means. This postmodernist take is not intended to be sophistry. It is intended to help show how the world actually works, in reality, in real life. It intends to show that some of what is sold to us as singularly produced is in fact "fake news".

## Denial or Unknowing Coproduction

There is a familiar saying that "where there's a hit, there's a writ", and we never have to wait very long to read in the press another case of artists (or their estate or their lawyers) laying accusations at the feet of another artist who they believe has stolen their creation. At the time of writing, there is a particularly interesting one that the press reported this week. It is interesting because the song in question has already been through the legal process of another artist laying claim to it itself and the song having to state that it was coauthored already (at first, it was coproduced without consent, but then consent was retroactively applied). In this case, BBC music reporter Mark Savage, reported the following headline "Lana Del Ray Says That Radiohead Are Suing Her" (BBC, 2018), and on her Twitter account, she let it be known that:

> It's true about the lawsuit. Although I know my song wasn't inspired by Creep, Radiohead feel it was and want 100% of the publishing – I offered up to 40 over the last few months but they will only accept 100. Their lawyers have been relentless, so we will deal with it in court.
>
> (Lana Del Rey, 2018)

The song in question is Lana Del Rey's "Get Free" (Lana Del Rey et al., 2017). The BBC article then notes that "Interestingly, Radiohead themselves were successfully sued by The Hollies over Creep's similarities to The Air That I Breathe. Albert Hammond and Mike Hazlewood are now listed as co-writers for the song, and split royalties with the band" (BBC, 2018). Lary Bartleet writing for the *NME* gave some clarity on the issue:

> Radiohead's publisher denies that any such lawsuit exists but explains that they are asking for Radiohead to be credited on the song. A statement from Warner/Chappell reads: "As Radiohead's music publisher, it's true that we've been in discussions since August of last year with Lana Del Rey's representatives. It's clear that the verses of "Get Free" use musical elements found in the verses of "Creep", and we've requested that this be acknowledged in favour of all writers of "Creep". To set the record straight, no lawsuit has been issued and Radiohead have not said they "will only accept 100%" of the publishing of "Get Free".
>
> (Bartleet, 2018)

Not surprisingly, those on social media responded in a variety of ways in support of or against each artist, but many noted the problem of owning a melodic sequence or a chord structure. So if Lana Del Rey's song "Get Free" is adjudged to have come from "Creep" and "Creep" from "The Air that I Breathe", then we can trace that song back to the one before it and that song to the one before that, and so on, until we get to the very first song. We have done the "research" and have found the *very first* song, the song from which all others come, and the very first song that was written before all other songs is called "This Song Sounds Like Another Song". If Warner/Chappell want an acknowledgment in favor of "all writers of 'Creep'", then it should hold that every song credits every contributor from the writer(s) of "This Song Sounds Like Another Song" onwards. Ridiculous, of course, but which bit is ridiculous? My example or the real situation?

Baudrillardian notions of simulacra in the recording are a firm part of the music production curriculum now for those of us teaching music production musicology in higher education, and Auslander, pulling on the work of Gracyk and Baudrillard, notes the simulacra of the recording as the copy that precedes the real (see Auslander, 1999; Baudrillard, 1994 [1981]; Gracyk, 1996). The concept behind this hypothetical first song "This Song Sounds Like Another Song", is of course to highlight the questionable link between origination and ownership. This system of ownership will have to crash at some point in the future, but for the moment, at least it is fair to say that the Lana Del Rey and Radiohead conflict will not be the last of its type and the crash will not be happening anytime soon. And so lots of songs (that sound like other songs) have been claimed to be owned, and the lawyers set to work. So we can legally own those things as writers and we can lay claim to putting one note after another in a particular sequence and own it (and defend it in court), so can we do the same with creative production? What if I creatively EQ, fix a specific compression ratio, and pan the guitars hard left and right? Can I own this? Maybe it is no different except that it can't be policed as well (yet) or maybe it's that no one cares as much. I should add that we are not advocating a "copyleft" ideology that "seems to argue for the reduction of, or extinguishment, of copyright altogether for an alleged greater cultural good" (Bargfrede, 2017: xi) or indeed are we affirming the copyright; rather, we are exploring the discourse. Perhaps, to retain our Deleuzian leaning, we should consider ourselves as the "copy *milieu*" or "copy middle" rather than the "copy centre".

## DEPRODUCTION

At a hypothetical point in the future, when collectively we have produced all that can be produced, the territory of production will begin to vanish. The process toward this is that of deproduction, and it might be said that this process has already begun. In reality, the final absolute point is unlikely ever to be reached, but the process toward that end will have a significant impact upon how we view production both by reframing the local and ultimately by accelerating output so that, if the totality of production is finite, output stops (the coal has been mined, the pits closed and dismantled).

The rise in technology that has driven the decentered Internet and globalization has produced some fine examples of distributed models of collaboration that Vera John-Steiner might never have considered even just two decades ago. Where there were territories that had borders and where maps remained fixed (at least for a while), there was some sense of security, but that has been replaced by the security of the omnipresent, that is, our ability to access information and experience anywhere, anytime. The authenticity that decided that the singer was only authentic if his or her singing accent represented his or her hometown (the physical territory) has given way to something much more local, namely, the global. The voice in one's room (in York, for example), that is the recorded voice emanating from the speakers, might be from New York, but that voice is nearer to the listener physically than his or her next-door neighbor and is more present in the house than the neighbor probably is. In this case, the territory, the social locality, has been reterritorialized by the global. Arguably, when a singer is in a Yorkshire studio in the flesh (the one from York but with the Brooklyn, New York, accent), we can confidently say that his or her accent is more authentic, because he or she relates to the voice that is *nearer* to him or her than the singer from York with the Yorkshire accent.

*Wikipedia* is a fine example of the global over the international and of a distributed model of collaboration. In one entry on deterritorialization, including a discussion on Anthony Giddens's notions of globalization in *The Consequences of Modernity* (Giddens, 1991), there is a particularly revealing intervention; the entry includes an edit in superscript in square brackets from a contributor who, in doing so, wittily demonstrates exactly how a distributed model operates:

> In the context of cultural globalization, some argue deterritorialization is a Cultural feature developed by the "mediatization, migration, and commodification which characterize globalized modernity". This implies that by people working towards closer involvement with the whole of the world, and works towards lessening the gap with one another, one may be widening the gap with what is physically close to them.
>
> (*Wikipedia*, 2018)

The global intervenes by becoming closer in proximity than the local (it is nearer to you than your neighbor). The intervention (the contribution of the phrase "weasel words") likewise destabilizes the local ownership of knowledge, just as it destabilizes and then re-stabilizes the authenticity of the recorded voice. As the world is at liberty to collaborate, to contribute, and to produce and technology increases the rate and quantity of contributions, we might consider what might result should it turn out that production is not infinite but finite. To transfer this to music production, can we reach a point where we have collectively produced every record that can be produced to the extent that not only does further production become impossible but the *idea* (and its apparatus) also fall out of use and hence out of existence? How could that happen? Here is one small way to start: Write every tune that ever did, and ever could, exist. Impossible? Perhaps, but we can have a go. We can begin step by step. Most tunes fall within certain parameters, maybe a few bars, normally within a key or mode that is limited; normally, it will be a tempered scale that is fixed, note lengths might be between a whole note and a sixteenth note, and so on. Suddenly, limiting parameters emerge that make it look slightly less than impossible, achievable even. And they could be written systematically and categorized, particularly if we can write the software to generate these. We would encompass every tune (at least within the chosen parameters) that has been written, and we would also then have written every tune that is to come. What then of ownership? Perhaps the money we will owe, because we will be sued for writing tunes that are already "owned", could be offset by the income from those future composers that we will sue because we have already written the tune that they will claim as their own. Answering the same concern from readers of Wired on his article "How many different songs can there be?" (Wired, 2015), Rhettt Allain reponded similarly, writing "If I sue more than I get sued I should be fine". Just imagine how that might play out in a court of law. As mentioned earlier, at some point, ownership will have to come crashing down, and that will be one of the stages on the journey of deproduction. It does not need to reach the final realization of all production for the process to have impact. Beyond this, we are not claiming a "death of art". Music will be there. In fact, *all* music will be there already. It is just that production will have stopped.

## OUTRO: THE RISING OF THE MANY

That will not happen, of course. This is conjecture and speculation, but it is not without actual purpose. Somewhere along the way to these (ludicrous) concepts, there lie a few shifts in our collective understanding of music production. And as we accept our collective approach to production and let go of the individual as producer, then we may find that we are freer to produce. Coproduction, in its reduction of the "I", ought also to reduce the grip on ownership that stifles creative acts and creative generosity. It may turn out that the field of study of coproduction is nothing more than how the known types of collaboration are manifest in the discipline of music production. If so, then that is fine; it is still worth the effort. And it would be hypocritical for Chris and I to lay claim to this territory. We are neither the first nor the last to research it. We are just another local calibration in the wider rhizome. However, it is time for the rhizome to spread. Coproduction may yet have much to offer as an activity and as a field of study.

## REFERENCE MATERIAL

Alix, C., Dobson, E., and Wilsmore, R. (2010) *Collaborative Arts Practices in HE: Mapping and Developing Pedagogical Models*. Palatine and The Higher Education Academy [Online]. Available from http://eprints.hud.ac.uk/id/eprint/12844/1/Collaborative_Art_Practices_in_HE.pdf [Accessed: 8 January 2018].

Allain, R. (2015) *How Many Different Songs Can There Be?* In Wired [Online]. Available from https://www.wired.com/2015/03/many-different-songs-can/ [Accessed: 3 November 2018].

Auslander, P. (1999) *Liveness: Performance in a Mediatized Culture*. New York and London: Routledge.

Bargfrede, A. (2017) *Music Law in the Digital Age: Copyright Essentials for Today's Music Business*. 2nd edition. Boston, MA: Berklee Press.

Bartleet, L. (2018) *Radiohead v Lana Del Rey: A Comprehensive Timeline of the Most Confusing Music 'Lawsuit' in Recent Music History*. In NME [Online]. Available from www.nme.com/blogs/nme-blogs/radiohead-v-lana-del-rey-creep-get-free-lawsuit-timeline-2216897 [Accessed: 18 January 2018].

Baudrillard, J. (1994 [1981]) *Simulacra and Simulation*. Trans. Sheila Glaser. Ann Arbor, MI: University of Michigan Press.

*BBC* (2017) BBC Commissions, New Works and Premieres, 2017. Available from www.bbc.co.uk/programmes/p05bbg38 [Accessed 5 January 18]

BBC (2018) *Lana Del Rey says radiohead are suing her*. BBC Music reporter Mark Savage. Available from www.bbc.co.uk/news/entertainment-arts-42602900 [Accessed: 18 January 2018].

Belbin, M. (N.D) *Team Roles* [Online]. Available from www.belbin.com/about/belbin-team-roles/ [Accessed 18 January 2018].

Bennett, J. (2011) Collaborative Songwriting: The Ontology of Negotiated Creativity in Popular Music Studio Practice. *Journal on the Art of Record Production*, Issue no. 5 [Online]. Available from www.arpjournal.com [Accessed 22 January 2018]

Burgess, R, J. (2002) *The Art of Music Production*. New York and London: Omnibus Press.

Deleuze, G., and Guattari, F. (1987) *A Thousand Plateaus: Capitalism and Schizophrenia*. Trans. Brian Massumi. Minneapolis, MN: University of Minnesota Press.

Giddens, A. (1991) *The Consequences of Modernity*. Oxford: Blackwell Publishing.

Gracyk, T. (1996) *Rhythm and Noise: An Aesthetics of Rock*. Durham, NC: Duke University Press.

Harding. P. (2010) *PWL From the Factory Floor*. London: Cherry Red Books.

Hepworth-Sawyer, R., and Golding, C. (2011) *What Is Music Production?* London and New York: Focal Press.

IFAI (2014) Conference paper at *The International Festival for Artistic Innovation*. Leeds College of Music, Leeds, UK.

Heidegger, M. (1962) *Being and Time*. Trans. John Macquarrie and Edward Robinson. Oxford: Blackwell Publishing.

John-Steiner, V. (2000) *Creative Collaboration*. New York: Oxford University Press.

Lana Del Rey. (2018) It's True About the Lawsuit, [Twitter post] posted 7 January 2018 [Accessed 8 January 2018].

Martin, G. (1979) *All You Need Is Ears*. New York: St Martin's Press.

Massy, H. (2000) *Behind the Glass Vol.1: Top Record Producers Tell How They Crafted the Hits*. San Francisco, CA: Backbeat Books.

Savona, A. [Ed.] (2005) *Console Confessions: The Great Music Producers in Their Own Words*. San Francisco, CA: Backbeat Books.

Sawyer, K. (2007) *Group Genius: The Creative Power of Collaboration*. New York: Basic Books.

Wilsmore, R., and Johnson, C. (2017) The Mix Is: The Mix Is Not. In *Mixing Music: Perspectives on Music Production*. Eds. R. Hepworth-Sawyer and J. Hodgson. New York and London: Routledge.

*Wikipedia* (2018) 'Deterritorialization' Entry in Wikipedia [Online]. Available from https://en.wikipedia.org/wiki/Deterritorialization [Accessed 25 January 2018].

## DISCOGRAPHY

Hammond, A., and Hazlewood, M. (1974) 'The Air that I Breath'. Polydor, Epic.
Hammond, A., Hazlewood, M., and Radiohead. (1992) 'Creep', Parlophone, EMI.
Lana Del Rey *et al.* (2017) 'Get Free'. Polydor, Interscope.
Unknown (N.D) 'This Song Sounds Like Another Song'. Mayditup Records.

# 16

# Narrative and the Art of Record Production

## Alexander C. Harden

## INTRODUCTION

A great deal of the charm of popular music recordings can be found in the stories they evoke for listeners. Throughout the mid- to late 20th century, notable experimentation has taken place in relation to storytelling in popular music, buoyed by the development of new music production technologies and practices that later led Eno to conclude that the studio is "a compositional tool" (Eno, 2004). The aesthetic importance of telling stories through recorded music can be observed, for example, in various rock albums of the 1970s, such as Camel's *The Snow Goose*, The Who's *Tommy*, or Pink Floyd's *The Wall*, each of which leverages the possibilities of the phonographic medium – or as others have put it, "the art of record production" (Frith and Zagorski-Thomas, 2012; Zagorski-Thomas, 2014) – in order to do so. As Frith and Zagorski-Thomas remind us, in the context of recorded popular song, all the materials that we hear are shaped through record production and the use of available music technologies. Indeed, even lyrics, which often invite the beginnings of song interpretations, are presented to us as performed by a singer and shaped through the production process in ways that Lacasse (2010) suggests offer "paralinguistic" interpretative cues. David Nicholls, recognizing the importance of the phonographic medium in musical storytelling, has coined the phrase "virtual opera" to acknowledge "the multiple ways in which certain works intended initially for purely aural consumption – *by virtue of their creation as recordings* rather than live performances – define for themselves a virtual dramatic space" (Nicholls, 2004: 103).

Although record production introduces new considerations in relation to how recorded music may evoke a story, there has to date been little work exploring the intersection between narrative and the art of record production, save a small number of publications which investigate particular instances of "sonic narratives" (Burns et al., 2016; Lacasse, 2006; Liu-Rosenbaum, 2012). Meanwhile, other discussions of narrative and popular song have addressed the role of lyrics and other media or contextual factors without reference to production, despite the important distinction which Moore (2010, 2012) makes between recorded *tracks* and *songs*. In this chapter, I aim to outline an analytical strategy of narrativity and the art of record production in reference to four areas in which the art of record production may support narrative interpretations. I will begin by establishing a working understanding of narrative before discussing several analytical concepts which will be used to address aspects of setting, characterization, event sequencing, and point of view or mood.

## NARRATIVE, NARRATIVITY, AND NARRATIVIZATION

In a traditional sense, the term "narrative" is used to refer to a text that consists of a particular rendering of a sequence of events in the form of language (Genette, 1976). Conjointly, "narrativity" is used to refer to the property of relaying a sequence of events. In this context of a classical understanding of narrative, events – and subsequently, narrativity – are taken to be encoded within a text, emphasizing matters of the text and authorial intention. As Negus (2012a) observes, though, such an understanding of narrative as a textual encoding of events restricts the agency of the listener in the act of interpretation. Indeed, even linguistic aspects of popular music in the form of lyrics or accompanying prose offer potentially competing possible narrative interpretations for different listeners.

More recent developments in the study of narratives have sought to reconceptualize narrative as a cognitive phenomenon. According to Fludernik, this body of work operating under the banner of "cognitive narratology" "demonstrates that readers do not see texts as *having* narrative features but read texts *as narrative* by imposing cognitive narrative frames on them" (Fludernik, 2010a). A cognitive position, then, holds that narratives are constructed by interpreters in response to information available to them in a given text. Two consequences of note arise from this redefinition of narrative: on the one hand, it broadens the range of media which may be investigated, which is especially useful, as recorded popular song involves a variety of musical and sonic information alongside the semantics of lyrics; on the other, the understanding of music as affording a narrative interpretation draws attention to the processual, emergent quality of narrative interpretation. To refer to such a process in which narrative is constructed by the listener in response to the affordances of a track, I will employ Fludernik's (2010b) term "narrativization".

## PHONOGRAPHIC NARRATIVITY

To illustrate this notion of the cognitive understanding of narrativity, I wish to consider The Beatles' "Yellow Submarine" (2015), which describes a simple sequence of events at sea set onboard the eponymous submarine. Whilst the lyrics are largely situational, details of sonic design are used to depict several moments described in the lyric and thereby support the narrativizing the track. The narrative use of record production in such a manner appears to have been aesthetically important, as the band reportedly added a series of Foley effects representative of ocean waves, the crew in the mess hall, and mechanical sounds following the initial recording session (Lewisohn, 2013). Following the second chorus, John Lennon can also be heard issuing orders to crewmates, which was captured from behind the door of the nearby echo chamber to simulate distance (ibid.). In this example, both the use of Foley and the different acoustical characteristics of Lennon's vocal timbre creates a distinction from the musical space of the instruments and other vocalists by affording a narrative space aboard the ship.

Although "Yellow Submarine" invites some discussion of the ways in which narrative is supported through sonic design, I would also like to reflect on how, in broad terms, such stories can be *evoked* for a listener rather than conveyed by *what* appears to be told. To do so, my approach will consider the combination of sound sources within the virtual acoustical space of the recording and the narrative associations or connotations afforded in response. This approach derives from a synthesis of several hermeneutical concepts: the "soundbox" and "phonographic staging", "landscape", and "anaphones".

Moore's (2012) soundbox is a heuristic model of the implicit acoustical space of a recording developed through the mixing process. It is four-dimensional, describing the spatial positioning of sound-events (in terms of stereo panorama, perceived distance, and perceived elevation) in relation to time. In conjunction with the soundbox, Lacasse (2000) proposes the concept of phonographic "staging" to describe the use of sound-processing that alters the spatial or timbral characteristics of sound events, positioning them within this implied space.

Whilst Moore's soundbox and Lacasse's staging together account for the distribution of sound-events within an implicit acoustical space, it is useful to employ the two concepts in reference to Trevor Wishart's concept of landscape. For Wishart (1986, 1996), the landscape describes the imagined sources of sound-events heard within a recording, such as the swirling of water in "Yellow Submarine". Wishart (1986) goes on, though, to make an important distinction between landscape and narrative anthropomorphism, i.e., the distinction between the sound events we *hear* and the *associations* with characters and so on that these evoke. To illustrate this, Wishart invokes the example of Tchaikovsky's *Manfred* symphony and writes of the final movement that

> we may be led . . . to associate the acoustic events with the idea or image of the distraught Manfred wandering through the forest. It may in fact be that there is some analogical relationship between our supposed experience of the events described in the programme and our experience of the acoustic events. . . . The landscape of Tchaikovsky's Manfred is however musicians playing instruments.
>
> (1996: 130)

In our example of "Yellow Submarine", therefore, whilst we might associate Lennon's voice with the captain onboard the ship, in the context of the landscape, the voice acts as a symbolic representation of Lennon shouting.

To narrativize a track, we should also account for associations made in response to aspects of the landscape. To account for this, the final theoretical concept which I would like to introduce is Tagg's anaphone. As Tagg (2013) observes, recorded music may connote extra-musical associations and phenomena. To account for such connotations, Tagg proposes what he calls an anaphone, i.e., a musical sound event that resembles non-musical forms of sensory perception. The most straightforward form of anaphone is what Tagg refers to as a sonic anaphone, in which a sound event resembles a source external to music. The swirling water in "Yellow Submarine", for instance, acts as a sonic anaphone that invites a listener to furnish his or her mental representation of the story with ocean waves. Tagg (ibid.) also outlines other forms of anaphone which resemble visual, spatial, kinetic, tactile, and social phenomena.

In the following discussion, I will consider the contribution of record production to narrativity in terms of four basic parameters of narrative. Firstly, I will discuss setting in terms of place and time. Secondly, I will consider the ways in which production can afford characterization through the representation of narrative agents using sound events in the sense I have just described (whether through the bonding of a voice to a narrative source or narrativizing instruments as representations of characters). Thirdly, we will explore the contributions of sonic design to the sequencing of events within a narrative. Finally, we will turn to the issue of how production can be used to evoke a point of view or "mood" that colors a narrative reading.

## SETTING

A popular and long-standing record production tactic to rapidly evoke a sense of space or place is the inclusion of ambient Foley effects, as we have heard in the example of "Yellow Submarine" through the use of water to simulate ocean waves, as well as the depiction of the crew in the mess hall. As a further example, take Bobby "Boris" Picket's 1962 "Monster Mash", a light-hearted piece that involves a series of monsters and the latest popular dance. The track begins with the landscape of a wooden creak positioned closely to our point of listening in the soundbox, followed by reverberant bubbling water and narrowband noise further back in the soundbox. Each of these sound-events acts as an anaphone which resembles the opening of an old wooden door, simmering fluid, and electrical sparking, collectively inviting us to narrativize the setting of a large tiled laboratory in which the story of the song unfolds. This establishment of a setting through the means of incorporating sonic anaphones signifies one of the most straightforward contributions of record production to narrativization as it invites the listener to situate his or her mental representation of a story within a particular geographical context.

Zagorski-Thomas (2009) and Eric Clarke (2007) propose that the use of staging or sampling to simulate particular forms of historical broadcast or recording media may connote a sense of temporal distance which through narrativization may situate a narrative setting. Throughout Robbie Williams's duet performance of "Dream a Little Dream" with Lily Allen in 2013, for instance, a subtle crackle has been added, which can be heard particularly at the start. The aesthetic use of noise here offers a reference to the distribution of recordings on vinyl, as we might associate with the mid- to late 20th century in which Mama Cass Eliott and The Mamas & The Papas popularized the song. Conversely, Zagorski-Thomas also proposes that "particular forms of clarity and audio quality are associated with modernity" (2009).

Through particular forms of phonographic staging or otherwise simulating historical broadcast media, a recording may offer chronological cues that inform narrativization. Throughout Gorillaz' "Feel Good Inc.", for instance, the lead vocal line is compressed and filtered throughout the first verse to give the effect of telephone transmission. A short beep, such as those used to separate messages on an answerphone machine, can be heard at the end of the verse, before a bed of quiet but noticeable hiss (associated with tape) and crackle (typical of vinyl) is added as the chorus begins. These staging cues suggest firstly that the protagonist is spatially dislocated from our point of listening whilst the use of noise suggests a temporal dislocation, as if the scene here is recalled by the protagonist.

## CHARACTERIZATION

In "Yellow Submarine", the staging of Lennon's backing vocals following the second chorus fulfills a dual narrative function: on the one hand, it consolidates the setting aboard the submarine; on the other, it supports the characterization of this voice as a commanding officer acting at a distance from the listener's implied position within the narrative space described by the soundbox. Here we see a case in which the staging of the voice offers us what Lacasse (2010) refers to as a paralinguistic detail. Other examples of phonographic staging to enhance the characterization of the lead vocals in this manner are common, including formant shifting, or equalization and compression (the "telephone effect"). Kraftwerk's 1978 track "The Robots" provides an especially clear case in which the lead vocal line is vocoded to evoke a cyborg-like singer (more recently replaced using a speech synthesizer).

In other circumstances, we might narrativize instruments in a manner described by Filimowicz and Stockholm, in which "sounds can be interpreted or experienced as characters in a story" (2010: 10). Such use of timbre to distinguish between or qualify characters has a long-established use in relation to narrative interpretations of music. In programmatic art music, the listener is often called upon to make associations between instrumental gestures and narrative agents without a determinant relationship between sound and associated narrative source. Sergei Prokofiev's *Peter and the Wolf,* for instance, represents each character with a different instrument of the orchestra and outlines a temporal development through the arrangement of musical material and motifs and the instruments which play them. To those aware of the program, for example, the clarinet is intended to afford ways of narrativizing the actions of the cat.

Following Filimowicz and Stockholm's assertion, Liu-Rosenbaum (2012) demonstrates an example of what he calls "sonic narrative" by anthropomorphizing groups of instruments. Echosmith's "Cool Kids" provides an example in which we can narrativize the guitar track in a similar manner. The first verse tells of a nonconformist outcast who is jealous of her peers' higher esteem whilst she remains largely overlooked. Accordingly, a muted electric guitar appears at the left- and rightmost extremes of the soundbox. Here, whilst the instrumentation is largely clustered within the middle of the soundbox, with the vocals and bass central and a moderate spread between parts of the drumkit, the guitar alternates each four bars between the extremes of left and right within the soundbox. One can narrativize the guitar as a musical representation of the lyrical subject, navigating the soundbox but failing to ultimately find a place to fit in. As we arrive at the chorus, the lyrics shift to describe the cool kids, and as they do so, the soundbox similarly reconfigures as the musical textures become more evenly distributed spatially, particularly due to the addition of multitracked voices. Here, the guitar is double-tracked, again placed either side of the soundbox, but switches to a loose strumming pattern and appears to recede, losing their clarity in the mix, now seeming to represent the new lyrical subjects: the unreachable cool kids.

## EVENT SEQUENCING

As illustrated through recorded music and earlier instances of radiophonic drama, a further area in which phonographic works may evoke narrativity is the ability to convey actions or events through associated sounds. So far, the uses of sonic anaphones in the other examples we have encountered are principally contextual and do not suggest a particular chronological sequencing of events. Elsewhere, though, the importation of concrete sounds into a musical discourse can evoke a sequence of events. In *The Marshall Mathers LP 2,* for instance, Eminem includes a 55-second track entitled "Parking Lot (Skit)" in which the landscape is constructed only from Foley effects and a shouted vocal line. In this case, the landscape depicts a driving car, running, and gunshots positioned close to the listener within the soundbox, whilst distant police cars approach throughout the track. Here, we may take this landscape to depict a police chase, whilst the chronology of sonic anaphones direct us toward a reading in which the protagonist attempts unsuccessfully to escape the police before turning the gun on himself. In the context of popular music discourse, Serge Lacasse

(2006) contends that Eminem's production similarly evokes a narrative reading in the particular example of his 2000 track "Stan". As Lacasse observes, the track can be considered in terms of three acts in which the protagonist, Stan, first writes to Eminem; Stan commits suicide having received no reply; and Eminem belatedly receives Stan's letters. Although the lyrics are highly descriptive of this event sequence, Stan's suicide in the second act by driving off a bridge is enacted entirely through the sampled sound of tires screeching, an impact, and large plunge into water in a passage from 4:48–5:00.

A further example is offered by Stevie Wonder's "Living for the City", a track which revolves around a protagonist and his family struggling to make ends meet within a socially unequal society. For the most part, the lyrics describe characters and feelings without a sense of temporal development. However, a passage during the middle of the track from 4:15–5:17 propels a narrative reading forward. At this point, the musical textures fade out to leave a landscape of speech and sonic anaphones that can be narrativised as a rapid movement through narrative time, leading to a penultimate verse appearing to take place some while later. This montage begins with anaphones of vehicle exhausts overlaid with a vocal exchange appearing to depict the protagonist getting off a bus and arriving in New York. However, the addition of a further voice, panned right within the soundbox ushers the protagonist over, as a siren can be heard distantly in the background. As the siren, acting as an anaphone denoting a police car, becomes more prominent and appears to move closer to our point of listening within the soundbox, the protagonist is directed to take a package across the street. Following this, the protagonist is arrested and further vocal and Foley excerpts, acoustically staged within the soundbox to complete the effect, depict his sentencing in court and locking in jail. In the verse which then follows this montage, the lead vocal style is notably altered and features considerably more vocal fry, pointing to a now-aged protagonist after being weathered by his incarceration.

## POINT OF VIEW / MOOD

Genette (1980) understands *mood* to be a property of all narratives resulting from the particular ways in which a story is relayed, particularly in terms of how directly it is relayed – that is, the degree of detachment from the narrator and the narrated events. In relation to recorded music, we might conceive of this detachment in relation to the extent to which the recording appears to enact an event, as though actively unfolding, or support the lyrics in the telling of a temporally dislocated event. To return to our previous example of "Living for the City", we can identify two approaches to mood by contrasting the manner in which, on the one hand, lyrics *relay* a narrative sequence and, on the other, the montage discussed above *enacts* one.

I would like to introduce two aspects of mood to which record production may contribute, the first of which we might otherwise think of analogously to point of view in visual media, which I will describe with the term "point of listening". The point of listening refers to the listener's implied location in relation to a narrative setting and/or event sequence, established in cases such as the "Living for the City" montage or "Parking Lot (Skit)", where the landscapes depict an extra-musical environment. The point of listening is informed by the spatial location of sound events in the soundbox and, in both these tracks, contributes to mood as it localizes our narrative encounter around a protagonist.

Elsewhere, other writers suggest that texture and the distribution of spectral energy in the mix offer a further way in which record production contributes to narrative mood and colors a narrativization. Tagg and Collins (2001), Liu-Rosenbaum (2012), and Burns et al. (2016) each associate the disposition of spectral energy with establishing a sense of mood, be it the prevalence of low frequencies to give a dark or dystopian feel or large registral gaps suggesting emptiness. With this in mind, consider Kylie Minogue's "Get Outta My Way", an example in which the disposition of spectral energy affords an attitudinal similarity with the sentiment of the lyrics. In the first verse, the protagonist casts the scene of an individual bored and feeling neglected by her unadventurous partner, joined by a four-to-the-floor kick and a rudimentary harmonic pattern. Aside from its unadventurous harmonic movement and rhythmic backing, the mix of the track may appear somewhat dull or deadened, caused by the limited presence of high-frequency elements other than the upper partials of Minogue's voice. By contrast, once the pre-chorus arrives, the protagonist resolves to take action and casts an altogether more positive light on her situation. Commensurate with this lyrical point, the harmonic pace and ambit increase, along with the mix, which becomes more animated and vibrant because of the inclusion of hi hats, guitars, piano, and additional synthesizer lines, creating a more even distribution of energy throughout the frequency spectrum.

## THE PET SHOP BOYS: "SUBURBIA"

So far, several of our examples have focused on the role of record production in evoking particular contributory aspects of a narrative independently of the musical and lyrical context. However, Nicholls (2007) proposes that the most complex instances of popular music narrativity involve a combination of media including music, lyrics, and album packaging or accompanying prose. To this end, I would like to close with a reading of The Pet Shop Boys' "Suburbia" and consider in greater depth how production acts in conjunction with the performed lyrics and music in this case to afford narrativization.

"Suburbia" was originally released on the Pets' debut album, *Please*, and appeared later in a remixed and extended edition produced by Julian Mendelsohn on the subsequent reissue, *Please: Further Listening*, as the "Full Horror Mix" (remastered in 2001), which I discuss later. The "Full Horror Mix" introduced additional lyrics beyond those of the original and made several distinct musical and production decisions. According to the co-writer and vocalist Neil Tennant, the song was conceived to offer a dark and troubled view of a suburban setting: "I thought it was a great idea to write a song about suburbia and how it's really violent and decaying and a mess . . . that the suburbs are really nasty, that behind lace curtains everyone is an alcoholic or a spanker or a mass murderer" (Tennant and Lowe, n.d.). Tennant goes on to further explain that the track was created shortly after discovering the duo's Emu Emulator sampler – used throughout *Please* – featured a car crash sample, which can be heard in "Suburbia" alongside sounds sampled from a film and self-produced sound effects. A tabular sketch of the track is included here and supports Tennant's description of the song in terms of a decaying utopia.

| Section | Description | Commentary |
|---|---|---|
| 0:00 – Introduction | Sampled dogs and percussion | Soundbox evokes a large open space. Landscape of dogs suggests a suburban setting, consolidated by the lyrics. Pitchshifting and harmonic language set a brooding, ominous mood. |
| 0:36 | Voice over introduced | |
| 1:14 – Instrumental refrain | Ionian electric piano refrain | Mix appears brighter whilst the ionian environment suggests a positive feel. |
| 1:31 – Verse 1 | Persona describes the scene, highlighting characters and police cars | Texture appears thinner with empty space around the voice within the soundbox. |
| 2:04 – Chorus | | |
| 2:18 – Verse 2 | Persona describes damaging local government building. | Cymbal crash evokes a smashing window. |
| 2:36 – Chorus | | |
| 2:52 – Bridge | | Anaphones of gunshots and violence; unquantized delay on brass evokes disarray and lack of order. |
| 3:40 – Instrumental refrain | Ionian electric piano refrain with added string pad. | Anaphones of police responding to the broken window which fade, suggesting a restoration of order. |
| 3:56 – Verse 3 | Persona describes media coverage of vandalism | |
| 4:13 – Chorus | | |
| 4:32 | Saxophone introduced | |

(*Continued*)

| Section | Description | Commentary |
|---|---|---|
| 5:03 – Instrumental | Collage of sampled dogs over percussion and bass | |
| 5:35 | Pitch-shifted voice over reintroduced | Saxophone begins to more closely resemble a wailing voice. |
| 6:50 – Instrumental refrain (play out) | | Fills accompanied with gating and delay suggest the positivity of suburbia is faltering. |
| 7:38 | Sampled Foley begins, music fades | The public revolt; the positive facade fades as the utopian ideal is lost. |
| 8:28 | Landscape augmented with rioting and gunshots | |

From the outset, the track evokes a dystopian suburban setting through the incorporation of sonic and spatial anaphones. It opens as tuned percussion, pans across the soundbox, and fades in, accompanied by an ensemble of sampled barking dogs with considerable added reverberation. Here, the breadth and depth of the soundbox acts as an anaphone for a wide, open space whilst the aggressive sampled dogs evoke a mood of trepidation. Anaphones also feature at several points later in the track. These include sirens (Doppler shifted to accompany staging movements), which resemble police car chases; crashes; and booming sounds resembling the destruction of buildings, and the use of gated reverberation on the snare, which gives it the more aggressive quality we might expect from or associate with a gunshot.

During the introduction, a string pad plays ascending chords in C Aeolian with various chromatic embellishments alongside Tennant's voice delivering a monologue describing the scene as the convergence of the city outskirts and paradise. However, the vocals are double-tracked with a pitch-shifted version of this monologue, lending an altogether more threatening message than the lyrics might offer by distorting the resonant characteristics of Tennant's voice. Following the extended introduction of string pads and spoken lyrics, the music enters C Ionian briefly with an instrumental statement of the chorus hook, before returning to the Aeolian mode for first verse.

Following the introductory passages, the lyrical material of the track is delivered across three verses. In the first verse, the lyrics paint an overall chaotic and dark image of the scene by incorporating themes of crime, being lost, and an uncomfortable independence. In the second, the protagonist refers to damaging the local government building, accentuated by a cymbal hit acting as an anaphone for the breaking of a window, and attracting the police in doing so. Both verses are followed by a chorus which, whilst lent a sense of positivity due to the major modality, features largely equivocal lyrics which draw from the idiom "run with the big dogs" to refer to keeping pace with the social elite, though the second stanza conversely refers to concealing oneself.[1] Following the second hearing of the chorus, an instrumental passage from 2:51 acts as a bridge to a restatement of the chorus hook, which leads to the third verse. After eight bars of this bridge, Tennant repeats a line twice, appearing to complain of boredom following sounds of gunshots. In this third and final verse, the lyrics address the media attention attracted by the vandalism mentioned previously.

Musically, the track repeatedly shifts from the Aeolian mode in the introduction, bridge, and verses to the Ionian mode during the chorus, reflecting the sense of ambivalence in the lyrics and the dichotomy that Tennant describes between perceptions of the suburbs on the one hand and the reality (in this scenario) on the other. Melodically, the verses are pentatonic and based around a four-bar phrase that pivots around the tonic before falling to the dominant. For Negus, speaking of narrative time in popular song, "Pentatonic melodies share characteristics with many other resources used in popular songwriting – they don't appear to go anywhere" (2012b: 490), suggesting, in this case, a sense of restlessness in conjunction with the lyrics.

The track ends with an extended outro from 5:03, which begins with a chorus of sampled dogs alongside the percussive textures before Tennant returns, delivering a spoken monologue with selective pitch-shifted double tracking joined by the opening string pad. A melodic line is provided at this point by a saxophone, which eventually reaches a particularly piercing G5, at which point its timbre evokes the emotional intensity of a straining voice. At 6:49, the instrumental riff of the chorus is repeated until the harmonic and percussive textures are left gradually fading alongside Foley effects of crashes and rioting. At this point, the repeated alternation between modes and Negus's observation of pentatonic melodies not appearing to "go anywhere" is also reflected in the production decision to fade out the music: both contribute to a sense of continuity without reaching a point of termination or advancement. Whilst this persistence is emblematic of the positive facade to Suburbia, the continuation of the sounds of rioting for some time after the music fades suggests that the suburban dream has become overshadowed by reality.

"Suburbia" combines each of the key elements we have seen in the other examples of this chapter and illustrates the role of record production alongside the recorded music and lyrics to offer ways of narrativizing the track as the decay of a problematic suburban setting. Firstly, the use of sonic anaphones and the construction of the soundbox establish a populated yet open setting augmented by the lyrics and accompanying music. Secondly, the pitchshifting of the spoken vocals provide a characterization cue, whilst we could also narrativize the closing saxophone as an anaphone of a screaming bystander. Thirdly, the anaphones of destruction and siren contribute to event sequencing by depicting particular moments of violence and the emergency response. Finally, different forms of narrative mood are evoked through the narrative detail of the lyrics, in addition to the production itself, which contributes to the changing mood throughout the track and establishes a perspective in relation to the narrative space – for example, through the smashing of a window, the police response, and eventual later rioting and destruction.

## CONCLUSION

Under a classical understanding of narrative as a telling of a sequence of events through language, song lyrics may offer a straightforward point of reference for identifying narrativity, though such a position neglects to acknowledge both the role of the listener in the process of narrative inter-pretation, as well as the phonographic nature of recorded popular song. In this chapter, using Fludernik's concept of narrativization, we have instead modeled narrative as an emergent cognitive phenomenon constructed by the listener in response to available information within a track. This recon-ceptualization of narrative invites greater reflection on the ways in which the phonographic medium informs narrativization and, using a synthesis of hermeneutical tools derived from writing on electroacoustic and popu-lar music, we have examined four areas in which the art of record produc-tion may afford narrativization.

In various examples, we have explored how the organization of the soundbox, landscape, and use of anaphones afford the narrativization of setting, characterization, events, and mood. As we have seen in "Subur-bia", and as advocated by Nicholls (2007), narrative cues derived from aspects of a track's production may act in conjunction with other informa-tion offered by the performed lyrics or music. Indeed, additional work to explore the contributions of other media – such as music videos, accompa-nying games, album packaging, and other prose – is necessary to develop a nuanced understanding of narrativity within musical (multi)media.

With the art of record production specifically in mind, a range of further issues relating to narrativity invite further consideration. Whilst we have briefly mentioned time here in relation to setting, we might for instance develop this to examine how the art of record production may afford the temporal sequencing of narrative events. On a related note, some consider-ation should be offered to ways in which production affords the evocation of diegetic levels as it might in film and the implications that follow for our relationship with the narrative space evoked by the track. Furthermore, whilst I have approached narrativity in relation to the internal world of a track, there is significant opportunity to reflect upon how the art of record production interacts with other forms of narrative in relation to listeners as well as producers or bands themselves. Together, these opportunities for further study highlight a range of exciting avenues for further inquiry in relation to the narrativity of recorded popular song.

## NOTE

1. In full, the idiom is repeated as "If you can't run with the big dogs, stay under the porch" (or some variation) to generally caution the addressee that he or she should act within his or her ability unless sufficiently capable.

## WORKS CITED

Burns, L., Woods, A. and Lafrance, M. (2016) Sampling and Storytelling: Kanye West's Vocal and Sonic Narratives. In: Williams, K. and Williams, J. (eds), *The Cambridge Companion to the Singer-Songwriter*, Cambridge: Cambridge University Press, pp. 144–158.

Clarke, E. F. (2007) The Impact of Recording on Listening. *Twentieth-Century Music 4*(1): 47–70.

Eno, B. (2004) The Studio as a Compositional Tool. In: Cox, C. and Warner, D. (eds), *Audio Culture: Readings in Modern Music*, London: Continuum, pp. 127–130.

Filimowicz, M. and Stockholm, J. (2010) Towards a Phenomenology of the Acoustic Image. *Organised Sound 15*(1): 5–12.

Fludernik, M. (2010a) Narratology in the Twenty-First Century: The Cognitive Approach to Narrative. *PMLA 125*(4): 924–930.

Fludernik, M. (2010b) *Towards a 'Natural' Narratology*. London: Taylor & Francis.

Frith, S. and Zagorski-Thomas, S. (eds) (2012) *The Art of Record Production: An Introductory Reader for a New Academic Field*. Farnham: Ashgate.

Genette, G. (1976) Boundaries of Narrative. *New Literary History 8*(1): 1–13.

Genette, G. (1980) *Narrative Discourse: An Essay in Method*. Ithaca, NY: Cornell University Press.

Lacasse, S. (2000) *'Listen to My Voice': The Evocative Power of Vocal Staging in Recorded Rock Music and Other Forms of Vocal Expression*. Dissertation, Université Laval. Availible from http://www.mus.ulaval.ca/lacasse/texts/THE SIS.pdf.

Lacasse, S. (2006) Stratégies narratives dans " Stan " d'Eminem: Le rôle de la voix et de la technologie dans l'articulation du récit phonographique. *Protée 34*(2–3): 11–26.

Lacasse, S. (2010) The Phonographic Voice: Paralinguistic Geatures and Phonographic Staging in Popular Music Singing. In: Bayley, A. (ed), *Recorded Music: Performance, Culture and Technology*, Cambridge: Cambridge University Press, pp. 225–251.

Lewisohn, M. (2013) *The Beatles – All These Years: Volume One: Tune In*. Reprinted Edition. London: Little, Brown.

Liu-Rosenbaum, A. (2012) The Meaning in the Mix: Tracing a Sonic Narrative in 'When the Levee Breaks'. *Journal on the Art of Record Production* (7). Available from: http://arpjournal.com/2216/the-meaning-in-the-mix-tracing-a-sonic-narrative-in-%E2%80%98when-the-levee-breaks%E2%80%99/ (accessed 28 October 2014).

Moore, A. F. (2010) The Track. In: Bayley, A. (ed.), *Recorded Music: Performance, Culture and Technology*. Cambridge: Cambridge University Press, pp. 252–268.

Moore, A. F. (2012) *Song Means: Analysing and Interpreting Recorded Popular Song*. Farnham: Ashgate.

Negus, K. (2012a) Narrative, Interpretation, and the Popular Song. *The Musical Quarterly 95*(2–3): 368–395.

Negus, K. (2012b) Narrative Time and the Popular Song. *Popular Music and Society 35*(4): 483–500.

Nicholls, D. (2004) Virtual Opera, or Opera Between the Ears. *Journal of the Royal Musical Association 129*(1): 100–142.

Nicholls, D. (2007) Narrative Theory as an Analytical Tool in the Study of Popular Music Texts. *Music and Letters 88*: 297–315.

Tagg, P. and Collins, K. E. (2001) The Sonic Aesthetics of the Industrial: Re-Constructing Yesterday's Soundscape for Today's Alienation and Tomorrow's Dystopia. *UK/Ireland Soundscape Community*. Available from: www.tagg.org/articles/xpdfs/dartington2001.pdf (accessed 4 March 2015).

Tagg, P. (2013) *Music's Meanings: A Modern Musicology for Non-Musos*. New York: Mass Media Music Scholar's Press.

Tennant, N. and Lowe, C. (n.d.) Suburbia. Available from: http://psb-atdead ofnight.net/mezi_radky/btl_please.php (accessed 15 September 2017).

Zagorski-Thomas, S. (2009) The Medium in the Message: Phonographic Staging Techniques That Utilize the Sonic Characteristics of Reproduction Media. *Journal on the Art of Record Production*. Available from: http://arp journal.com/621/the-medium-in-the-message-phonographic-staging-tech niques-that-utilize-the-sonic-characteristics-of-reproduction-media/ (accessed 1 November 2014).

Wishart, T. (1986) Sound Symbols and Landscapes. In: Emmerson, S. (ed.), *The Language of Electroacoustic Music*. London: Macmillan, pp. 41–60.

Wishart, T. (1996) *On Sonic Art*. Amsterdam: Harwood Academic Publishers.

Zagorski-Thomas, S. (2014) *The Musicology of Record Production*. Cambridge: Cambridge University Press.

## Discography

*The Beatles* (2015) "Yellow Submarine (2015 Stereo Mix)". *1 (2015 Version)*. [Digital Download] Calderstone Productions.

Camel (1975) *Music Inspired by the Snow Goose*. [LP] Decca.

Echosmith (2013) "Cool Kids". *Cool Kids*. [Digital Download] Burbank: Warner Bros.

Eminem (2000) "Stan". *The Marshall Mathers LP*. [Digital Download] Interscope.

Eminem (2013) "Parking Lot (Skit)". *The Marshall Mathers LP2 (Deluxe)*. [Digital Download] Aftermath Records.

The Gorillaz (2005) "Feel Good Inc.". *Demon Days*. [CD] Parlophone.

Kraftwerk (1978) "The Robots". *The Man-Machine*. [CD] Los Angeles, CA: Capitol Records.

Minogue, K. (2010) "Get Outta My Way". *Aphrodite*. [Digital Download] Parlophone.

The Pet Shop Boys (2001). "Suburbia (The Full Horror) [2001 Remaster]". *Please: Further Listening*. [CD] Parlophone.

Picket, B. 'Boris' (1962) "Monster Mash". *Monster Mash*. [Digital Download] Master Classics.

Pink Floyd (1979) *The Wall*. [LP] Harvest.

The Who (1969) *Tommy*. [LP] Track Record.

Williams, R. and Allen, L. (2013) "Dream a Little Dream". *Swings Both Ways (Deluxe)*. [Digital Download] Farrell Music.

Wonder, S. (2002) "Living for the City". *The Definitive Collection*. [Digital Download] Universal Island Records.

## FURTHER READING

Herman, D. (2009) *Basic Elements of Narrative*. Chichester: Wiley-Blackwell. (Herman proposes four fundamental properties of narratives from a cognitive perspective and gives a general introduction to a variety of narratological concepts.)

Ryan, M.-L. and Thon, J.-N. (eds) (2014) *Storyworlds Across Media: Toward a Media-Conscious Narratology*. Lincoln: University of Nebraska Press. (Ryan and Thon's collection provides insightful perspectives on how the affordances of different media support narrative interpretation and the imagination of possible worlds.)

# Part Four

# The Context of Production

# 17

# "Clever" Tools and Social Technologies

## To What Extent Does Wider Access to Technology Necessitate a Step Change in Approaches to Teaching Music Production?

## Alex Baxter

Music is undeniably important, a universal language that forms part of every culture. Everyone engages at some level with music and has at least one piece of music that evokes memories or feelings. It has the power to relieve stress and be therapeutic but can equally help push people to train and work harder. It can draw crowds, whilst its presence can equally let individuals cut themselves off from the natural sonic environment. This, coupled with music's ability to exist seamlessly within both foreground and background space, makes it a unique art form. It is no surprise that being involved in the production or creation of music is therefore appealing to large numbers.

Interestingly, the uptake of non-compulsory conventional music education appears to be challenged. Whilst there are a range of study routes, such as national vocational qualifications (NVQs), and the Business and Technology Education Council awards (BTECs), which aim to focus more on practical training, Gill and Williamson (2016) point to music GCSE uptake rates of only 7.4 percent. The percentage of those studying at the higher level of A-Level (either music or music technology) drops significantly further. In 2015, The Department of Education announced the intention that *all* pupils starting year 7 (11-year-olds) would take English baccalaureate (EBacc) subjects – English, mathematics, history, or geography – the sciences, and a language when they reach their GCSEs in 2020. Music (and indeed every arts-based subject) is noticeably absent, with Daubney (2015 cited in Ford 2017) suggesting uptake rates will likely fall further and that by 2020, music as a taught subject "could be facing extinction".

## INSPIRE, MOTIVATE, AND ENGAGE – THE COMPLEXITIES OF MEETING (AND IDEALLY EXCEEDING) EXPECTATIONS

Strategies to attempt to avoid a disconnect between people wanting to engage in making music and problems that can arise within formal music education if they are not carefully mitigated against have been well documented over a large time frame. Elliot (1989) highlights that "music (including one's conception of what music is) divides people as much as it unites them" (1989: 12). Equally, Green (1988) points to "cultural expectations" and "dissonance" if resources fail to match expectations. Whilst this originally related to metallophones and tambourines being poor substitutes for electric guitars and drum kits in schools; 30 years on, from a teaching music production within a university setting point of view, this concept of cultural expectation (and avoiding dissonance) still very much relates, especially so at the point when a student is choosing *where* to learn music production. The key areas that arguably cause dissonance are access to studios and digital audio workstations (DAWs), the amount of time available to use these specific resources, whether whoever is doing the teaching will be teaching the things that the student wants to learn about, and whether the resources available are perceived "capable" of enabling the creation of a desired "sound". The current high cost of higher education is now arguably a further factor on *whether* to learn music production formally at all or instead to take a more individualized learning route.

The original method to learning the trade of a music producer was of course becoming a "tea boy" (or girl) in a studio. This was the beginning of the apprenticeship model for those wanting to learn how to produce music. The role was actually one of the hardest to get right. As an "unknown", there was the need to show competence and understanding but with no arrogance, developing quickly the ability to be invisible but always there and implementing a solution (before anyone else) and just at the point when others could see it was needed. Crucially, the tea role was where the ability to problem solve and troubleshoot started – having a strategy to make sure everyone got the right drink let alone the right number of sugars in their coffee/tea. Failure to do this was a surefire way to instantly annoy people en masse and ruin a "vibe" before even getting the chance to demonstrate any audio or production competencies. Progressing on from this role, often to that of an assistant engineer, was the sign that people actually trusted you.

The number of studio complexes, however, is now substantially smaller than it was, resulting in an equally substantially reduced number of opportunities to enter the world of music production via this route. Additionally this route was usually only available for one, or a very small group of people, at a time in a specific studio, and the need to be "useful" was important. Davis et al. (2014) point to growth in popularity of music technology degree programs at the same time as the "decline in informal apprenticeship systems [such as the old 'tea boy' route] that have typically provided a gateway to employment in the recording industry" (2014: 1). Universities are often required to recruit much larger numbers to their music technology, production, and audio engineering programs than the one or two from the older more informal routes, however, and to this extent, it highlights an imbalance – small numbers of students studying music at school (and Daubney's stance that music could be facing "extinction" in this area) feeding universities that require healthy numbers wanting to study on their programs in order to make courses viable to run), feeding a diminishing number of professional studios. Interestingly, pressure is applied to universities following the implementation of the UK government's Teaching Excellence Framework (TEF), whereby participating higher education providers receive a gold, silver, or bronze award in order to reflect the excellence of their teaching learning environment and student outcomes. The framework was developed by the Department for Education in England, but individual providers in Scotland, Wales, and Northern Ireland are also able to take part. Factors such as recruitment and employment outcomes (along with other areas such as value-added returns and student feedback) impact on the overall TEF rating of an institution. Small numbers of applications and limited employment options are therefore not a good thing – fortunately there are other more unconventional routes in and equally there is still demand for capable music technology, production, and audio engineering graduates within what must now be considered the broader music industry.

Successful businesses are very aware of the need to continually innovate and evolve provision of their products, and as the numbers of high-end studios have diminished, so too has the need for the same level of very high-end and often large format studio products with their associated high purchase costs. The Focusrite brand is one such example of this evolvement; originally established in 1985 to serve high-end professional recording studios, its products were focused on high-quality recording and production equipment for professionals. More recently, its focus has switched to broadening its range of products to serve a wider customer base, to include professional, commercial, and hobbyist musicians – one of the company's current key straplines being "*We make music easy to make*". This specific targeting of the novice market and the creation of products that are easy to learn how to use yet importantly enable users to craft a "professional sound" has led to a range of unconventional routes that budding music producers and potential university students now take to learn about music production and indeed how to produce music. "Bedroom production" is arguably the new apprenticeship model, and companies that make these products know this. Whereas previously, profits were made from the sale of a few large-format "pro-end" products, now the cheaper product, with "expertise built in" that can be billed to help a music producer (or aspiring music producer) achieve a desired sound is where sustainable music product design and development is at. Substantial demand for these well-designed "clever" technological tools exists – there are of course substantially many more bedrooms than professional studios.

A "clever" technological tool is therefore a product that can automate certain production strategies or tasks. Expertise is built into the product in order to enable functions that are designed to enhance music or sound quickly, as well as be simple to apply, and this helps with creating the perception that they are important to creative flow and production realization. If the functionality of a "better button" can be designed in at the product-development stage, then as long as it does succeed in the achievement of a goal, without significant time investment (this could be learning how to achieve the goal conventionally or implementing something conventionally), it can actually be highly beneficial to all users, offering convenience to professional or commercial users and a better sounding result for the hobbyist or novice.

Learning from others both formally and informally is well established as a practice within the area of music, but with the rapid development of social media avenues and technologies, so too is the ability to access a much broader range of learning opportunities from increasingly widening settings. Motivation and learning as a consequence of social interaction and imitation relates back to the social learning and cognitive theories put forward by Bandura et al. in the 1960s. The concept being that people learn from one another, via observation, imitation, and modeling. Social media technologies offer flexible learning opportunities and also a rich range of media in which to learn new knowledge and techniques – video sharing, wikis, online courses, blogs, and hosted chats to name but a few. Liu (2010) points to the fact that more and more students commute to university rather than live on campus and more and more need to juggle studying with part-time or full-time jobs. This is aligned with my own experience; the image of the traditional student is rapidly changing. Liu continues by highlighting that whilst not necessarily originally developed for learning purposes, social media tools are "wonderful communication tools" with social engagement, options for direct communication, speed of feedback, and results among the main reasons driving social media adoption. The demise of older models of music production training has on one side left a void, but arguably the learning landscape is much broader now – people have much easier access to music production software and hardware tools at hobbyist level prices, enabling these tools to exist in the bedroom or the project studio. If an element of "automated experience" is also built into these tools in order to simplify the music production process, effective results are achievable without years of training. This coupled with videos shared on social media platforms, especially those platforms where video sharing is the core functionality, means that users have a place to go when they want to find a solution to a problem they face. These videos tend to support the "how to do X" model of learning but crucially allow for a highly individualized learning approach, especially relating to pace, enabling watchers to go at a speed that is comfortable for them – rewinding sections that they may need more time to comprehend. McGoogan (2016) reports that in 2015, more people visited YouTube than Google, making it the second most visited website in the world (after Facebook – another social media platform). Commonplace now is the enhanced video learning model, where a paid subscription to a video training series includes the ability to send in individual questions for answering in a later video or resource. Interestingly, the option to ask a question was often not available in the old "tea boy" training route (a session could not be stopped or slowed down to facilitate the asking of questions), and in larger classroom learning environments, there can be a nervousness to ask questions. Where does this leave the formalized academic context for learning to be a music producer, if devices and tools are designed to be "clever" and an abundance of informal learning resources via social media platforms are available to provide the training?

## TEACHING MUSIC PRODUCTION AT UNIVERSITY – THE COMMON STUDENT ENTRY POINT AND THE DESIRED EXIT POINTS. . .

Consideration of the typical student entry point and desired exit point(s) enables the approach to teaching music production to be better scaffolded. Students who decide to follow a formalized academic route to learning about music production at the degree level often start their studies with skills and experiences drawn from a very broad range of areas. Sometimes these experiences are gained from the more traditional music learning opportunities catered for in school education, but most often, they are either blended with or are solely derived from the experiences gained from producing music in a bedroom and learning technique through a variety of social media platforms. Comfort zones exist, and often students feel the perception that they are reasonably well versed with producing music in a particular favored musical style. Commonly, prior experience is centered around "in the box" tools – students will often state a preferred DAW and plugin set used in the creation of their music and can often feel uncomfortable if forced to switch to different tools that they are not used to – "I could make this sound amazing if I was doing this in X" being a common phrase for example. Plugins are often used following word of mouth / social learning discovery – either from watching a video of someone achieving a specific result, which motivates the student to want to achieve a similar result, or from attempting to find out which plugins a favored music producer uses to again attempt to emulate a similar sound. Plugin settings are often shared informally and interrogation of chat boards and wiki spaces often reveal questions around what compression settings to use for vocals (or other instruments) with often very little detail as to how the material was recorded, its dynamic range, or its level. Interestingly, many students describe themselves as either a music producer or wanting to be a music producer, but unpicking what this means often leads to many differing responses. Shepherd (2009) identifies seven distinctive types of record producer – the producer as the engineer, the mentor, the remixer, the musician, the artist, or the visionary. The seventh trait is much less easy to quantify – that of the producer with the "magical touch" or secret formula that guarantees success, the "genius" who can take techniques, expand on them, and invent new ways of doing things and inspire others to be interested in these new ways and adopt them at the same time. The term "music producer" is therefore blurred – it means different things to different people, and this certainly needs to be considered in order to appropriately plan a teaching approach and curriculum.

In terms of the exit point, in addition to the expected overarching attributes that universities expect their graduates to have – traits such as professionalism, employability, and enterprise; learning and research skills; intellectual depth, breadth, and adaptability; respect for others; social responsibility; and global awareness to name but a few, there is the obvious need for music production graduates to be "industry ready". However, the fact that technical resources and "clever" tools are within almost everyone's grasp (the same tools that students were using before entering university), there is much competition and choice, which can make it very hard to cut through as a music producer. The void that the demise of the older models of production training has left is a challenging one to fill within an academic context. Real-world projects certainly help to galvanize and provide focus and often instill a belief, especially when starting out, that there is the potential to be noticed and that future work may be forthcoming. This is largely removed when studying in an academic context, and as such so is the energy and adrenalin that comes with working with unknown people or recognized (famous) people who are clients. In an academic context, recording and/or producing a band where everyone is your friend is simply miles off being anywhere near the same experience. Rather than attempting to fully fill such a void though (a challenge not least because of complex organizational structures and increased student numbers), there is a place for an education experience that is much more routed in exploring music production in its widest sense, including equipping students with skills and traits that the older models of production training perhaps just either expected to be there or did not require at all. Williamson and Cloonan (2007) point to the fact that "there is no such thing as a single music industry" (2007: 320). The step change needed in teaching music production in a formalized academic context should be to fully acknowledge this by ensuring that graduates understand how they can fit (and importantly be able to fit) within a broader industry with broader career routes running through it. Ensuring students are exposed to areas connected to music production and technology that they might not have previously known existed is just as important; this should be one of the hallmarks of a good degree course and indeed is arguably easier to simulate in the formal academic context than it is in the commercial environment where "time is money" and a specific output is required. A series of eight areas should be fostered in order to shape the learning of music production and bridge the typical entry points to higher education with the desired exit point(s) – these eight areas also interrelate by impacting and reinforcing each other as the areas are developed over time.

## TECHNICAL UNDERSTANDING

The scope available to developing technical understanding broadens significantly over a three-year period of study. Often students embarking on a program of study at the very beginning believe this to amount to being able to use a specific DAW well – perhaps even everything there is to know about using a specific DAW. In reality, whilst being able to use a DAW (or range of DAWs) is important, there are numerous alternatives to degree study if this is all a student really wants to achieve from a course. The proliferation of tutorial videos available on the Internet (both free and also paid for content) can do this at a tiny fraction of the cost of higher education. Some universities even embed this type of content into their teaching approach, and this can be beneficial in order to provide effective starting points for discussion, further exploration, or connection to the area of technique development. In a way, this is one way that students can be exposed to observing a broad range of professionals and differing approaches, and can, if the content is high quality, lead to a more enriched experience. Technical understanding must go much deeper than just this, however, for those wanting to engage further in music production and must aim to craft a knowledge and understating that, being almost second nature, can be applied to different scenarios within a number of spaces. Technical understanding includes understanding the typical engineering practices of sound capture and microphone choice and placement, taking into account acoustically how sound behaves in the space where it is being captured, created, manipulated, or enhanced; and methods needed to mitigate room-related problems along with appropriate signal flow, routing, and structure decisions. When working with others, solid technical understanding and knowing how all the equipment works (whether this be in the studio or on location) enhances the flow of a session and enables the focus to be fully formed on the job in hand – that of producing music. This area of technical understanding naturally draws a similar alliance to high-level technical competencies required and found in the more traditional ways of learning music production in a studio, but in a formal academic context, they can be constructively scaffolded.

## EXPERIMENTATION

The bedroom production route is all about experimentation, and as one of the more typical initial individualized routes that students follow before formalizing their study within an academic context, it needs to continue to be fostered. Unconventional ways of working allow for freedom, a sense of no shackles, and a number of producers have achieved great success via this route. A potential shortcoming of teaching music production within a degree setting can be the typical structure of investigation leading to a piece of summative assessed work at the end of a period of study and the fact that students understandably want a clearly defined task and mark scheme, so as to know exactly how to achieve marks. Tasks which are more formative or which receive little or nothing in the form of an allocated grade, even if they are marketed as being highly beneficial to developing practice, tend to be a harder sell to students who are often needing to juggle studying with other commitments. Time experimenting, using the studio as an instrument, investigating, and discovering new approaches to recording and sound creation, exploiting ambiences, layering material, using equipment (even broken equipment) in unconventional ways, and generally exploring the breaking of established practices is, for a developing music producer, crucial and connects very well into the "easier to grade" areas of technical understanding and technique development. The best courses offer a sustained scope for experimentation and ensure that personal experimentation exists, is encouraged, and is appropriately rewarded. Assessed tasks need to have opportunities for thinking outside of the box built in, in the same way that cohorts also need to be continually exposed to the experiments that producers and budding music producers have explored and the results that have been achieved. Easy access to resources (whether these are studio, lab, or portable resources) aids in experimentation. Davis et al. (2014) calculate that a typical student may amass 500 studio hours over a three-year degree period, compared to 7,500 hours a busy professional might amass (working 40-hour weeks) over the same period. Five hundred hours over 90 study weeks (one year typically representing 30 weeks of study) works out at around 5.5 hours of studio time a week, and this figure may fluctuate depending on the number of studios an institution has and the number of people needing to access them. Music production students should be encouraged where possible not to just think in study weeks (blocks of time that are punctuated by vacation periods) and should also be encouraged not to see the submission of a summative piece of assessed work as the closing of a set of learnt skills. Outside of term time, there is good opportunity to continue to access studio resources (which may be significantly underused during these time periods) to continue developing an individual's music production craft, and exploiting this also goes some way to reducing the differential hours gap.

## PROBLEM AND TROUBLESHOOTING

The systematic solving of problems and troubleshooting requires the blending of understanding, creativity, and analysis. Constrained tasks can be useful in order to attempt to push students down more experimental routes of working and extended thinking, often with the rationale being an attempt to support an area of technical understanding. However, such tasks need careful thought so as to ensure that the result is worth the process in the eyes of the student and is not just an un-authentic experiment. Often in both the academic context and the traditional industry approach to learning music production, the working space is already defined and often maintained by technicians. Much can be gained from a problem-solving and troubleshooting perspective, however, if students are actively included in the building of a space. This can involve both the spec'ing of a physical room (acoustics measurements, sound treatment placement, and position, AB testing, and so on) through to the connecting up of equipment, speaker placements, software, outboard configurations, and so on. Involvement such as this becomes highly authentic and real – the student knows that this is a space that they will be using when they are producing their music. Only accessing and using spaces that are up and running leaves a substantial void when a student no longer has access to that space to work in, perhaps after completing a program of study. Involvement with studio design concepts therefore provides opportunities for students to hone their own spaces, and take steps to improve their own critical listening position. Simulating this in the more traditional industry method of learning music production in a commercial studio is naturally problematic, would result in studio downtime, and therefore is invariably never an option.

## INDUSTRY CONNECTIONS AND THEIR INPUT

In addition to the skills, research, and professional links of both academic and technical staff connected to a program of study, the input of external industry professionals and organizations through master classes, focused sessions, and supporting curriculum steering is vital in order to further endorse course relevance, quality, and innovation. A benefit of the older studio route to learning the craft of music production was the number of people that would flow through a studio and the clear variety in the approaches that could be observed in order to achieve results. Course accreditation, which promotes the supporting of links between industry and education, such as that provided by Joint Audio Media Education Support (JAMES), is one way to ensure that a broad consortium of music, entertainment, and media industry organizations feed their expertise in and that students have opportunities to be exposed to a variety of approaches. Interesting outcomes often happen in panel discussions (perhaps hosted by a course lecturer), where the panel members are drawn from differing areas, for example, a producer, a mastering engineer, and a re-recording (dubbing) mixer, because often in situations like these, the panel members will also ask questions of each other in order to understand the differences in approaches taken between them – this can be very powerful for students because they see that even very established professionals, making a living from the broad industry that they want to move into, still ask questions. Practical learning experiences with industry professionals working with small groups of students provide rich learning opportunities and also the opportunity for students to ask questions around technique and reasoning – often harder to do in the old industry route because of the need to remain "invisible". Bringing a number of industry professionals in at a time helps to simulate a sense of the outside industry "inside" and makes the industry feel closer to the students through immersion in time-based projects, such as recording and producing an EP over a week, heightening the adrenalin experience. Both the process and the results can then be reflected on with respect to developing professional practice.

## AWARENESS OF THE BROADER INDUSTRY AND OPPORTUNITIES AND BEING MORE THAN JUST AN END USER

The broadening out of the music industry means that it is now even more important to expose students to the huge range of connected areas which make up music production or music technology as a whole – including those areas that students may not even know exist when they start their course. As music production tools have become "cleverer", so too have tools from other disciplines. Just as the broadening out from the studio into the bedroom has opened up accessibility for engaging in producing music, computer coding is no longer solely for the computer scientist. Tools that bring coding closer to the musician, such as Cycling74's Max/Msp, and DSP Robotics' Flowstone are further enabling new ways to produce, shape, manipulate, and perform, as well as to create new plugins and software applications – the potential creative outcomes from these being almost limitless. Additionally, these tools also bring connections to the world of open source electronics platforms for rapid, easy-to-use hardware and software prototyping for creating and engaging with interactive projects. Other areas that also connect in relate to the world of sound design, forensic audio, audio archiving, and corporate audio. The proliferation of bedroom recording and novice production in non-specialist spaces arguably enhances the need for producers with audio restoration skills in order to mitigate problems that have been inadvertently incorporated in to the recording and production process.

## TECHNIQUE DEVELOPMENT

On one side, technique development is the fusion of technical understanding and ideas formed through experimentation opportunities and observing other peoples' approaches and strategies. At baseline, this is about being able to apply and channel knowledge and understanding in order to achieve an appropriate result. There are, however, an additional number of other techniques which need to be fostered, and these should align to developing all of Shepherd's (2009) previously identified seven types of record producer roles. Fostering learning strategies and techniques for working with others, developing strategies for communicating with different people, balancing personalities, being easygoing, and being able to stay focused and keep a project to time (under pressure), as well as having a developed confidence in order to coax or support clients appropriately to get the best out of them and achieve the desired results, also fall under the crucial area of technique development. The three-year period of degree study provides time within which to develop, alongside the skills needed to work as a producer or engineer, these "softer skills" and a sense of overall maturity – traits which are good to have and highly transferable across all industries. They are also especially important for securing the first job.

## ABILITY TO PROMOTE ONESELF AND COLLABORATE

Collaborating with others, effectively making friends with other creatives in connected industries, helps to provide vital interdisciplinary experiences. The formal academic context is good for this where often these other creatives (film and TV students, animation students, and so on) are only a staircase away and studying in the same building. They provide excellent opportunities for working out how to work with others in a melting pot where it is important to be able to get your ideas across well. Links such as these also help to plant seeds for the future – that film and TV student that you worked with one day might be in a place where he or she needs a music producer, therefore providing the building up of a contact book of people who are all roughly at the same level.

Whilst integrating detail relating to the music business is important, so too in the 21st century is the development of entrepreneurial skill. With music technologies and "clever" products able to exist more in non-standard spaces and be used by a very wide range of people at varying skill levels, the ability to promote oneself above the competition is vital. It is no longer satisfactory to just have a website – now there is also the need to really understand the wealth of rich analytics available that provide detail on how many people are engaging with your content and in which ways. Being able to translate this detail into developing a personal career strategy is one of the keys to increasing the chances of success, and therefore, the scope for this needs to be fostered into the planning and delivery of a degree course. Tying in to this are the benefits of connected media and social medias and exploration of other methods for direct targeting of a defined audience. Because the informal route of learning how to produce music uses the communication tools of social media, options exist for savvy students to be part of the training of others through a dedicated channel of high-quality created content.

## CRITICAL REFLECTION AND EVALUATION OF APPROACH, CONTEXT, AND RESULT

Key to supporting the interrelation to the previous seven areas is the development of higher-order thinking through evaluation. This is synonymous with what is generally expected from degree study and features in many guises – on one side supporting the individual music producer's journey via opportunities to evaluate successes and processes, to learn from mistakes made, and to connect these to adjusting practice, and on the other side, promoting the critical reflection and discussion of musical works related to specific aesthetic criteria as well as the broader music industry context – this includes reflection on where musical works sit in relation to comparable artists, the historical lineage of music production practice, and current trends. This wider context analysis supports the development of the "producer vision" and also enables the student to be able to identify just who the audience is, including how to access them and how this audience specifically accesses its produced music.

## CONCLUSIONS

The ease of access to "clever" tools and social technologies has certainly seen the numbers of budding music producers rise in recent years. At the same time, the term "producer" has broadened and blurred in order to become more of a "catch-all" phrase with various sub-roles being able to be defined. The old industry route to learning the craft of music production is still a viable route. It provides focused entry within a very specific area, but competition for places is high and roles still tend to need to be mastered before a progression route becomes available. However, embracing music production in its widest possible sense offers a huge range of opportunities – arguably more so now than ever before. The appropriate course of action here is to acknowledge the strengths that the various routes can offer and play to these strengths – avoiding the temptation to want one route to simply fill the role of the other. The formal academic context very much still has its place as one potential route and can certainly embrace from a broad base the connections to the numerous areas that relate to music production over the typical three years of degree study. The need to maintain currency should be the focus by blending skills and technique development with reflection, in order to enhance practice and approach, whilst also keeping an eye on current and future developments of the wider industries and integrating these – embracing all the connected potentials that such developments can offer.

## BIBLIOGRAPHY

Bandura, A. and Walters, R. (1963) *Social Learning and Personality Development*. London: Holt, Rinehart and Winston.

Davis, R., Parker, S. and Thompson, P. (2014) 'Preparing the music technology toolbox: Addressing the education-industry dilemma', *Journal of Music, Technology & Education*, Vol. 7, Issue 3.

Elliot, D. (1989) 'Key concepts in multicultural education', *International Journal of Music Education*, Vol. 13, Pages 11–18.

Ford, L. (2017) 'Teachers blame EBacc for decline in Music student numbers'. [online] 9th March. Available from www.sussex.ac.uk/broadcast/read/39525. [Accessed 17th August 2017].

Gill, T. and Williamson, J. (2016) 'Uptake of GCSE subjects 2015. Statistics Report Series No. 107'. Cambridge Assessment. Available from www.cambridgeassessment.org.uk/our-research/all-published-resources/statistical-reports/

Green, L. (1988) *Music on Deaf Ears*. Manchester: Manchester University Press.

Liu, Y. (2010) 'Social media tools as a learning resource', *Journal of Educational Technology Development and Exchange (JETDE)*, Vol. 3, Issue 1, Article 8.

McGoogan, C. (2016) 'YouTube is now more popular than Google.com on desktop computers'. [online] 1st February, Available from www.telegraph.co.uk/technology/2016/02/01/youtube-is-now-more-popular-than-googlecom-on-desktop-computers/. [Accessed: 2nd July 2017].

Shepherd, I. (2009) 'What does a music producer do, anyway?' Available from http://productionadvice.co.uk/what-is-a-producer/. [Accessed 15th June 2017].

Williamson, J. and Cloonan, M. (2007) 'Rethinking the music industry', *Popular Music*, Vol. 26, Issue 7.

# 18

# Contemporary Production Success

## A Roundtable Discussion With Mid-Career Music Production Professionals

## Russ Hepworth-Sawyer

## PARTICIPANTS

Three colleagues, all known to the author personally, were chosen for their similar ages and different experiences and routes to the industry.

British-born, award-winning producer and engineer Adrian Breakspear, now working from Sydney, Australia, provides an insight into the industry on both sides of the globe. Adrian started assisting in studios such as Abbey Road and more latterly the Strongroom, London. He then made the brave move to Australia to soon become the man behind the desk at Sony studios there. Adrian has just won an ARIA for his work with Gang of Youths.

Andrew Hunt brings insight into modern-day record production through his experiences of being at the heart of the London production community as a long-standing board member of the Music Producer's Guild. He deservedly also won the MPG Breakthrough Producer of the Year 2017 with his work with Jonathon Holder & The Good Thinking.

Finally, Mike Cave is a Liverpool-born mixing and mastering engineer, having started his professional career after a qualification in London at Liverpool's Parr Street studios working with the likes of The Charlatans and many other acts. Mike now works from his Loft Studio mixing and mastering.

Adrian and Mike were interviewed live either in person or by teleconferencing. Andrew was interviewed by questions and answers using email.

The process of music production has changed considerably in the past 20 years as these professionals have risen through the ranks of their industry. One of the aims of this roundtable was to ascertain any perception these professionals obtained of both the change in the industry since they started and any historical change that had preceded their careers that they had heard of from their peers. Another aim was to seek the underlying philosophies to their work and professionalism. Attitudes to production have also changed, and capturing this was a key aim also, for example, the roles within the studio and record-making have dynamically altered in recent years.

The questions posed were all, essentially identical, with slight subtleties adjusted to each interviewee's specialism. For example, questions that relate to recording were more applicable to Adrian Breakspear, whereas mastering and mixing questions were more appropriate to Mike Cave. As a musician, now producer, Andrew Hunt's questions were tailored accordingly. The core questions are posed in the following, interspersed with their responses. Not all the answers are represented within this chapter but are selected for their appropriateness to the narrative.

## HOW WOULD YOU DESCRIBE MUSIC PRODUCTION?

*Andrew Hunt:*      Today, being a producer, it's probably a "creative project manager". Just like a film director. For me, making a record is making something more incredible than real. What film is to a theater, a record is to a live performance. Hence the film director similarity. It certainly combines organization, technical skill, and lots of creativity.

*Adrian Breakspear:*      It's helping to realize the artist's material in the best way possible that I can. To translate the artist's ideas into the most representative outcome possible. Not everyone can articulate what they want to hear. It's my job to translate that to reality. Take those early albums from The Strokes or Arctic Monkeys or Kings of Leon. Those sorts of songs would be ruined if they were overproduced. They were intended to have a few guitar parts and a vocal and they translate. Obviously as the band develops, they add new layers and become further produced, but those early songs were written with that in mind. That's production.

*Mike Cave:*      Production is a team effort. Even if you're totally driving the session, it's still a team effort. For example, Digitalism, they're out DJing most nights and I'd send them a mix and then they'd play it out overnight and in the morning in my inbox I'd have a few suggestions for revisions. They can hear it in the music's native place. Working like this is how it really works for me. It's all about working together.

## COULD YOU OUTLINE THE PATH YOUR CAREER HAS TAKEN THUS FAR? WHAT ARE YOU MOSTLY UNDERTAKING NOW (PRODUCING, RECORDING, MIXING, MASTERING)?

*Mike Cave:*      I basically split my time these days between mixing and mastering. My background is very much of the traditional route through commercial studios as the tea boy, tape op, assistant, so that route really. I've now been freelance for the past 15 years just cracking on making records. When ProTools first came out, commercial studios used to hire it in, as most sessions were still using tape. However, for those sessions that a producer wanted a ProTools rig, the studio would hire it in for £300 a day or whatever. I jumped on that pretty quickly and ended up being someone who knew the software and therefore became useful in sessions. A role I took on if you like. It was hard then, as you didn't have access to the system unless the rig was in front of you. It's not like these days where you could download a copy of ProTools native and at least understand the infrastructure and what not. So for the studios, this was dead money, but they weren't ready to invest. This was at a time of TDM cards and rigs were sort of £30k–40k to buy. I soon realized that if I got my own rig, I could be charging for me and the rig, so I sort of get that sorted by a hire purchase agreement with a company. This was the kit that let me set up ultimately. Mixing nowadays is my thing, and whilst I've produced quite a few records, I tend to have fallen into what I enjoy which is mixing mainly out of my own facility called The Loft in Liverpool.

## WOULD YOU CLASSIFY YOURSELF AS AN ENGINEER-PRODUCER OR A MUSICIAN-PRODUCER (USING RICHARD BURGESS'S ORIGINAL CATEGORIZATIONS FROM ALL THOSE YEARS AGO), OR SOMETHING OUT OF THIS?

*Andrew Hunt:*    Well that's broadened as I've progressed. When I started, I was, without doubt, a musician-producer, but as I progressed from project to project, I took on more elements of the process (often due to budgets actually). Engineering became a love and a passion. I obsessed over it, really honing my craft. As such, it was a jaw-dropping honor to be shortlisted for Breakthrough Engineer of the Year in the 2017 MPG Awards. Don't get me wrong; I was over the moon to win the Breakthrough Producer Award that night, and if I'm really honest, at heart, I probably think of myself as a producer, so that was the one I really wanted.

*Adrian Breakspear:*    I guess I'm one that comes as much from the engineering side as much as the musical side. I don't think I really separate the two roles that much. I mean some people I've worked with have never been able to engineer themselves, but for me, I'd never want to separate the engineering side from the production side [when producing]. I guess my philosophy overall and I'm working with a new band or artist is to not put my own stamp on it, so to speak, but give them the best version of what they want or what I feel they should sound like, or enhance what they are already doing. With some bands, it's quite obvious they know how they should sound and want to sound like and it's helping them achieve this by whatever means necessary and sometimes that involves almost doing a "Frankie Goes to Hollywood" and replacing everything, or getting them to play to loops, or copying and pasting parts around, but perhaps it's just nips and tucks. More often, it's just me making suggestions and hints that enhance what they're trying to achieve. In most cases, the band knows how they want to work, and my role is to make nips and tucks and to clean out the deadwood, keeping the things that are clear and removing the things that are not working, by replacing or adding to them to give them the best example of their songs I can. Generally, I'll be thinking texturally, but if needed, I'll get into the music too. I don't really want to be one of those producers that people come to for "that" sound. To me, it's the music that speaks, and I facilitate. It is nice to get known for doing a certain thing [especially if successful], but it's dangerous and can date you. It pigeonholes you, and then you don't get a lot of other gigs you might have done. For example, I've obviously done the Gang of Youths stuff [Adrian won awards for this in Australia], but perhaps less known I've been mixing EDM and pop tracks and recording acoustic projects. Right now, I'm working on Muriel's Wedding, the musical, with Kate Miller-Heidke, which is quite a random thing to do. So I've been recording that but not producing that, which is quite a challenge. I quite enjoy getting all these random jobs and not living in one pigeonhole. So I guess that's my philosophy. "Try and make the band sound the best they possibly can do".

*Mike Cave:*   I think those coming into the industry today really need to consider being an engineer before saying they're a producer. It would be a touch shortsighted not to have those skills and play just that one role. There are, having said that, producers that are totally hands-off, and I still work with a few. It's rare though and for all the reasons of budgets, it becomes very difficult financially. I think professionals these days need to wear several hats to remain competitive.

I believe that engineering needs to become like driving a car, second nature. That frees you up then to be creative on the production. It's like you use both sides of your brain. The technical is on the engineering, although second nature, whilst you're concentrating on the creative. That's why a session is so tiring. After 10 hours in the studio on a busy session, you're a bit dazed. Today though, you've got to have all these areas covered to be competitive.

## WHAT APPROACHES OR PHILOSOPHIES DO YOU TAKE TO PRODUCING OR UNDERTAKING YOUR WORK? IS THERE A PATTERN, A STANDARD? IS IT SOMETHING THAT IS UNIQUELY "YOU", OR DO YOU KNOW OF OTHER PRODUCERS OR HAVE YOU OBSERVED FROM OTHER PRODUCERS WHAT THEY DO AND ADOPTED BEST PRACTICES SO TO SPEAK?

*Andrew Hunt:* My philosophy to production [for example, engineering, production, the people, the message] is it's everything. For me, the purpose of making a record is to convey the emotion of the artist's intention through the song. This has to encompass every element, from the writing, the recording, the mixing, and the mastering, the people, the environment, and so on.

*Adrian Breakspear:* Unless it's a project with a low track count stuff where I know there's just a guitar and a vocal and that's about it, I tend to avoid the highest sample rates. On larger projects, you lose processing power and you double disk space. As the projects develop, it can hamper flow and the operation of the project. The Muriel's project starts with 40 to 50 tracks from the live show, which we're building on. So to try and do that at high res, which would impede the workflow. I'm capturing stuff that is good in the first place. None of that should really suffer at 48kHz and 24bit. It still should work through great conversion. It's not practical at the moment with the technology. Clients aren't really asking for any more than 48k. And to be honest, lots of people are just upsampling to 96/24 for delivery to MFiT (Mastered for iTunes). I hear lots of bad high res work, which should have been sorted before worrying about the tiny bit of improvement high res can give over an excellent captured at 48/24.

*Andrew Hunt:* I always have to really get to know the artist. I need to build a level of trust where we both/all feel so relaxed and open, there's no holding back. Trust in both directions. They need to feel completely comfortable with me. Then you're open to push the creative boundaries bearing their tastes in mind (because you know them) and equally to know when and how to push the performance to evoke the right emotion. It's even more important when you're then working on the records on your own. You need to be confident in the decisions you're making and if you have a full understanding of the artist, you know if those decisions are right or not.

*Adrian Breakspear:* [Speaking of current times] I'm not really very big on the slating of current records and lauding the '60s and '70s records. You know that people sometimes say that no records after 1995 were any good or "why can we not make albums like the Beatles anymore?" My answer is that if you want to do that you've got the tools to shape the sound to achieve that. You just need to shape the frequencies, such as the tops and bottoms, and be careful as you go, you need to bounce things down and not be too precious if the balances aren't quite right. If you want to go back to those bounces and make changes, then you can't. That's what makes the shape of those records, not the esoteric tape machine. The esoteric gear doesn't just make it the way they think, and with plugins and so on, we can get close with the equipment these days. It's just the vibe and the approach to the production. The hip-hop kicks that actually appear on those Amy Winehouse record when people try to compare her to The Daptones, make it of its time. They really just want flavors of things.

If I were to submit a mix sounding like the Beatles with a more mid-range bias to the sound and so on, or Otis Reading where the vocals are distorting at the top and the bass drum is swallowed in the mix and the snare off to the right slightly, the label wouldn't want that at all. We perhaps listen with rose-tinted hearing aids a little too much probably. We listen with our hearts and not our heads too much. I'd quite like to try one day, for experiment, to make a record using those techniques where I'd have to submix things down and not go back to them. I could use the ATR plugin on UAD with a tape emulation and so on, I could perhaps get close, especially at 7.5ips.

It might be fun to deliberately limit myself and the production to just capturing with just a tape emulator and then use the TG channel stuff from Abbey Road, where you've only got the brilliance control or treble and bass and see how it comes out. Continue that restriction with two mics on the drums, such as a 47 on top and 47FET on the bottom. To not allow to stray outside of what they had in the early 60's. I might just do that!

*Mike Cave:*    My bugbear is quality control. I approach every job that comes in as though it's going to be the "best record ever made". I know that's not going to happen, but my approach to it would be the same. Obviously, I get jobs in that are not ready and still need work before they can be mixed and mastered. I need to be careful as these are not the quality we're after, as I'd need to do a lot of work to fix it (additional production) before it can be done. I need to be honest with labels and the like. For example, stuff comes in from labels that I was convinced was the demo. I asked them for the mix files, and they told me that this was it! I had to decline the work. It's not worth putting my name to something that's not going to sound as good as it needs to. The adage "you're only as good as your last job" is still the way things are. The problem with things like this are that if I do the job, it would never get approved anyway, and as a result, the phone would stop ringing for work. So I need to be careful. It's important when to say no too! Quality, on time and on budget – that's my philosophy.

## WHOM HAVE YOU LEARNT YOUR TRADE FROM AND HOW? DID YOU ASSIST, AND WITH WHOM AND ON WHAT SESSIONS?

*Andrew Hunt*

Andrew highlights his self-taught route: I'm completely self-taught. Also, everyone I've ever worked with, be that other engineers or artists. You always keep learning. I read a lot of interviews and books too. The information is always out there. Primarily though, it's about experimenting and listening. I've not really ever assisted. I worked on a project with Jimmy Hogarth and Cameron Craig once. That was fun. It was location recording in a big church, so Cameron asked me to help out, as the job would be too big for one person. He was definitely the "engineer", but we had an assistant too. I sort of filled a "second engineer" role. I also did a similar job for Mick Glossop in an old music hall. Apart from that, I've always been the producer in the studio. Ooh, I tell a lie. Shortly after I did the job with Jimmy and Cameron, Jimmy called asking if I had a number for a good assistant, as he was stuck. It turned out that my assistant was ill that day, so I'd moved what I was doing. I was at a loose end, so flew over in the car and tape op'd for him. Another fun day.

*Adrian Breakspear:*

I wouldn't say I've had one specific mentor. I've been lucky enough to assist and work with a bunch of different guys. But probably I've worked most with Mike Crossey, who's now in LA, but started off in Liverpool and then bounced about back and forth from London and then moved to LA, where he is now. He's done The 1975, Wolf Alice, Nothing but Thieves, and when I worked with him, I worked on some of Razorlight, a bit of Arctic Monkeys, and a whole bunch of other bits for artists that didn't make it. Then I worked with him on the B-side album for The Kooks. I got a lot to watch over his shoulders for both recording and mixing. I also worked for Chris Sheldon, where I assisted on his mixing session. He's the guy who did things like Foo Fighters and Biffy Clyro. I worked with Guy Massey a little bit, who was at Abbey Road. I was a runner at Abbey Road for six months in the early 2000s. The studio was recording things like Lord of the Rings part 2, Harry Potter part 2, and Gangs of New York. Starsailor's second album Silence Was Easy. It was an exciting place to be.

There's also Jamie Lewis, who's a friend now, who we've done a lot of stuff together. He was Mike's (Crossey) assistant but was thrown into being a producer before he'd been his own engineer, which was great for me as a guy ready to make the leap into engineering. Then I'd be working with him eventually as an engineer assistant and ProTools guy. It was good for him to learn how to rely on me, and we learnt our way together. We did several albums together with the Dutch band Direct and The Tunics and a bunch of other sessions. It was a good period for me.

Markus Dravus, who did Coldplay, was a good person to work with for a few days. I did several sessions with him and worked with Rory Kushnan, who was a great person to work with. I've also done odd days with Mike Hedges and Ian Grimble as well as Nick Terry, Pedro Ferrer, so there's been lots of people I've worked with who have been really useful to get a feel for production. It has stopped me being a "mini-me" of another producer or engineer, so that helps me not being pigeonholed.

What's changed a lot, when I was learning, people used to hire the room to get their hands on the SSL to mix. So they'd come into all this hardware, and I'd not really assist them in the mixing, and it's rare I'd get involved when they were using tape, but when they'd come in with a hard drive for ProTools, and there'd be a conversation about what plugins they'd want in advance and they'd know what the spec was and they'd come in and install their favorite plugins they'd want to install. They'd then mix on the console, and I'd be there as an assistant to make sure they've saved the recalls, and they'd go away, and in a couple of weeks, they'd come back and they'd ask me to help them recall those mixes. They'd make their changes, but now if they go into a commercial studio, they tend to print all the hardware processing back into ProTools and then don't need to come back to do the mix revisions with you on the console. Obviously working for Sony as I do, I'm not in that commercial studio environment at the moment, but talking to my friends at Miloco, this is what I'm told is the norm in London currently. Alternatively they now bring their own rigs, stick it up two channels of the desk, and just use the monitoring, which is obviously a good way to work. Recalls are just not necessary anymore.

What happens when the crop of maintenance engineers retire, as many of them are in their sixties? There are very few guys that can service a console or do proper wiring work. There are amp techs and guys that can create and design their effects units, but those people to fix a psu in the field . . . is not something I am capable of. But we need people like that at the moment coming up. I can fault find, but I can't tell you which capacitor has popped.

*Mike Cave:*     The most influential producers would be those who have taught me by me watching them work with artists and getting the best out of them through the etiquette, diplomacy, and psychology – people skills basically. I learnt more through observing them more than any of the artists I'd worked for.

Ken Nelson, who did all the Coldplay early records. I was his assistant at Parr Street for many years. He's a super cool guy who's so chilled with artists. He was never short of showing me tips and tricks. He can work with anyone, and even the most awkward artists, he could deal with them with expertise. I'd learn more with him in a couple of hours than a day on my own. I learnt so much. I'm still mixing for him today, and we've a great working relationship.

Jeremy Wheatley is another I did a couple of records with. He's a good mixer these days, and I learnt a few things off him.

The best bits I learnt being an assistant at Parr Street was learning people skills. As stupid as it sounds, making tea, and getting the measure of everyone in the room. Watching producers deal with clients and vocalists who are not getting the take. It's how to get the best out of someone before the session collapses. How do they deal with that psychology and getting the best out. There's real value in that. There's really only two things that make a good traditional record: performance and the song. Performance is so key to it. So watching the seasoned producers getting the best out was one of the best things I ever observed. I also learnt how not to work from some producers who did not work as well! I can do better than that!

Learning the studio etiquette was a really key thing as an assistant. The studio can be quite tense at times, and every person in the room is going to add a certain dynamic to the session whether they are talking or not. It's how the session works in the room, rather than just letting it happen. So it's finding that balance, and as an assistant, it's important to know how to fit in. My view was to not talk unless spoken to and to be watching and learning the whole time. Anticipating the problems in advance of them happening, and if, for example, the mic choice was wrong and you'd get mics ready because I've been pre-empting what they might need and have it on hand (i.e., a patch cable ready in my hand) for the engineer when they need it. Efficiency.

I'd grown up at Parr Street where I was for six or seven years. I'd been employed as an assistant engineer and an engineer, and I had to take a leap of faith and go freelance. It was The Charlatans who became my catalyst for doing so. They'd been coming into the studio, and then they asked me to go away with them and record [engineer] an album with them, and it turned out that I ended up sort of coproducing it with them. As an assistant in a main commercial studio, you can't really go off for six weeks and do an album with someone. You can't really come back from that. It took something as big as that for me to make the move and sacrifice that assistant's job. So I guess The Charlatans took me under their wing. At that time, they were a big band, and it was quite nice to be validated by them and gave me the confidence to go freelance. Luckily since then I've been busy thankfully.

I was quite an early adopter of ProTools and especially earlier than anyone in my vicinity. So I picked up a lot of work as a ProTools operator and editor. Also as a programmer too. People hired me in to take control of the ProTools system, and like with The Charlatans, I ended up running both ProTools and the desk and a touch of production too.

## SO DID YOU REALIZE AT THE TIME THAT THIS WAS THE FORERUNNER TO HOW MANY STUDIOS AND PERSONNEL WOULD OPERATE IN THE FUTURE?

*Mike Cave:*

The other leap of faith was mixing in the box. That was not easy for me. I was so used to a tape machine, a desk, and speakers and a few bits of kit – and that's what I'd learnt on. The move to the mouse was terrifying. I jumped in and made the best of it I could. It kept me quite busy in the end, as I was able to offer all-in deals for the labels earlier than many, as I could do instant recalls on mixes. And using my own facility, I was able to manage the costs better. In previous times, a producer would not be able to be as certain on the budget for an album; I could be more certain of the costs, and at a time when budgets were shrinking, this became attractive to them to keep costs down. This is the way all projects go these days.

There is a benefit to this in so much that I'm always working out of my room, and I know it's sound rather than being thrown into another room and having to learn it quickly. I can be more efficient and more successful working in my space.

## SPEAKING OF ASSISTANTS, DO YOU USE ANY REGULARLY, OR MIGHT YOU SHOULD BUDGETS PERMIT?

Speaking about the role of the assistant engineer, the myth is that this is a dying role in the production process, **Andrew Hunt** explains the times he uses an assistant: I often use assistants. I need someone, as I'm often producing and engineering myself. In some cases, playing too! The problem is always the same, when someone is good enough to fill the assisting role perfectly for me, they're too good to be an assistant and are ready to engineer! Then they're off!

*Adrian Breakspear:*   I regularly get people asking whether they can sit in on sessions, with a view to becoming an assistant or just for the experience, but unless I know the band really, really well, and it also has to be a session busy enough for them to watch and not a session where I'm too busy, otherwise they'll be on their phone before you can say anything when comping a vocal. You want something busy enough to get a bit of experience. I've had one or two sessions like that, but they're rare really. The opportunities are less these days.

I have a couple of guys that regularly assist me, but it's pretty hard. When I say regularly, I mean once or twice a month rather than once a week. I'd love to have a full-time assistant, who could cover the things I don't want to do and do the voice over before the main session. But there's just not the budget, and as house engineer, it's my job to do this, so I do.

*Mike Cave:*   I'm pretty self-sufficient, as I'm mostly mixing and mastering these days. I've a couple of people who I can rely on to prep mixes for me. If there's a lot to do and I need someone to help out, I infrequently call upon people to help, but it's quite rare these days.

## IF/WHEN YOU HAVE ASSISTANTS/APPRENTICES, WHAT WOULD BE THE METHOD BY WHICH YOU'D TRAIN, BUT ALSO THE MAIN MESSAGES YOU'D PROVIDE THEM?

*Andrew Hunt:*   I always work on the same philosophy, one I adopt myself. If you do something, just do it well, and then you'll be known as the person who does things well. When something needs doing, who do "they" ask? The person who does things well! I don't need or want an assistant to show me what they know and try to prove themselves. That will come in time. I have a job to do and a goal to achieve. I need them to help me perform that task. I need them to be invisible but always there. I need them to be in the background but have my back, ready, second-guessing. I need them to be silent until I need an opinion from them, when I need them to tell me quietly and professionally, so I can assess the best option. That one's important. I think it's really difficult, as everyone wants to show that they have valid opinions which give them worth, but there are already lots of opinions in the room, and I will have listened to them all and been on the project for some time already. Recording is only a part of the project, and I know exactly what I'm trying to achieve by that point. Their time to speak up and have their voice heard will come in time, but day one, two, or three is not the day to voice it. It will most likely just hinder the process.

## OF THE CONTEMPORARY BREED OF FELLOW UPCOMING PRODUCERS, WHAT DO YOU THINK IS DIFFERENT ABOUT TODAY'S PRODUCERS FROM, SAY, 25 YEARS AGO? WHAT STARK DIFFERENCES DO YOU OBSERVE, BOTH IN PRACTICE AND HOW YOU MANAGE YOUR WORK?

Asked if they know of any changes in their roles in the production process from days gone by, *Andrew Hunt* explains:

> I wasn't working as a producer 25 years ago, but from what I hear, the budgets were large, there was a lot of disposable cash, and this was reflected in the working practices of the time. Of course, that's different from most people working today, but if we look further back, to the '50s, I think it's almost gone full circle. In the really early days, it was the producer who was often the person who found the artist, developed them, and made the record. I don't think that is so different to today. A lot of my peers have achieved what they've achieved by hard work and belief. Most by taking a risk and working with an unknown artist, eventually getting the break with them.

One aspect producers today are having to face is more personal marketing and PR. Andrew Hunt explains how he develops relationships with his clients and none more so than *Jonathan Holder*:

> We're like two strange cogs that fit. Jonathon is an exceptional talent as a musician, a writer, and a producer in his own right. We are very different but have a complete trust in each other. If neither of us agree on a point, it means we haven't found the best idea yet! We keep working until we do. It's striving for these high standards and respect of each other's talent that keeps our working relationship strong. We trust each other's opinion and know our fortes.

*Adrian Breakspear:*   Being someone that's come from the engineering background, I find it a little hard to separate the two roles, and if I want to get a particular guitar sound, it's probably better for me to try and get that sound rather than articulate it through several people such as the guitar player and the engineer. Unless, of course, I'm working with someone who knows me so well, it's already done. Unfortunately, I've not had the situation where I can train someone up at every session I do, and if I had, then I'd probably be more willing to delegate things. However, I'm also a bit of a control freak, as we all tend to be. If I'm working on a production from the ground up, where there are parts to be played and parts to be programmed, I'd not be averse to bringing in a drummer, because I simply cannot play the drums, but I'd feel a bit shitty if I got a guitar player in, as I consider myself a reasonable player. Similarly with engineering, I feel I can do that, so it would seem weird for me to get someone else in to engineer when I can just do it myself. I've not managed yet to get someone on board I can rely on. The reason I guess I've not had the assistants through the route I had in London is really because of shrinking budgets. Once upon a time, I would go and assist and engineer for people, and in those days, you'd have a producer and engineer, a Pro-Tools operator and the assistant(s), and that's four roles. Nowadays, that's been boiled down to two or perhaps one role these days. I don't tend to have an assistant unless it's a really, really big session, like the one I've got on Sunday. I'll have an assistant on that, but the rest of the time, I'll not ask, as it's not necessary, and I can't justify the cost these days.

*Mike Cave:*   Record sales and budgets have been the things that have changed. Take the Napster thing, and when that all kicked off, the labels perhaps got complacent. Record sales and the way people consume music has been changed forever, outside of the labels' control. They thought they were so powerful, and they thought they'd never lose control of distribution. They dropped the ball, and given that, they're now playing second fiddle. They can't go back now and charge the value they wanted, but people don't value music now and are used to it for free. The question is now how to monetize music for artists and the industry. The music industry needs to discover new revenue streams. The budgets are so tight now, and as such, the quality of music has taken a touch of a dive. Artists are making music at home now and in less-than-ideal environments. If you wanted a mix done, you'd hire a mix engineer, and that's how it was. For a reason! Now people are doing their own mixes and have a view that the mastering engineer will make it good. So cumulatively, the quality drops at every stage a little. As it hits the market, it's noticeable. For example, a lot of labels have their in-house mastering now to save budgets.

Ken Nelson used to say to me, "If you do cheap jobs, you become a cheap producer". It's so easy to say, "Leave it with me, and I'll to the extra elements to make this record work", but you'll find that you'll not be charging what you should for the extra week it's taken you to do the music to the level it needs to be. It's important to get paid and in some ways stand your ground. It's important to say it the right way to people so you don't get a bad reputation. You can't deliver great records without charging for your time as I've a facility to pay for.

## WHAT NEW AND UNIQUE CHALLENGES DO YOU, AS AN UPCOMING PRODUCER, FACE TODAY?

> Income streams. In order to "earn" from your work, are you writing with your artists, or are you seeking points on contracts today? Are there other methods/tactics used to ensure your income?

Asked how he monetizes his work today, *Andrew Hunt* outlines that he'd never ask for income streams he doesn't deserve:

*Andrew Hunt:*     Income was always difficult. The most interesting projects are often from independent artists who have little, if any, budget, but you have to find a way!

Often, I write with the artists, sometimes a little to push the song to its full potential, sometimes a lot, sometimes not at all. If I have written, to whatever degree, I only think it's fair to agree on a writing split. Otherwise, I'll ask for a fee and a royalty on sales and streaming. Although the latter is almost non-existent, it's still important to have it, as a small amount of lots of records builds. This goes for PPL too. You have to look after every corner, as it all counts. However, if the artist or label can't afford to pay you what you need up front, you have to come to some kind of arrangement, as then you're gambling. Everyone has faith in their own project and will assure you that it's going to be huge, but that hope doesn't pay the rent, so I have to be realistic. Don't get me wrong, I'm also a hopeful dreamer!

## MIXING – WHAT ARE YOUR VIEWS AROUND THIS IN THE CONTEMPORARY CLIMATE? THERE'S INCREDIBLE CHANGE HAPPENING AT THE MOMENT WITH MACHINE LEARNING AND ARTIFICIAL INTELLIGENCE IN MIXING (CITING IZOTOPE'S NEUTRON). HOW MIGHT YOU USE THESE CHANGES TO YOUR BENEFIT, OR DO YOU BELIEVE YOUR SKILLS AS A PRODUCER WILL BE NEEDED STILL AS YOUR CAREER PROGRESSES?

Asked about mixing and the technological advances, such as iZotope's Neutron or some online mastering offerings, including artificial intelligence and machine learning, **Andrew Hunt** explained:

*Andrew Hunt:*     I haven't used any of these pieces of software or services yet. I know of them though and think it's amazing technology but have a view. . . . For me, making a record is about conveying an emotion. It's not about what is, by comparison, "right" but what is right for the emotion and intention of the song. These are very unique and human qualities. A lot of the reason behind why I'm employed or chosen is for my tastes, not because my technical abilities are capable of making a mix match the "industry standard for the genre" (whatever that is). I make records because I love it. Not because someone is paying me. So I hope I can continue!

*Mike Cave:*     From my perspective, I don't think there's a chance of automated mixing anytime soon. I think of it like driverless cars. Would you want to get in one and go up the M62 [motorway/freeway]? I think the automated "mastering" services that are out there should not call it mastering as it's not! Similarly, we're not going to get there with automated mixing anytime soon. There's just stuff that's better done the old way. Why do people like tape emulations? It's because we liked the way things sounded then, and they sounded great. There are people who want a great record but don't want to pay for it to be mixed. There's that triangle with the axis saying Good, Cheap, Fast, and you can't have all three! If you want something great, you, a lot of the time, need to hire great people (and facilities).

I'm concerned about upcoming professionals having to work for free. Spec mixes are always thought of as a way in, but I'd try to, wherever possible, to avoid this. To the people asking you, they'll not value your work or respect it. If they had paid for it, they'd want to fix it with you and get it right. They'd perceptively respect and value the work involved and therefore the output. It's very difficult for those coming into the industry now. Getting a reputation is hard work. Sometimes, if it's a job I really want to work on, I'll say to the label, look my rate is X, but I really want to do this, so I'll do this one for free, but I let them know anything in the future will be at the rate. Also I'll do a deal – well, I'll do this one for free if you give me that other gig at full rate. That can be tricky, and it's rare I'll do it. Sometimes it does pay off, and I get work from the same A&R afterward that does well, so it can gain respect.

An adage I live by is "I'm not here to make money, but I need to make money to be here". I can't pay all my studio overheads and equipment without money, so it's about being able to be here for people, and to do so, I need to charge and earn a reasonable living.

## MASTERING – WHAT IS YOUR VIEW OF THIS IN THE CURRENT TIME? WHAT ARE YOUR EXPECTATIONS OF MASTERING AS A CONTEMPORARY PRODUCER?

*Andrew Hunt:*     I value mastering greatly. It's my last set of independent ears before the world hears the art we have created. I work really hard at a mix, and I'll have always put hours of consideration into it. I hope I've got it bang on. The best compliment to me would be for the mastering engineer to say – "It's great. I don't have to do a thing". I don't want a mastering engineer to do something because they feel they have to.

I usually use the same one, and I've developed a great relationship with her. I trust her. I trust her to let me know if I've missed the mark, and I've not conveyed the right emotion through to the mix. I trust her to say, "You know what, I think you should do x, y, and z". Thankfully, that's never happened, but we have that kind of relationship.

In my role as producer, I take raw ideas, blank canvases, or demos and help the artist make that into records. Through the process, I'll have got close to it. I'll have spent hours molding it. The mastering engineer will dress it for the outside world. I want them to let me know where we're at. Have I done what I set out to do? Also, I mix track by track without too much regard to the other tracks. That consideration will have happened much earlier in the process. I want the mastering engineer to take all my individual tracks and help them to sit as one piece of work without losing what we set out to do or what I achieved in the mix.

*Adrian Breakspear*:   My aim is that my mix goes to the mastering engineer, and they say, "All I need to do is to put some limiting to make that loud", then I'd be happy. That rarely happens, but I want my mastering engineer to be my quality check, but also the technical person who managed the PQ codes and DDP and so on. The mastering engineer is the person to make the project unify things. That's why I'd probably not be a mastering engineer, as I hear the minutiae, and like most, I still hear things however well I've done.

I have to let go, especially when the mastering engineer hasn't heard the constant issues I hear in all my mixes, but I hear those same issues on other people's released mixes. We all work at that level and can hear the things we'd want to change. The client's approved it, and the mastering engineer has made an album from it, and the client's still happy. I've got to eventually let go.

Chatting to mastering engineer Leon Zervos who is ex-Sterling Sound and he's back over in Sydney now, and I had an EP, which I mixed on the move with changes in the studio, and there were some brighter tracks, some darker mixes, and he'd unify the whole thing to a really good standard.

*Mike Cave:*   When I was at Parr Street, we had what we called the edit suite. We had a Neve room, an SSL room, a demo room, and an edit suite working on Sound Tools and a TC Electronic Finalizer. It wasn't really there to be a mastering room; it was really just to be able to do radio edits and also have the ability to get our clients get their audio approved by the labels. It was the inroads to digital editing to me but also a touch of the mastering process. It's what I call today "mock mastering". Obviously, if my mixes go to another mastering engineer, I'll send them the non-compressed mixes but also give them the reference of the "mock master" to work from, as that's what everyone, and the label, have got to listen to and know. The mastering engineer could think of this as the starting point.

Anyway, as I started to mix a lot of records later on, I would sit in on the mastering sessions, or as many as I could. Sometimes, I could choose whom I wanted, but at other points, I'd go out and sit in on as many sessions as I could. More often than not, they'd want me to take the tapes over anyway. I'd be a little bit annoying and answer questions and watch what's going on. Eventually, I started to learn a bit about what they were doing.

One day, a mastering engineer listened to my mock master and said, "I don't know what I'd do better than that". That experience gave me the confidence to go out and consider doing it myself. To hear that from a mastering engineer was a real confidence boost for me.

As I came up the ranks, a lot of it is confidence with ability. I've got to appear confident to get the work. The A&R and artists need to see I'm confident to make them confident I can do the work for them.

## THE INDUSTRY IS GOING THROUGH A LOT OF STRUCTURAL CHANGE AT THE MOMENT IN THE WAY MUSIC IS CONSUMED, THE WAY MUSIC IS MADE (AI, ETC.), AND WHAT IS EXPECTED OF PRODUCERS. WHAT WOULD BE YOUR THREE MAIN PREDICTIONS FOR THE FUTURE?

*Andrew Hunt:*     Ooh, I'm not so sure about all of that. Right now, I'm concentrating fully on the record I'm making at the moment. Every day, the record I'm working on is the most important. Then the next record will be the most important, and I will adapt and change to try to make it the best record I can. I'll just concentrate on that for now. Let's see where we end up.

*Mike Cave:*     I'm getting less work with labels but more directly working with the artists, which can be exciting. It's sometimes better having fewer chefs in the kitchen, so to speak. What I'd really like to see coming up is an industry standard for delivery, level matching, and different digital platforms that play their audio. Some form of standardization that stops me having to second-guess what my work will sound like coming out of one streaming service or another, as

they all use different codecs. I have to deliver a whole load
of folders for each platform when I master, as it provides
me with so many headaches. So at the moment, some uni-
fication would be good for creativity.

With regards to high-resolution audio . . . in some ways,
Mastered For iTunes makes sense in terms of a bit of
future proofing, and to be honest, the "downsizing" sounds
pretty good. The problem is the record-buying public are
not really after more, like we are. So we have to master so
the audio sounds good on the phone!

## CONCLUSION

In the pursuit of exploring mid-career producers as they develop, all three
interviewees here are of a similar age and have begun their careers in UK
studio complexes. They, like many of their peers, have diversified in their
roles and have specialized as their careers have developed, either out of
choice or from necessity. Their starting points have not all been traditional.
As we have learnt, Mike Cave and Adrian Breakspear have both benefited
from in-house training from a major studio complex, whereas Andrew
Hunt, a musician initially, has developed his skill set to production, bringing a different experience and approach. Despite these different routes, it
appears that many of their approaches and philosophies can be noted as
similar and converging. One might consider that the industry has shaped
their views and actions in similar ways.

The interviewees' experience of being "trained" has also been noted
and discussed. Many soft skills and client-conscious skills dominate the
conversation. It is also expressed that those trained within the studio paradigm receive their training very much by observation, rather than any traditional educational technique. One must consider whether, in pedagogical
terms, current educational establishments explore these aspects as being as
important as the interviewees consider them in their own training.

Interviewees look to the future and, on the whole, do not appear to be
greatly bothered about the forward march of technological change that
could, through machine learning and artificial intelligence, potentially
change their more skilled roles of mixing and mastering forever. The interviewees did, however, comment on and infer the considerable change in
the industry and the way in which they and their peers are paid and try, like
we all may, to consider the impact on the future.

Much more could be made of these interviews than we have room to
explore here. It is hoped that the interviews provide an interesting insight
into those professionals working at the middle of their careers in music
production in different locations and performing differing roles. It is hoped
the experiences presented here will be of interest to those studying the
changing role of music production and the music producer in the 2010s.

# 19

# Conversations With Women in Music Production

## Kallie Marie

Popular culture in Western society has seen a recent rise in public con-sciousness around ideas focused on equality and women in the workplace, and many are asking, "Where are all the women?" in fields like STEM, business, and government but also in the arts. For example, Kim Gittleson (BBC, 2017) reported on the lack of women in economics, Ruth Marcus (*Washington Post*, 2017) raised questions about where all the women are in the current presidential administration of the United States, Adi Ignatius commented (*Harvard Business Review*, 2013), on the struggles of work-life balance for working mothers and the disparity in the concerns of what should be an equal concern for working fathers, and Megan Erickson also wrote on the subject (Big Think, n.d.), asking where all the women sci-entists are. Both the research and curation of the interviews conducted led to the emergence of patterns that were not anticipated on the behalf of the author. These collected conversations shed light on the nuances of shared experiences and differences across generations of women who have worked at various and high levels as music producers and audio engi-neers. Each of the women interviewed shared unique sets of experiences, perspectives, and outlooks. Their collective voices are referenced with the utmost respect, weight, and gravity that their own narratives carry. Some of the key aims of this research, alongside offering voice to these women's narratives and experiences, were also to show what sorts of challenges women still face and how this impacts both the people making the music and the music being made. During the course of these conversations, each of the women viewed themselves, their roles, and their careers thus far, both uniquely and collectively, in different respects. However, an unantic-ipated by-product of the research was that many were reluctant to discuss a few areas relating to these topics, or to even discuss them at all, which really spoke to the present state of women in music production because of the uneasiness that many felt about these issues and the impact that speak-ing about them publicly could have for their careers.

The interviews were carried out in a variety of ways: some in person, some over video conferencing software, some via voice over IP systems, and some others via email. All interview subjects were asked the same set of questions, in the same order. All questions were grouped into categories that had specific themes and goals, and this was mentioned to the interview subjects as they progressed through the questions. Those responding to the questions over email could see them outlined in such a way on the document they were sent. It was not possible to display all responses here. The categories of questions were organized around the following themes: Gendered Perceptions Influencing the Recording Process (of Women as Artists and/or Producers), Women as Producers and Engineers, The Importance and Impact of Women in Music Production, Accessible Technology – The Great Democratizer?, Thoughts on Shaping the Future, and Barriers Faced by Women in Music Production: Obstacles Yet to be Overcome. These categories were devised to provide an investigative and thought-provoking look at several facets of the lived and shared experiences of the interviewees. They were also devised to evaluate and help us understand what has contributed toward the positive advances that have been made and what needs to be done to continue making more strides in these areas, as well as what is as of yet unaddressed.

The women interviewed were all producers, recording engineers, or audio engineers, each with at least ten years of professional industry experience, working at the highest levels of their specialities. Women who were DJs, beat makers, self-producing artists, or mastering engineers were not included in this round up, as those are separate specialities that will have their own nuances and are likely to be covered in other volumes dedicated to those specialties within the broader scope of music and audio. The women quoted herein were also chosen with as much variance in age, musical style, race, geographical location/nationality, marital status, and orientation as possible.

An unanticipated signifier that really acted as a barometer for the present climate and how much ground still has to be gained was the overarching reluctance many women demonstrated in discussing these topics at all. Some said flat out no, while others were only willing to do so if they could remain anonymous. One could draw conclusions about a specific group of people, if they are uncomfortable with sharing their experiences about what it is like to work in their field as themselves. If a group of people fear losing work for having discussed their experiences of their work environment, their challenges, observations, and success stories, they are not yet on equal footing with others in their field. This then acted as a signifier, or barometer, that provided a frank look at issues that the interviewees raised, namely that, as an industry, music production can, and should, work toward something as an effort of continued and shared self-awareness.

As is the case with many societal issues, elevated consciousness is often the first step toward change. Raised societal awareness can create knock-on effects that often have a lasting impact in many permeable areas. For example, in order to see changes in many workplace environments, some argue that equality must start at home, both in what children see patterned in the division of work among parents and caregivers but also in how labor and tasks are divided up between adults in the home. Arlie Hoschild (1989) noted in his book *The Second Shift*, details of work life, home life, and the balances and inequities experienced by women and men in the home between the 1970s and 1980s. As long as women working outside the home are still doing more than half of the housework and child rearing, many women are working a second shift. Gloria Steinem, in 2009, commented that society cannot expect that women will be equal in the workplace until men are equal in the home.

Initially, the focus of this research was to uncover insights into the patterns that created these successful producers and engineers who have all come from varying backgrounds but share the commonality of their gender and succeeded in varying circumstances and socioeconomic climates. Evaluation of possible ways forward, which might ensure the continued successful inclusion of women and thus the fostering of variance in the types of producers and engineers who participate in the music industry in the future, was also explored. It is considered that the inclusion of more diverse producers and engineers might impact the types of music that may be created; for if more diverse portions of society are involved in music production, then so too it could be anticipated that the types of music produced and the methods of production may also diversify.

To gather insights into the successes, experiences, and perspectives of each of these women, as producers and engineers, key topics and experiences were explored. One of the topics discussed was their early access and exposure to technology and science, their perceptions of how their early careers were shaped, and the challenges that they may have faced along the way. It is important to start at the beginning and look at the relationships these women had with technology at an early age. Evidence suggests that the types of exposure a child receives in early developmental stages crucially affects the child's interests and skills later in life. Many of the interviewees shared an early exposure to not only music and technology but also to science and in some cases, the mechanical aspects of recording. Another important distinction to make, in our contemporary digital mindset, is that technology doesn't necessarily mean ubiquitous digital technology for these women. For many, this meant electronics, early computers, building-block-type toys, or scientific gadgets and experiments. The producers and engineers were all asked to reflect on this.

In the interview, Sylvia Massy, a multi-platinum producer and engineer, reflected on her early exposure to technology and said,

> I was fortunate as a kid to have a father who built tape machines when I was a baby, so he gave me a microphone to play with when I was two, and he recorded me playing with a microphone. So I was very familiar with the equipment very early. When cassette tape recorders were first introduced I would do these silly little productions with the neighbor kids and we would pretend like we were in a theatrical show and we'd each have a role, and then I would put in the sound effects and the music and everything in the background, and that was super fun. I did have opportunity to play around with the equipment when I was young and it was consumer grade equipment, but when I was about 18 or 19 I learned how to play

an instrument and joined a band and we went into a professional studio; it was my first experience in a professional studio, and the engineer was kind enough to say, "Here's some faders. You can play with these, and it will do certain things. Like you could make an echo here, you could make a delay here", and I was fascinated with that. It was something that I wanted to do because I was familiar with the equipment early.

Abhita Austin, an engineer and studio founder, upon reflection of those early days, had this to say:

Yeah, I had tape recorders, and VHS tapes, and I would record. . . . In hindsight I guess that's the beginning, but I would always record my voice, record my friends, and I was young. I was like 5/6, very young. I would always take things apart. I was always just fascinated in a way with these gadgets, and eventually I got like those little Casio keyboards, where you could record a second.

Johnette Napolitano, film music producer and artist, reflected similarly: "I bought right into every innovation there was the minute it came out".

While producers, male and female alike, may share the same innate curiosities, early exposure affects all children in the same way. Importantly, all the interviewees had similar experiences and curiosities around technology and recording. Meanwhile, society perhaps has coded to them that these activities are not for their gender. It is a unique combination of their individual home environments, drive, personalities, and the fostering they received from home that possibly led to their rejection or ignorance of, in part, what society might have had them do instead.

Perhaps as long as society continues to nurture young girls in ways that focus on exposing them to interests deemed feminine, with the frequent socialization of young girls with toys and sociological cues that imply women are non-technical, it would continue to lead to fewer women participating in certain career fields.

Dr. Elizabeth Sweet gave a succinct TEDx Talk (Sweet, 2015) that touched on the multifaceted impact of gendered toys and how they have historically gone through periods of being less gendered and then more gendered and how this correlates to childhood development. Dr. Cordelia Fine (2010) notes in her book, *Delusions of Gender*, that both early cognitive stages, as well as the social cueing, help children learn about themselves and what are deemed as acceptable interests for their gender. Let Toys Be Toys, a UK-based organization, raises awareness on its website (Let Toys Be Toys, 2014) on why play matters and the lasting impacts and inequalities that are created during early childhood education. Furthermore, the George Lucas Educational Foundation, as cited on the Let Toys Be Toys website, mentions studies showing that the future of who participates in STEM shows a direct correlation between early childhood play and future career paths saying,

Gender-typed toys are not effective in fully developing children's academic, artistic, cognitive, musical, and physical skills. In addition, they perpetuate stereotypes that can limit a child's scope of possibilities for future education and professional aspirations. Toys are children's tools for exploration and discovery. They are not only a fun part of childhood, but necessary in a child's development.

Judith Elaine Blakemore, professor of psychology and assistant dean at Indiana University-Purdue University Fort Wayne | IPFW, Department of Psychology, in 2014, conducted two studies that evaluated the perception of toys and their effectiveness in child development. The toys were categorized as strongly feminine, moderately feminine, neutral, moderately masculine, and strongly masculine. Blackmore found that "strongly gender-typed toys appeared to be less supportive of optimal development than neutral or moderately gender-typed toys".

Of the women interviewed, each exhibited curiosities for taking things apart, liking technology and music, and in some way getting exposure or access to things that helped further these curiosities through toys, family environment/makeup, or encouragement. The more women are educated in these roles, through exposure and now through computers being a regular part of education, we can and should expect to see a continued number of women taking interest in these career paths, so long as their early developmental environments not only offer access to such modes of play but also send strong messaging that these types of play are acceptable for all children. The women interviewed for this book were not all from generations that had this mindset, so it is quite lucky that some were nurtured in environments that encouraged this exploration but also that they were confident enough to follow their interests into adulthood and develop their careers from them, however out of the "ordinary" it seemed. It is not only early development at play here but also a common preconception that women's brains aren't technical. This plays into a loop of causation through lack of early exposure and will be discussed later, when the burden of proving oneself to be technologically capable is discussed.

The observed patterns that emerged during interviews are noted here, but beyond this, they illustrate a picture of commonality experienced by women in their field. It brings to realization that in their experiences, as isolated individual practitioners of their craft, women audio professionals seldom have the opportunity to reflect on or share with others in their field. Many echoed sentiments of being unique, not like others, "an odd bird", or not like other girls, or of "always being one of the guys".

Many of them expressed viewing themselves as somehow being different, which raised questions around the idea that if they knew how like each other they were and how alike they are to countless other women in varying technical fields, they might feel less othered. Perhaps this illustrates that their sense of other is rooted in societal conditioning and cues, received in early childhood education and in present-day media and marketing. One example might be the marketing of Father's Day gifts of tech and gadgets, while Mother's Day primes society to give women flowers, chocolate, and trinkets.

The messaging of the classic construct of masculine as active and feminine as passive can feed into these subliminal assumptions about the career choices and roles. Part of the isolation of having few peers or mentors to interact with is the fact that isolation is often part of the job. Working long hours in studios is commonplace and can lead to an isolated lifestyle, both in the work environment as well as the social. This isolation, coupled with a lack of visible representation and role models in the field, may also lead to even deeper, sometimes unconscious, feelings of isolation.

This research is presently in flux, due, in part, to organizations and media coverage of prominent women in music production. One example would be the Audio Engineering Society's promotion of the UN He for She campaign, particularly championed in the UK Section. The rise of social media has also had an impact on allowing otherwise isolated individuals to see themselves differently and connect with others. Having accurate representation not only impacts the mindset of the women in the field presently but the aspirations of other girls and young women in the future. This is starting to change, perhaps with the help of social media influence; however, one might argue that previous patterns and mindsets live on in the generations of people who are already working in the industry now. It could, perceivably, take a long time for deeply rooted, highly coded social norms to change.

Normalizing young girls' interests, whatever they may be, the same being said for young boys, could help remove future stratifications and rigid roles along gender lines in employment and interests. Sending messages early on to children about what is "normal" for their gender has lasting impacts, both in self-awareness and in the activities that they seek out. This impact is prevalent among adults on a deep sub-conscious psychological level; for example, Fine (2010) documents that even among women at Stanford, seeing other women engaged in mathematics, science, and engineering (MSE) led to them having more interest in the subject areas. This is significant because today one of the leading responses to the lack of women's involvement in music production and audio engineering is the classic "Well, maybe they're not here because maybe they just don't like it". There's substantiated research to suggest the opposite and that what it really has to do with is a variety of nurture, social cues and conditioning, and stereotype threat, for those who are faced with stereotype threat not only underperform but also grow disinterested in the topic as they wrestle with and view themselves as "other". As Fine (2010) wrote, "[W]omen who are invested in masculine domains often have to perform in the unpleasant and unrewarding atmosphere created by 'stereotype threat'. Anxiety, depletion of working memory, lowered expectations, and frustration can all ensue". Fine also notes that, "Stereotype threat can do more than impair performance – it can also reduce interest in cross gender activities" (Fine, p.42 "I Don't Belong Here"). Stereotype threat also has noted cognitive effects and can impact learning and, as mentioned earlier, a participant's working memory. Having this collective self-awareness helps shed light on things that one might otherwise decide to dismiss, blame themselves for, or internalize. When asked if they had ever worked with another woman who wasn't an artist, just merely had team members in the studio, whether engineer to assistant engineer, producer to engineer, or vice versa, the majority of women lacked any experiences or had very few experiences of working with other women as part of a production team, and for many, it had never occurred to them that they had not had such an experience. The fact that they had never even reflected on what it means to never have had any instances of this in their memory is interesting enough in and of itself, as this will likely have had a psychological impact on their self-image and social grouping, which in turn will likely impact their performance at work, whether or not they are aware of it.

In a role like producer, how will this impact their creative and technical output? The interview subjects were asked to reflect on their time in studio environments and whether they had had opportunities to work with women, besides artists, and if so how it was received by clients to have a woman or two on the session.

Andrea Yankovsky, an audio engineer formerly from Power Station/ Avatar Studios, reflected,

> When I was on staff at a studio, it was very rare to have outside team members come in, such as a producer, engineer, to have more than one woman aside from myself. I was really lucky, in that I was working at Power Station/Avatar. Zoe was managing, and one of her strengths was that she just made gender a non-issue, for running the business, and so there were female assistants and female PAs. When projects would be staffed with multiple assistants, there were women. But in terms of outside independents coming in, I can count on one hand the number female engineers and producers.

Hillary Johnson, a music producer and recording engineer, spoke of her similar experience of having the opportunity to work with other women, "The place where I worked in the '90s, where I was the manager, two out three assistants were women, because I hired them".

It would appear that while many women found it rare to have other women in technologically related roles on their teams, either as engineers, producers, assistant engineers, or interns, these two women did have the opportunity to work with other women, and in those instances, they were women who were hired by women.

One interviewee, who wished to remain anonymous, who has produced and engineered for 18 plus years on the West Coast of the United States, had this reflection on the topic:

> I think when you are the only woman on the team, sometimes you get treated like the token, and there's a bit more pressure because you feel you're representing all women in that space. So whether you make a mistake or do something great, it can feel like it will be interpreted as a reflection of the competency of all women in that role.

When there is a whole team of women, your gender is not as noticed and there is not as much pressure to uphold or shatter a stereotype. When there are other men around and you're the only woman, the client often assumes you are not in the technical role and will automatically approach or defer to the male coworker. When there is a team of women or several women, they have no choice but to talk to a woman and there is less resistance.

Abhita Austin also had reflections along these lines about her experiences in the studio:

You know I've been on sessions where I assisted, and the clients would come in and greet everybody except me, because I was the only woman, and then when I went to the patch bay to plug stuff [in], they were like "OH! WOW she's assistant engineer!?" So they just thought I was somebody hanging out.

These experiences are not unique to music production and are shared in other areas such as STEM, business, government, and manual labor. For example, Kristen Schilt in her research "Just One of the Guys?" (2006) noted, "Similar patterns exist among blue-collar professions, as women often are denied sufficient training for advancement in manual trades, passed over for promotion, or subjected to extreme forms of sexual, racial, and gender harassment that result in women's attrition". In presenting these experiences, it potentially removes it from being an isolated experience that the person may internalize, and also it possibly sheds light on the phenomenon that is experienced by women, across varied fields.

Perhaps when a social grouping of people are all experiencing similar phenomena in the workplace, while also being veterans in their field, it does suggest that there is something more than anecdotal collusion afoot. Many interviewees did comment that things were worse for them in this regard when they were starting out and that it had to do with the level of the clients or the studio that they were in. On the face of it, one could say that it's not that big of deal, but as the patterns emerge, one may start to ask what is the impact of having most of a professional life in isolation and of being second-guessed and "othered" in the workplace? The impact this has on the output of the people who experience this could potentially be immeasurable but no less important. Creative and technical output are likely impacted and impeded, not just for the individuals, but for the field and industry as a whole.

Some areas are so subtle, ingrained, and pervasive among the general society and culture that they aren't often recognized as such by those who experience them, because of the frequent internalization of these social norms. For example, women in STEM as well as other fields like politics and finance are still reporting similar experiences, as alluded to earlier. The self-blame and self-doubt are part of a larger social construct, a consistent undercurrent, where for a variety of reasons, more complex than there is space for here, young girls and women are taught to question themselves and their experiences and to doubt their abilities. This leads to regular self-blame and internalization, with mindsets like, "I must have imagined it" or "I am overreacting" and sadly even, "It's my fault. I shouldn't have . . ." This internalization, or fear of being at fault, was deeply rooted in the working experiences of the engineers interviewed, and many expressed concerns over distracting colleagues and carefully choosing clothing. Sylvia Massy mentioned, "For years I wore a T-shirt and jeans to the studio, now I dress up, and wear heels, and I will wear a skirt whenever I can, but it may not be appropriate".

Kerry Pompeo, a seasoned audio engineer and producer in New York City, echoed similar thoughts:

> [W]hen I was coming up, that's how I felt I had to be, and I felt that I had to always wear jeans, and always wear a T-shirt, and always have my hair in a ponytail, and have a crew neck up to here so that I wouldn't distract anybody.

Andrea Yankovsky also echoed similar sentiments in passing, "[Y]ou start to wear turtlenecks in the summer . . ."

These engineers and producers have brought with them society's unwanted gift of self-doubt and internalized views of themselves as distracting. The context of the issue this presents is beyond the scope of the discussion here but nevertheless was a theme that repeatedly came up, without prompt. Many of the engineers and producers interviewed commented that confidence was key for young women in the field and something that was severely important to a young woman's success in the recording studio as either a producer or engineer.

Still there are those who may either experience or not perceive these instances and may unknowingly perpetuate them without realizing it. These subtle areas of subtext and ingrained behavior might be the last frontier of change. These subtleties are apparently not unique to the music business but are part of a larger and more pervasive and insidious global, historical, cultural, and socioeconomic context. These sentiments are echoed through many areas of society globally, with initiatives like Slut Walks in urban neighborhoods of the United States aiming to tackle some of these perceptions, while the sale and trafficking of women and girls worldwide persists, all the way to the covering and hiding of women's bodies or the selling and commodification of them. If society cannot divorce itself from notions that what women look like or wear is more important than what they say, do, and are capable of, it will likely continue to affect women's perceived value (whether internalized or projected onto them) in any workplace, not just the music industry. Whilst not a main aim of this research, personal appearance does illustrate one small micro facet that evidently impacted the interviewed engineers and producers, as during the interviews, this was not asked, as the topic is too complex for the space allowed. It is noteworthy that this sentiment was raised independently of the thread of questions. These issues do permeate many industries and affect people working and living within these constructs. In turn, this shapes aspects of the industries that they work in and invariably their collective output and the music being created.

Demographically speaking, these experiences will be shaped by each person's unique set of circumstances. Although each of the interviewees' career paths varied, many faced similar challenges. Some felt that gender was a factor, while others said they couldn't know for sure or suspected that it was a mix of other factors like their race or age. For example, Johnette Napolitano said, "I will say that, yes, I've had to get my bitch on to some soundmen who assume I don't know shit, but it was much more ageist than anything else".

While Ebonie Smith, an award-winning Atlantic Records audio engineer commented:

> I cut my teeth making records in Africa. Working in studios and going around, so the gender issues are way, way more pervasive, and I had a lot of people say, "Hey, you know, you should be cooking. You should be taking care of your man. You shouldn't be out at the studio, late at night. You should be home by a certain hour". My perspective is, I don't hear you. I'm out here, and I'm going to do what I've been called to do with my life, and that is to make records, and nothing is getting in the way of that.

For many of the women interviewed, a mindset of putting "blinders on" and ignoring naysayers was a key coping mechanism early on and ongoing, if needed.

How each of the interviewees experienced and handled these situations is very different. For example, overcoming stereotype threat is likely a common challenge, and many of these women will have dealt with it, sometimes in several layers (both race and gender or both gender and sexual orientation for example) often without being conscious of each of these factors. It is also well documented that the extra cognitive load of such a phenomenon existing can cause those under stereotype threat to unknowingly underperform, as well as having to deal with microaggressions simultaneously. Andrea Yankovsky summed it up by saying,

> It's like the kids at the school that are in the flight path of the airport; they learn how to tune it out, but it still has an effect. I would generally just ignore it, that was my coping mechanism. Even if it bothers you, don't show it, don't let them see that it bothers you, because then they are just going to do it more.

Many mentioned that they had received little to no mentoring, had to learn early on to ignore microaggressions and comments, and that they each worked diligently at being prepared to prove themselves beyond the average level expected of their male peers, while also making serious life choices about family, friends, and relationships in order to have their careers, functioning with little to no sense of community and combating, often daily, the task of either establishing their credibility despite being a woman or carefully presenting themselves (almost curating a persona on a very high level) from the types of music they chose to produce or the types of clothing choices they made. Seemingly, there are similar character traits that many of the women shared, among these being resolute in focus, being able to ignore everything else around them and forge ahead, being passionate about music, and being deeply curious about technology. While their male counterparts, we conceive, share some of the same hurdles, uniquely the engineers and producers interviewed here appear to have also faced the task of proving themselves as credible to an extra degree because of society's views of women. For example, something that Ebonie Smith termed so perfectly as "the burden of proof" is the common experience of proving oneself professionally as a woman. It is often a harder task, as well as being seen as credible, believable, or even to be believed to be who you say you are or that you are actually employed in the field you say you are in. Overall, many women expressed not being believed to be who they said they were or that they had to work harder to prove that they could do the job, because overwhelmingly they were either not believed to be technical, not believed to be physically strong enough, or not believed that they are actual technicians and producers. Sylvia Massy recounts such an instance:

I was talking with a band called Avatar, a metal band from Sweden, about producing a record for them. They said, "Come see us at this show, we can meet afterwards", so I said, "Ok I will be there." The show was in Baltimore, Maryland; they were fantastic. We had never met before, and they didn't know what I looked like. After the show, they were at a table signing autographs, and there was a huge line of people waiting. Mostly girls. I said to the security, "Hey, I'm supposed to meet the band", and you know the bouncers said, "Yeah, girl, yeah. Get in the back of the line", you know? I was a bit miffed and said, "No. I'm a producer. I am supposed to be meeting the band", and they were like, "Yeah yeah yeah, sure, get in the back of the line". So I stood there for a few minutes, and I thought, well, what the heck, I'm just going to push my way through and go talk with them. So I busted through the line, through the security, and went up to the band, banged my hand on the table, and I said, "I'm Sylvia! I'm supposed to be meeting you here!" All the security rushed in and grabbed me and dragged me out of the venue. The band was like, "No no no no, she's supposed to be here!" And we had our meeting, and I got the job.

Maybe the guys, the bouncers think I am one of those girly girls or whatever . . . a fan. You see, I had a similar, previous experience with the band Limp Bizkit, where I was supposed to meet the band after the show. As I stood in that same "meet-the-band" line with all the girls, I just got frustrated, and I thought, "I don't want to be like one of these girly girls here". You know, what do you call them? Groupies. You know there are girls who just wanna see the band, go meet the guys, and I'm not that – so I got frustrated, and I left. I just walked away from the Limp Bizkit project, whereas I could have probably done that record that sold six million copies.

Ebonie Smith weighed in with a similar experience:

> I still have, outside of the label system, people who don't believe that I'm the person I say I am. I went to holiday party last year, down at a pretty big studio Downtown Manhattan, and I was put on the list by an engineer friend of mine, who works down there, and I get down there, and I am not on the list so I'm like, "Dude, you know I told you I was coming down. What's going on?" So try texting him, but because he's running around, like he can't text me back, and so the guy at the front desk is like, "You're not on the list", and I am like, "Well, can you call this guy?" and he's like, "You gotta text him or something, so he can come down and let you up". So I wait for a while and then somebody comes down, and I heard someone go, "Atlantic Records engineer is here". You know, Atlantic Records engineer, who else could that be? So, I go up to try to get in, and I'm like, "I heard somebody say Atlantic Records engineer, that's me". And he goes, "Really? You are?"
>
> At that point, I just left. So I still have these experiences. . . . Part of it, I think does have to do with my gender, and it may have to do with my race as well. It may have to do with the fact that I look younger than I am. There are lots of reasons why people discriminate against you; they don't think you fit the look of what they're expecting. . . . But that has happened to me, and you know before I was even working there, at Atlantic, you have the burden of proof.

A common thread, in nearly every interview, was that the general consensus was that more readily accessible technology, now cheaper than ever before in music recording history, as well as the advent of the Internet, has really helped level some of the playing field and has helped to dismantle gatekeeper influence and importance, remove some of the entry barriers, and provide opportunities for exposure, as well as access to knowledge bases online in the areas of training and support. Hillary Johnson felt that:

> Because of the fact that there's so much music being produced on computers, and computers are something that women are taught, just like men, in school [now], so maybe women are just more comfortable with a computer, and they only have to learn about microphones, and music and that sort of thing. . . . So maybe because we are taught to not be technical, maybe it's just easier for women to start something and to continue it because there's less technical challenges.

Kerry Pompeo also had similar thoughts on the topic:

> Because the technology became accessible to anyone, behind their own closed doors; I think where people would have been a little bit

more shy to step into a male-dominated room, they could learn for themselves at their own pace and explore in a comfortable way. So I think that that shift, in the business, and that's like what happened a lot to my career. People used to have to come to me to record everything, then I was getting a lot of, just mixing jobs, because it was like, "Hey I recorded this at home. Can you make this, can you use your ear, to make this sound better and nicer?" Now I'm seeing more female producers coming back to the studio, because they have that foundation, and they've been accustomed to it. So, the studio, as a business, is becoming more of, "Let's get back in the studio and create". Where it used to be like, "Here's the stuff from home". But I think that that really kind of helped the shift in the business of recording music and really helped with democratizing who had the ability to do it.

Of course, not everyone has the same access to technology, and while many have the Internet or some access to it, an anonymous engineer, who was quoted earlier in this chapter, pointed out that:

Still about one-fifth of Americans don't have Internet access, and many people underestimate the number of people who are computer illiterate or have smartphones but no computers. Phones are being marketed as doing the same as a computer, but the differences are huge. So while many people are getting more and quicker access to tools for music production, low-income populations, which disproportionately include women of color, are actually falling farther behind faster.

So assuming we're talking about middle-class and wealthy women, they have better access now than ever before and can realize their artistic visions on their own or in their control. This is due to cheaper equipment, more software available, and access to education online, through community colleges, and through YouTube. But to truly hear all voices, they have to remember or learn how to share the resources and knowledge with low-income women of color so they can also project their ideas, visions, and stories into the world.

In keeping with continued progress, areas that still provided some challenges were also considered. Many spoke to shared experiences of microaggressions, and though micro, the focus really is that of the cumulative effect it may have on an individual's creative/technical output, as well as her performance in her field. While some said they rarely experienced any such microaggressions at this stage in their career or couldn't recall previous experiences anymore because they had become so accustomed to ignoring it, others said they had been fatigued by them, and still others had internalized them to an extent feeling that it just comes with the territory of being a woman. Leslie Gaston-Bird, an audio engineer, educator, and AES Board member felt that:

[Y]ou know I don't know of anybody who's going to say, "you don't know what you're doing, woman." I don't think any of the stuff we have to deal with is overt because, fortunately no man is that much of an idiot, to say something so overtly sexist. So it's almost like any sexism we face is going to be this underhanded stuff.

Ebonie Smith gave an example of what she found frustrating as a form of microaggression:

The assumption of a lack of competence in areas, which can come across in a passive-aggressive way; an example would be assuming that I don't know how to use certain tools or just assuming that I can't learn. I think the most frustrating thing is not being given an assignment, because there's a fear that I don't know how to do it. You can still give me the assignment, and I will figure out how to do it. Trust that I can learn! That's something that is frustrating too as a microaggression. Like the assumption that you can't learn something . . . I say all the time, I just want the privilege of being able to fail. I learn more from failing than anything else, and I would like the privilege to fail. I think there is this pervasive need or desire to keep women from failing. You know, I think generally in society, men want to be able to rescue or solve problems for women. "You don't have to fail, because I can just do it for it you" or "[Y]ou don't have to fail, because I can just solve the problem". It's kind of like, no, I want to be the best human being I can be, so let me fail. Let me fail, especially if it's something that is not life or death. And failing at putting something together, or failing at doing something technical, if it's not the end all be all, then it's ok! Because, with my interns, I train them, I give them a real good briefing, but I like for them to fail.

Hillary Johnson weighed in saying,

I think that we are in a time where what you're talking about is probably all there is. In other words, there's no blatant sexism. It's all subtle. Unless, it's just really obnoxious, and you're like, "This is ridiculous, get the hell out of here." But I mean again, when I started, I keep referring back to when I started, because when you start that's when you get all your confidence and experience, and you don't have those two things, then you're gonna have career fatigue really quickly. I think if you have confidence and experience, because any sexism is sort of like, well, it's like just par for the course. You either ignore it, or you address it, and you move on.

Andrea Yankovsky discussed microaggressions while working in the studio, noting that, "Whether it bothers them or not, it doesn't matter. But it's there. It's like the kids at the school that are in the flight path of the airport; they learn how to tune it out, but it still has an effect".

Some felt that the best way to combat these instances was to work with other women and like-minded individuals, as well as with more inclusive organizations. Some women optioned that segregation could lead to problematic "separate but equal" causation and that continuing to lead by example rather than to separate was the way to handle the situation. Because this is a complex and multifaceted issue, many women seem to be wrestling with these notions and how to address them. Kerry Pompeo explained it like this:

> Where I see the risk, and this is where I sometimes have struggles with, and I totally appreciate it, I'm not knocking the women's groups, that are meant to give us a place to vent, and bond, and empower, and come into a collective, but I feel like we do detriment to ourselves, if it has to be a "female thing", because then we're putting ourselves in a box, that is going to make it very easy for people to say it's *that*. I think that it's not a realistic depiction of the world too, because it's not a hundred percent men, it's not one hundred percent women, so I think that when it does go, "The Lilith Fair" route, that you really don't give people a choice but to say, "It's Lilith Fair", or say, "It's a chick group", or say "It's chick rock", because at that point, they're not lying. It is chicks; they are playing rock music, or chick hip hop. They're not lying if [it's an] all-female group, and this stereotype, even though they might have more balls than any dude, the stereotype might follow them because it's all women, and that is something that needs to be broken down, because what is "chick" and what is "girly"?

Still, some of the women interviewed even felt uncomfortable for anyone having noticed that they were women. Again, this very layered issue, beyond the scope of music production, is something that men in music production will not be navigating, while marketing themselves to achieve certain types of status or prestige in their careers. The fact that so much extra energy and headspace is given to these nuances shows a very palpable and yet insidious side of prevailing outdated cultural norms that while outside of the recording studio influence the women at the helm of them. Hillary Johnson made mention of navigating this in her own career, saying:

> Well, for years, and probably still to a certain extent, it's not that I fought against being represented as a woman, you know a female engineer, but I kind of tried to shy away from it, because I felt that was not helping me. I felt that by people focusing on the fact that I was a woman doing this, was actually taking away from any credibility, and so, this is how I have felt for years, and only recently thought, "Maybe I should have been using this to my advantage?"

How is it to be a veteran in your field and to feel that your gender might discredit you? What is so non-credible about women? What does that say about society's messaging about women's worth, and how is that then filtered into the workplace and the recording studio? That is the key, not that the music industry puts that on these women engineers, but that as a society, these constructs exist and are brought into the music industry, through deeply ingrained mindsets, cultural mores, learned social cues, and women's internalized debased self-esteem. It is evident, from the interviews conducted here, that some light has been shed on certain aspects of those challenges that appear to remain. They are seemingly small, but they are cumulative and make a lasting impact on the careers of the people they affect as well as their creative output. These pervasive assumptions and mindsets are deeply rooted in many cultures globally. Why are women not credible? (For example, in Pakistan, Iran, and Saudi Arabia, women's testimonies only count as half in a court, and in other countries, they are not allowed to drive. In the USA, women were not allowed to have a credit card until 1974.) It could be argued that if women are not credible, then why would you hire one to be in charge of recording or producing your record? No one would likely admit to it, but even women themselves grapple with the idea of women being credible. In the interview, Kerry Pompeo, for example, mentioned:

> Because you have to prove more, that's a fact! That's one thing that I will not deny, is that women have to prove more. I am guilty of it. You know when I hire or interview, I am guilty of it, and I don't like it, and I'm like, "Damn! I just did something that I would have hated if I was sitting in that seat". But it's such a subconscious thing that's ingrained in us.

Several of the women interviewed made mention of adopting tactics to make sure they were deemed credible. These tactics ranged from being sure to only work on or create certain types of music, dressing differently to make sure they weren't seen as feminine, or overpreparing to make sure that they appeared infallible in their area, because the minute they faltered, they knew they would be perceived as not capable and instantly be discredited. Sylvia Massy noted that her love of hard music worked in her favor to help ensure she wasn't boxed in by stereotypical gendered views of the kinds of music a woman might create; "[M]aybe that's the best thing I could have done, was to start with hard music, so I don't get pigeonholed into, 'Oh, she just does girl music'. I've never had that problem, ever".

These nuanced experiences are consistent threads that weave through the commentary shared by these women, which illustrates patterns of thinking and behavior in and outside of the music business and its dynamics. In many instances, when a pointed question about a specific nuance or experience was raised, their responses to the questions unearthed a lot of other inferences, along with their intended answer. Most did agree, though, that without changes in society, there would be no further changes made to the prevailing climate and that the remaining progress to be had would be that much harder to attain, if at all. Hillary Johnson succinctly reflected on the topic saying, "I think that anything that happens out in the world, or in our immediate society, just sort of trickles down into the studio. I don't think that only the studio will change".

In other words, it could be argued that until women are viewed and treated differently by society as a whole, one couldn't expect that the music business itself would be isolated from such attitudes and habits. Of the women interviewed, all seemed to feel positively about the future generation of women coming up in the music business, as they have much more access to mentors, role models, and supporting organizations. Hillary Johnson again astutely noted:

> If these bigger companies, whether they be manufacturers, that are forced, that are asked, to not use women in their ads, and they don't because they feel a pressure, so when they get the pressure, and they don't put the women in the ad. Then you also have an organization, like Women's Audio Mission, supporting them. That sort thing, of showing the rest of the industry that there's a problem, and we are addressing it, and we are making it better, and I think the idea of showing that we are trying to make it better will make it better in the end. Rather than saying that "there's no problem", I think that the people that don't know that there's a problem, they're not going to see it. They are never going to see it, or it's going to take a really long time for them to see it, and I don't think that they're the ones who should be targeted. Eventually, these people will come around or they'll be dead . . . and I think people's individual pressure to not get left behind is pretty powerful.

It speaks highly of the women who have carved out paths for themselves without the benefit of so many of the resources that the current generation can benefit from. Going forward, it is hoped that the newer cohort of producers and engineers will have better access to resources and communities to find mentorship and peers. There is an increased amount of visibility online for many of these women, and this in turn will hopefully help reduce things like stereotype threat and keep younger generations inspired and aware of the different career paths available to them within the music production industry. The potential collective output from more diverse types of people will potentially lead to more diverse types of music being created.

These conversations shed light on the existence of a number of women who have been working as producers and engineers for several decades already. They can and should be named, remembered, hired, and heard. It is hoped that they may continue to lead the conversation, however they choose. To have been entrusted with the stories of these select few women in music production is very valuable, and it is just the beginning of a larger conversation that can continue as music, its production, and culture's intertwining relationship with music production continues to evolve through both cultural shifts and technological advances.

## BIBLIOGRAPHY

Anon. (2014) 'Let Toys Be Toys'. [online] 16th October. Available from http://
    www.lettoysbetoys.org.uk/what-do-toys-have-to-do-with-inequality/.
    [Accessed 8th November 2017].
Erickson, Megan. (2011) 'Big Think'. [online] 27th September. Available from
    https://bigthink.com/think-tank/where-are-all-the-women-scientists. [Accessed
    8th November 2017].
Fine, C. (2010) *Delusions of Gender: How Our Minds, Society, and Neurosexism
    Create Difference*. New York, NY: W.W. Norton & Company.
Gittleson, Kim. (2017) 'BBC'. [online] 13th October. Available from https://
    www.bbc.com/news/business-41571333. [Accessed 8th November 2017].
Hochschild, Arlie Russell, & Machung, A. (1989) *The Second Shift: Working Par-
    ents and the Revolution at Home*. New York, NY: Viking Penguin.
Ignatius, Adi. (2013) 'Harvard Business Review'. [online] 4th April. Availa-
    ble from https://hbr.org/2013/04/where-are-all-the-women. [Accessed 8th
    November 2017].
Marcus, Ruth. (2017) 'Washington Post'. [online] 15th September. Available from
    https://www.washingtonpost.com/opinions/where-are-all-the-presidents-
    women/2017/09/15/f5bb1ff2-9a59-11e7-82e4-f1076f6d6152_story.html?no
    redirect=on&utm_term=.f61d07b007f0. [Accessed 8th November 2017].
TEDx *UCDavis Talk*. Available from https://www.elizabethvsweet.com/tedx-talk.
    [Accessed 14th November 2017].

# Index

Note: page numbers in *italics* indicate figures and in **bold** indicate tables on the corresponding pages.

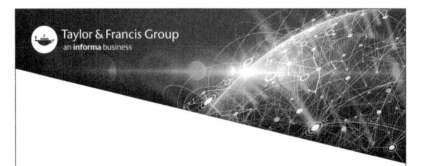